HPBooks®

How to Rebuild Your HONDA Car Engine

By Tom Wilson

Another Fact-Filled Automotive Book from HPBooks®

NOTICE: The information contained in this book is true and complete to the best of our knowledge. All recommendations on parts and procedures are made without any guarantees on the part of the author or HPBooks. Because the quality of parts, materials and methods are beyond our control, author and publisher disclaim all liability in connection with the use of this information.

The cooperation of American Honda Motor Co., Inc., is gratefully acknowledged. However, this publication is a wholly independent production of HPBooks.

CVCC® is a registered trademark of Honda Motor Co., Ltd.

Publisher: Rick Bailey; Editorial Director: Tom Monroe, P.E., S.A.E.; Senior Editor: Ron Sessions, A.S.A.E.; Art Director: Don Burton; Book Design: Paul Fitzgerald; Book Manufacture: Anthony B. Narducci; Typography: Cindy Coatsworth, Michelle Claridge; Drawings, photos: Tom Wilson, Myron Hemley, American Honda Motor Co., Inc., others as noted.

Published by HPBooks, Inc.
P.O. Box 5367, Tucson, AZ 85703 602/888-2150
ISBN 0-89586-256-5 Library of Congress Card Catalog No. 85-60113

1st Printing

Contents

ACKNOWLEDGMENTS

Writing a book is one of the best ways to discover who your friends are. Listed below are the many people who went out of their way to help.

Standing head and shoulders above the crowd is Myron Hemley of Fallbrook Camera. Myron is directly responsible for the excellent photos throughout the book. Whether it was staying at work until 2 a.m., or suggesting a different angle or lighting, Myron was there. He snapped more than half of the photos, and processed, proofed and printed nearly all of them, including the cover.

Steve Beegle wrenched on yet another car for the camera. Randy Lyles of British Masters Racing provided tools and experience. Scott Mergle towed incapacitated Hondas hither and yon and provided parts and experience.

Much of the hard-to-get information and machine-shop expertise came from aftermarket Honda specialists. Terry Haney of Haney Brothers Honda in Tucson, AZ, Serge Harabosky of A/T Engineering in New Milford, CT, Oscar Jackson of Jackson Racing in Huntington Beach, CA, Tom Nelson of Dragmaster Co., Mike Paar of Hondaworks, Inc., in Mesa, AZ, Larry Scott and Les Shofstall of Lesco Engineering in Escondido, CA, and Jesse Waddell of San Diego Motor Machine & Supply all went the extra mile.

More pro help came from Pat Garrett of Toyota Carlsbad, Brant Geilow and Pat Morgan of All Foreign, Irene Jeremice and Robert Morris of Fel-Pro, Bob Lopez and Ron Thompson of Federal Mogul Corporation, Janine Sine and Terry Davis of TRW and Barry Williams of SCID.

Honda's help came from several sources. Dave Jenkins and Larry Wroblewski of Needham, Harper & Steers helped get the extensive Honda lineup into sharp focus. Bob Murphy of American Honda Motor Company helped with information and illustrations. At the dealer level, Charlie Baladez of Cush Honda and Phil Escalante of Hoehn Motors answered questions.

Extra thanks to Tom Monroe and especially Ron Sessions of HPBooks for being the best editors going.

Special thanks to Jan for her help and patience.

Introduction

Introduced in 1976 as Honda's "large" car, the Accord has been Honda's best-selling model in recent years. Beginning in '82, Accord production has been supplemented by an assembly plant in Marysville, Ohio. Shown is '83 hatchback with 1751cc CVCC engine.

Now that Honda cars are commonplace, it's difficult to remember a time without them. But, before the introduction of the Civic in 1973, there was but a handful of Honda cars in North America. First in were a couple of S500, S600 and S800 sports cars. These small, rear-wheel chain drive, highly tuned cars were never officially imported, but some found their way here anyway. In 1970, Honda began exporting N600 sedans to the U.S., followed by the Z600 coupe in '72. The 600s were shopping-cart size, front-wheel-drive cars that used motorcycle engines. Neither the N or S series cars is covered in this book.

Honda's more thorough approach to the North American market—the Civic—was an instant success. Without being too small, the early Civic made the most of a compact exterior and light weight. Part of the weight savings was attributable to the all-aluminum 1170cc engine. It relied on a crossflow cylinder head and rpm for efficiency and power. About a year later, the 1170 was replaced with the larger-displacement 1237. The 2mm-bore increase helped offset the negative performance aspects of added safety and emission controls.

In 1975, a totally new engine was introduced in the Civic, the 1488. The new engine used a cast-iron block and boasted Honda's highly acclaimed Compound Vortex Controlled Combustion—*CVCC*—cylinder head and induction system. Both the 1237 and 1488 were offered in the Civic, with the 1488 being the clear winner in the sales department. The 1237 was discontinued after the 1979 model year. The '75 through '83 1488s are covered, along with 1170 and 1237 engines.

Honda introduced the Accord in '76, and with it, a 1600cc engine. The 1600 is essentially a 1488 with a 6.5mm-longer stroke. It also has a cast-iron block and CVCC cylinder head.

1979 marked the introduction of the sporty Honda Prelude. Like the Accord, the Prelude brought a new engine with it, the 1751. This time, bore was increased 3mm to 77mm; stroke was increased another 1mm to 94mm.

In '80, the Honda lineup was restyled. Replacing the price- and mileage-leader 1237 was a new Civic powerplant, the 1335. The 1335 was interesting because it used an aluminum block similar to the 1200s, but with a CVCC head like the iron-block engines.

In keeping with its position as the Honda flagship, the Prelude was the first redesigned model of the mid-'80s. In the '83 redesign, the

1751 was dropped and a new 1830cc engine added. This new engine is not covered in this book.

In '84, the Civic and Accord were also redesigned and got new 1332 and 1488 engines in the process. Like the 1830, the 1332 and "new" 1488 are significantly different than earlier engines and are not covered.

To sum up, this book covers two 1200s—1170 and 1237—and three CVCC engines—1488, 1600 and 1751. The 1335 defies neat categorization because it's half 1200 and half CVCC. Think of it as a marriage of the two engine families.

Unfortunately, rebuilding Honda engines is a job many people shy away from. They've heard enough stories about auxiliary valves and color-coded bearings to scare them away. That's too bad because working with Honda engines is easy. At worst, the auxiliary valve is merely another part to replace. The coded bearings are certainly no extra work. In fact, they provide a ready-made means to fine-tune oil clearances. With the 1200s, you don't have to consider auxiliary valves because they don't have them. And all Honda engines are relatively lightweight and easy to maneuver. Rebuilding these engines doesn't mean pulled muscles and sore backs from wrestling a bulky, 200-lb block around the garage.

More importantly, this book provides the information needed to quickly and accurately rebuild your engine. The step-by-step procedures, photos, graphs, drawings, lists and captions I've included will guide you through the rebuild with minimum effort.

In the course of researching and writing this book, I talked with many pros—both factory-trained and garage experienced. I spent hundreds of hours scouring junkyards, rebuilding engines and talking to experts in dealerships, volume-rebuilding factories and independent shops. In short, I've done the legwork so you don't have to.

Read through the book and study the pictures before you start to work. You'll be better able to handle the upcoming jobs with the proper tools and planning. Don't skip steps because they seem unnecessary. Your engine will not run as long or well if you do.

Always double-check your work and the work you farm out. Machine-shop and parts people make mistakes, too. The time to catch a mistake is *before* the engine is back in service, not after the damage is done.

Finally, work safely. Your work area should be as clean as possible. Keep tools, rags, cables, lines and parts out of the way when not needed. Use common sense when it comes to lifting heavy parts. Don't smoke around gasoline! If you get tired, take a break or quit for the day. The risk of injury isn't worth rushing the job. Take your time and do the job right the first time—you won't have to do it over.

With that in mind, let's get to work.

1800 = 1751?

Honda model designations, such as 1300, 1500 and 1800, are based loosely on engine size. I choose to refer to each engine by its exact displacement, not its model designation. Following is the actual engine displacement for each model:

Model	Engine Displacement (cc)	Chassis	Year
1200	1170	Civic	73
1200	1237	Civic	74-79
1300	1335	Civic	80-83
1500	1488	Civic	75-83
1600	1600	Accord	76-78
1800	1751	Accord	79-83
1800	1751	Prelude	79-82

In the process of rounding off engine-displacement calculations, you may see variations of these numbers in other publications. For example, you may see 1169 for 1170, 1487 for 1488, 1599 or 1602 for 1600 and 1750 for 1751. Even Honda shop manuals don't agree on displacements! The point is—there are only six different engine displacements covered in this book, not 10.

Time To Rebuild? 1

This trusty Civic has gone way past the century mark without a rebuild. Owned by an aircraft mechanic, it has received regular maintenance since new.

Rebuilding a Honda engine should begin with a thorough diagnostic checkup. Why? First, a low-power, gas-wasting engine may not need a complete rebuild. Many times, a valve job or simply cleaning the auxiliary valves will suffice. Second, if the engine does need rebuilding, you can begin the project with a good idea of what's needed. This allows you to budget the rebuild more accurately. Finally, the checkup may reveal that the engine doesn't need rebuilding, just a thorough tuneup, or carburetor or distributor overhaul. Therefore, a complete examination of engine condition is the first step in restoring lost efficiency, whether it's just a tuneup or a major rebuild.

Don't think that just because a car engine has gone 100,000 miles it might blow up tomorrow, even if it runs quietly, uses a minimum of fuel and oil and has good power. This outlook can cost many trouble-free miles of service from the original engine. Accumulated mileage alone is *not* a good indicator of engine condition.

Obviously, some engines have seen their best days before 75,000 miles. Others are still going strong way past the century mark. The two biggest factors in engine longevity are maintenance and operating conditions. No engine can *give* good service if it doesn't *get* good service.

How an engine is run affects how long it will last. Giving the powerplant time to warm up before extracting full power from it can measurably lengthen its service life. Over-revving an engine is particularly hard on the rings and valve train, resulting in a *smoker* long before its time. Dirt roads also take a toll by relentlessly bombarding the air filter until some grit finds its way into the induction system. Broken gaskets and loose hoses can mortally wound an engine under dusty conditions. Frequent stop-and-go driving is harder than mile after mile of freeway cruising. Worst of all, dirty oil and a dirty oil filter will do an engine in quicker than anything. In short, you must account for the duty and maintenance an engine has seen when starting diagnosis.

EXCESSIVE OIL CONSUMPTION

Oil consumption is determined by an engine's internal clearances and is an excellent gage of engine condition. When an engine is new, its internal clearances are easily bridged by an oil film, resulting in tight seals at the piston rings and valves, and high oil pressure at the main and rod bearings.

As miles pile up, these parts wear and dimensions increase. When this happens, oil cannot bridge the gaps between rings and cylinder walls and is sucked into the combustion chamber and burned. You see the wispy trail of blue smoke from the exhaust pipe. At the crankshaft, increased dimensions provide less resistance to oil flow allowing oil to work its way to the ends of the journals where it's flung off. This causes more oil to try and fill the void, thus dropping oil

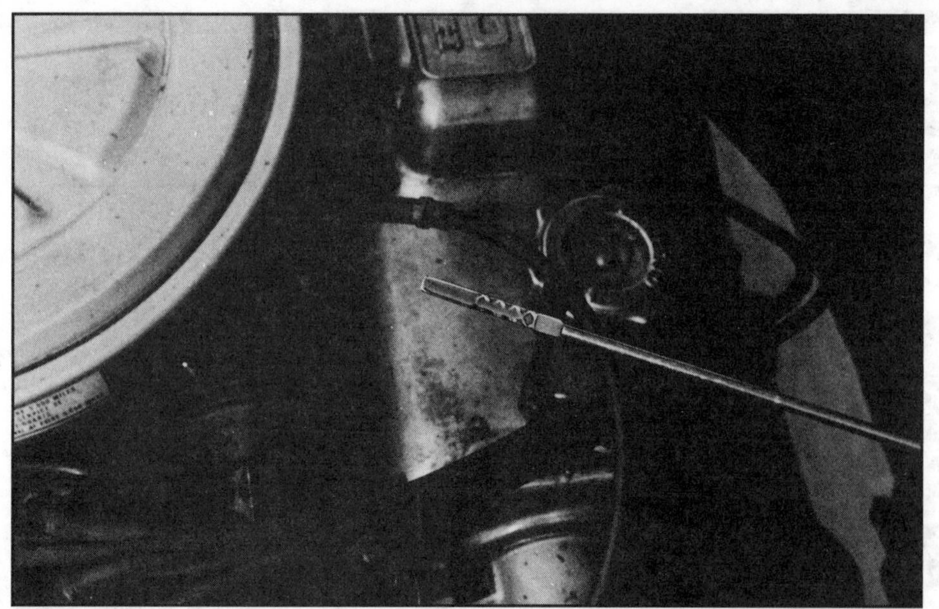

Oil consumption is an excellent indicator of an engine's internal condition. To check, make sure oil level is at top of dipstick crosshatch area. Write down odometer reading and read dipstick regularly. When oil level drops to bottom of crosshatch area, note mileage and subtract first reading from second. Difference is oil consumption rate.

pressure. It also causes increased oil consumption because more oil is splashed on the cylinder bores. When oil consumption is high and oil pressure low, a rebuild is required to correct excess clearances.

How Much Oil Consumption is Excessive?—Certainly, there's no harm in burning a quart every 1000 miles or so. Only when you get less than 1000 miles to a quart should you worry. And if a new quart is needed every 500 miles, your engine definitely needs attention. If you think your engine is burning oil, monitor its consumption carefully. Your memory is a poor tool compared to a written mileage record. Also, if oil consumption is a problem, you should be able to see blue smoke under certain conditions. Read on for details.

Don't let an oil leak fool you when computing oil consumption. Usually, the oil-consumption figures above account only for oil *burned* by an engine, not that lost due to leaks. Actually, most oil leaks won't count for much compared to that burned due to worn rings or valve guides. Nevertheless, fixing a big, obvious leak may drop oil consumption considerably!

There are two things that allow an engine to burn oil: worn rings and valve guides. Both allow oil to enter the combustion chamber where it is partially burned and sent out the exhaust. A single puff of smoke immediately upon startup after sitting overnight usually means worn guides and valve-stem seals.

Another good test is to find a long hill to coast down while in top gear. When you reach the bottom, glance in the mirror as you open the throttle. A puff of smoke indicates worn guides *or* rings. But, if the engine lays down a smoke screen, you can bet the rings are at fault.

CVCC Oil Consumption—On 1335, 1488, 1600 and 1751cc CVCC engines, faulty auxiliary-valve O-rings or prechamber gaskets can cause excessive oil consumption. O-ring leaks occur as the O-rings dry out from heat—common on 1976—'79 1488 and 1600cc engines. Prechamber-gasket leaks can occur if auxiliary-valve-holder torque drops below 50 ft-lb, or if old gaskets are reused.

The best way to find faulty O-rings and prechamber gaskets is to remove the auxiliary-valve parts from the cylinder head. This is possible without removing the head. Be suspicious of the auxiliary-valve O-ring and gaskets if oil consumption is high, but a *wet/dry compression test,* page 14, shows ring condition to be OK.

Blowby—Just as worn rings and cylinders allow oil to enter the combustion chamber, they also let combustion gases pass in the other direction, into the crankcase. These *blowby* gases pressurize the crankcase, robbing horsepower and contaminating the oil. Telltale signs are blowby vapor blowing out the dipstick hole and oil filler. Really bad cases can blast a dipstick out of its hole!

Modern engines, including all the Hondas covered in this book, use a positive crankcase ventilation (PCV) system to vent the crankcase. The PCV system routes blowby back into the induction system through a hose and metered orifice, where it is subsequently drawn into the combustion chambers and burned.

Valve Guides—Some oil passage past the valve stems and piston rings is normal. After all, if the rings and guides were sealed oil-tight, they would wear out in less than 10 miles from metal-to-metal contact.

Excessive oil loss through the guides occurs when guide-to-valve clearance is excessive. This allows oil to practically run down the valve stem and into the port. Then, next time the intake valve opens, the oil is sucked into the combustion chamber. Exhaust valves can pass oil in the same manner. But because an exhaust port is a hot, mostly high-pressure area, excessive clearances there result in blowby into the rocker cover.

POOR PERFORMANCE

Perhaps when I say *performance* I should say *efficiency.* This is because when an engine yields poor *fuel economy* and *power,* it is inefficient. Diagnosis should determine if the engine is using the right amount of fuel to produce the expected amount of power. Remember, when I say performance, I mean both engine power *and* fuel consumption. If power or fuel economy drops, engine internals may or may not be the cause. The diagnostic tests later in this chapter are designed to systematically uncover the problem.

To locate the source of poor performance, start with a professional tuneup. All aspects of engine performance should be checked using an engine analyzer until you are satisfied the engine is as well-tuned as it can be. Listen carefully to what the mechanic says about your engine because he sees many, many engines and has learned to spot troubles quickly. If your engine responds with new-found pep and fuel economy, odds are that its internals are still serviceable. But, go ahead and do the following diagnostic tests as a double-check.

If a tuneup doesn't restore lost

power or fuel economy, perform the following diagnostic tests. However, don't go straight to the tests without first doing a tuneup. If you do, some test results will not be accurate. For example, a cylinder won't have full compression if the valves are incorrectly adjusted. Vacuum testing will also be affected.

Before an accurate engine diagnosis can be made, engine must be in top-notch tune. Plugs, timing, carburetion and emission controls have to work correctly. A *dyno-tune* is especially helpful when trying to determine how many useful miles are left in your engine.

CAUSES OF POOR ENGINE PERFORMANCE

A quick look at the most likely internal engine problems will help put them in perspective before you start testing for them individually. Keep in mind that internal-combustion engines such as your Honda are nothing more than air pumps. They perform work by inhaling air, compressing it and expanding it, harnessing the expansion and exhaling the byproducts. So, anything that hinders an engine's breathing reduces its efficiency—both power and fuel economy. Burning fuel only makes the air expand.

The key to an engine's pumping efficiency is the tightness of the combustion chamber—that area formed by the piston top, rings, cylinder wall, head gasket, head, valves and sparkplug. If any of these parts allow air to escape from the combustion chamber, engine performance will drop. Additionally, engine breathing will suffer if the valves and valve mechanism are in poor shape.

Generally, a worn-out engine will perform poorly and use a lot of oil. This is usually caused by worn rings and cylinders. But it is also possible for engine performance to be low and oil consumption to be normal. In this case, the valves, camshaft, valve springs or head gasket could be at fault.

Blown Head Gasket—Unfortunately, head-gasket problems are familiar to many 1976—'77 Civic CVCC and Accord owners. When a head gasket *blows,* it allows hot combustion gases to escape into the cooling system or another cylinder.

If the gasket blows into the cooling system, the combustion leak is confined to one cylinder—you might not notice the power loss if the blown portion is small. Not at first at least. The hot gases will heat the coolant very rapidly, causing the engine to continually overheat, or at least run very hot. You should suspect this type of blown head gasket if your Honda warms up within the first few miles of driving and continues to heat. The compression and combustion gases will also pressurize the cooling system, causing continual coolant loss.

To check the head gasket, perform either a compression test or leak-down test. And, for this particular type of blown head gasket, a color-changing fluid test of the cooling system is a good idea. The compression test will reveal the bad cylinder with a lower-than-average reading. The leak-down test will have a noticeable rise in leakage, coupled with bubbling in the radiator, particularly if the thermostat is open or has been removed.

The color-changing fluid test is a little different. A clear jar containing a column of colored fluid is held against the radiator-cap opening. Once the engine is fully warmed, so the thermostat is open and coolant flow heavy, a squeeze bulb atop the clear jar is worked to force air from the radiator tank through the fluid. If there are any combustion gases in the radiator tank—and hence the coolant—the fluid will change color. On my test kit, it changes from yellow to green. The speed and severity of the color change indicate the extent of the combustion leak.

Even worse is a head gasket blown between cylinders. An absolute power thief, this type of blown head gasket is painfully obvious because power is lost in two cylinders. A compression or leak-down test will show two adjacent cylinders with poor readings.

For all you pessimists, yes, it is possible to have a head gasket blow both between two cylinders and into the cooling system, with all the symptoms just described.

The 1976—'77 Civic CVCC and Accord were notorious for blowing head gaskets. The problem stemmed from the different expansion rates of the CVCC engine's cast-iron block and aluminum head. As the engine warmed up, the block and head expanded differently—the aluminum head "grew" much faster than the iron block. During the heating and cooling cycles the head gasket was literally torn, or *sheared,* between the two parts—combustion heat and pressure finished the job. Honda cured the problem in early 1978 by introducing a new molybdenum-disulfide-coated (MoS_2) head gasket and redesigned head bolts. The slippery gasket lets the block and head expand without taking the gasket in all directions. The redesigned bolts feature a helical groove cut into their shanks. The grooved shank weakens the steel bolt so it will stretch with the expanding aluminum head. It sounds scary, but it works.

Usage affects head-gasket life on these early CVCC engines more than others. Basically, an early engine must be fully warmed before extracting full power from it. This allows the head and block to expand more slowly, permitting the head gasket to more-easily conform to the expanding cylinder head. The worst case is in cold climates, when the engine is fired up and immediately driven off at full throttle. Combustion gases cause a rapid heat buildup in the cylinder head. This fast cylinder-head temperature rise causes rapid cylinder-head expansion on a cold block, usually resulting in a torn head gasket.

Burned Exhaust Valves—When a mechanic says a valve is *burned,* he means that some of the valve's *face*—sealing surface—has been eroded away or cracked by the blast of hot combustion gases. Think of combustion gases as an inefficient cutting torch and you'll understand why valves burn.

A burned exhaust-valve face cannot provide a gas-tight seal against its seat, causing a large drop in power and compression. In extreme cases, a small chunk will be missing from the valve, allowing all compression to escape out the exhaust. The engine can then only run on three cylinders. Such extreme cases show immediately on a compression test because engine-cranking speed doesn't change on that cylinder's compression stroke and the gage reads little or no compression.

One major cause of burned valves is incorrect valve adjustment. Unlike many modern engines, Honda engines don't have hydraulic valve lifters, which automatically maintain valve clearance. Instead, Honda valves need adjustment every 15,000 miles. Skipping this service or maintaining insufficient valve clearance can easily lead to burned valves—a valve can cool only when it is fully seated. Some cooling takes place through the guide, but not much compared to the cooling at the seat when the valve is closed. If a valve stays open longer due to tight clearances or whatever, it has both less time to cool and absorbs even more combustion heat—thus the burning process is begun. Even well-maintained valves can burn if a piece of carbon gets caught between the valve head and seat as the valve closes. This holds the valve partially open and can start the gas-erosion process. Once it starts, the valve-burning process is rapid.

Exhaust valves are much more prone to burning than intakes. This is because exhaust valves are exposed to combustion heat on both the combustion-chamber and port sides. Intake valves, on the other hand, are cooled by each passing intake charge and are heated only on the combustion-chamber side. Because intakes run so much cooler than exhausts, they are much less apt to burn.

CVCC auxiliary valves experience such a rich, cool mixture that they rarely burn. By contrast, the main combustion-chamber and exhaust-valve temperatures of the 1976—'79 1488 and 1600cc engines are very high. As a result, exhaust valves on these engines are usually completely worn by 100,000 miles. It's rare to find a reusuable exhaust valve when rebuilding one of these engines.

Cam-Lobe Wear—Worn CVCC-engine camshafts is the number-one problem reported by most Honda mechanics. Wear is always most pronounced on the lobe at the bellhousing end of the engine—number-4 intake on most engines. Usually, it can be detected during a routine valve adjustment.

No one seems to know why this lobe wears so quickly. One theory is the close proximity of an oil drain hole to the lobe. Oil around that lobe drains quickly into the sump instead of puddling on the head. Thus, there is less oil whipped up by the whirring camshaft—*windage*—to lube that lobe. Another thought is that the bell-housing end of the engine is the hottest. Coolant passing through the thermostat picks up heat from the block and other end of the head. Still another theory involves *shrouding.* The valve train surrounds that cam lobe with rocker-arm-shaft pedestals, the wide auxiliary-valve lobe, rocker arms, springs and the like. Oil has a hard time reaching that lobe. Of course, all three factors could contribute to that lobe's early demise.

Whatever the reason, it's almost guaranteed that the end lobe will be worn on high-mileage CVCC engines. Of course, cam wear is accelerated by dirty oil, and infrequent oil and filter changes. However, even with frequent oil and filter changes, I've seen this lobe go "bye-bye" in less than 75,000 miles. Once wear is through the surface hardening, the lobe and rocker arm wear like a pillar of salt in a rain forest.

With a worn cam lobe, performance suffers because of decreased engine breathing. If an intake lobe grinds down, the valve won't open completely and the cylinder won't completely fill as the piston moves down. If an exhaust lobe wears, a similar thing happens. Not all spent combustion gases are exhausted. Consequently, when the intake valve opens again, there's less space for the new mixture and the sparkplug fires a diluted air/fuel charge.

Unlike many engine problems, a worn camshaft usually goes unnoticed by most drivers. Performance drop is so gradual that it is difficult to detect. Worn lobes don't cause a visible or audible trace like oil consumption or a rod knock, so detecting a worn cam is usually found by a mechanic as he performs a valve adjustment.

Oil-Pump Gear—Another type of cam wear occurs at the oil-pump drive gear. It's common on 1170 and early 1237cc engines—sometimes to the point of gear-teeth *breakage.* If the gear fails, the oil pump stops pumping and severe engine damage can occur. Also, metal particles sheared from the gear can damage the engine, especially the oil pump. Visually inspect the camshaft oil-pump drive gear.

If worn more than a light polishing, *replace the cam and oil-pump driven gear as a set.* They are a *matched* pair. When worn, drive-gear teeth will be concave. Many times, the driven gear will not show wear, even though the drive gear is worn out. Replace it anyway, or the new drive gear will quickly be destroyed due to a mismatch.

Carbon Deposits—Although carbon deposits don't fall under the category of engine damage, and a rebuild is not necessary to remove them, a few words about carbon will help engine diagnosis.

Carbon is a solid byproduct of incomplete combustion, some of which sticks to the combustion-chamber surfaces. Both gasoline and motor oil are hydrocarbons, so burning them in the combustion chamber in the wrong amounts causes excess carbon deposits. The most common source of harmful carbon deposits in an emission-control engine such as the Honda is excessive oil consumption, although a rich air/fuel mixture can be just as bad. Prolonged idling and slow driving can also cause carbon buildup. So, carbon deposits are a symptom of a problem, not the source. Merely ridding your engine of carbon will not cure the problem, only delay the symptoms. Therefore, while there may be ways to get rid of carbon buildup without overhauling an engine, curing excessive oil consumption may mean an engine overhaul.

Auxiliary intake valves on CVCC engines are especially prone to carbon deposits. The 1975 1488cc engine is the worst offender because it runs cooler and richer than other CVCC engines. See the example on page 98.

When auxiliary valves clog with carbon, engine performance and driveability suffer because the resulting overall air/fuel mixture becomes too lean. Other symptoms include hesitation, rough idle and poor gas mileage. To test for clogged auxiliary valves, completely warm the engine. If the car has a manual choke, pull it out to the first detent. On automatic-choke engines, prop the choke partially closed with a screwdriver. If the engine smooths out and has improved response, clogged auxiliary valves are likely culprits. If pulling the choke makes no difference, the problem is elsewhere. Check the exhaust system for obstructions.

Carbon deposits cause trouble in two ways. First, they may *shroud* the valves. Carbon deposits build up on the back side of a valve and restrict air/fuel-mixture flow into the cylinder. This is what happens with auxiliary intake valves.

Carbon deposits in the combustion chambers can also cause damage. Carbon easily heats to incandescence, causing *preignition* and *detonation.* These types of abnormal combustion can damage an engine by placing a heavy load on engine internals. Imagine red-hot carbon in the combustion chamber. When a fresh intake charge is compressed on the compression stroke, the hot carbon *preignites* the mixture. A moment later, the sparkplug fires and the mixture also starts burning near the plug. The two flame fronts collide, sometimes producing an explosion—*detonation*—rather than even burning. The resulting sudden pressure and temperature rise is more than the engine was designed for. Piston, valve and ring damage can result if preignition or detonation is prolonged. Although not as severe, preignition *without* detonation causes excess combustion-chamber pressure and temperature, but without pinging or knocking.

Detonation is very similar to preignition, but the second ignition source—glowing carbon—lights the mixture *after* the sparkplug has fired. Again, combustion-chamber temperature and pressure exceeds engine-design limits and damage occurs. Audible signs of detonation are pinging or knocking, sounds akin to billiard balls striking one another.

Admittedly, severe engine damage from preignition or detonation isn't prevalent, but severe cases can burn or blast holes in pistons, break rings, and deform main-bearing caps. So, prompt attention to the cause of abnormal combustion is wise. They are usually associated with low-octane gasoline.

Recent research indicates that a small amount of knocking or pinging is not harmful to an engine, but does reduce fuel economy and power. Nevertheless, be concerned if your engine is knocking heavily. Besides carbon buildup, detonation can be caused by *stale* or low-octane gas, over-advanced ignition timing and engine overheating. Check for these problems if your engine detonates.

There are other problems associated with carbon. Earlier, I mentioned how a valve could start burning if a piece of carbon got lodged between the valve and seat. The same sort of thing can happen to a sparkplug. If a piece of carbon sticks between the plug electrodes, the sparkplug will short out and the cylinder will misfire or go totally dead. Plug replacement or cleaning cures that problem.

A carbon-aggravated problem most people are familiar with is *dieseling*—the engine *runs-on* after the key is turned off. A hot piece of carbon acts like a diesel-engine glow plug by providing an ignition source other than the sparkplug. Besides ridding the engine of carbon, slowing the idle and reducing spark advance a few degrees will help reduce dieseling.

If your engine isn't showing signs of carbon—dieseling, detonation and the like—resist the temptation to pour a "mechanic-in-a-can" into your engine. Internal engine cleaners work, loosening carbon throughout an engine. This loose carbon then bounces around the combustion chamber doing its mischief—burning valves, shorting plugs, scratching cylinder walls and so on.

If your engine shows signs of carbon buildup, investigate the cause. Are you constantly idling the engine or putting around town—perhaps on a delivery route? Taking the car on an occasional *double-nickle* freeway drive might help. Maybe the carburetor is out of adjustment and in need of a professional tuneup. Or, perhaps the engine never dieseled before you noticed blue smoke when accelerating. If the latter is the case, the culprit could be worn rings or auxiliary-valve parts. You'll have to fix these problems before preventing carbon buildup.

DIAGNOSIS

Now that we've looked at the more common engine problems, let's start in on how to find them—without taking the engine apart. Your engine may or may not be exhibiting problems, but do the tests anyway. If it has a problem, you'll find it. If not, you'll have established a baseline of your engine's condition. From there, you can decide whether to rebuild now or later.

NOISE DIAGNOSIS

Internal Noises—Diagnosing engine noises is a difficult and imprecise art. Many factors influence the way sounds are perceived, not the least being the human factor. When inves-

VALVE JOBS AND HIGH-MILEAGE ENGINES

Some old-mechanic's tales are true. One you may have heard concerns valve jobs on older engines. According to this tale, perform a valve job on a older engine with tired rings and cylinders, and *increased* oil consumption will result.

Well, it's true. A valve job helps seal the combustion chamber, which results in higher combustion pressures. These higher pressures easily push past worn rings, thereby increasing blowby. Also, more oil is drawn into the combustion chamber during the intake stroke too, resulting in more burned oil. It all adds up to increased oil consumption.

Internal noises can be transmitted through wooden dowel. Place dowel against skull, just forward of ear. Other end should go against a *solid metal* part of engine: head, block or bolt head. Rubber-mounted accessories and gaskets absorb sound. Likewise, don't "listen" through rocker cover. Air space and gasket between head and rocker cover muffles sounds. If you have access to a stethoscope, use it. A stethoscope—below—amplifies sounds, and its ear plugs block background noise.

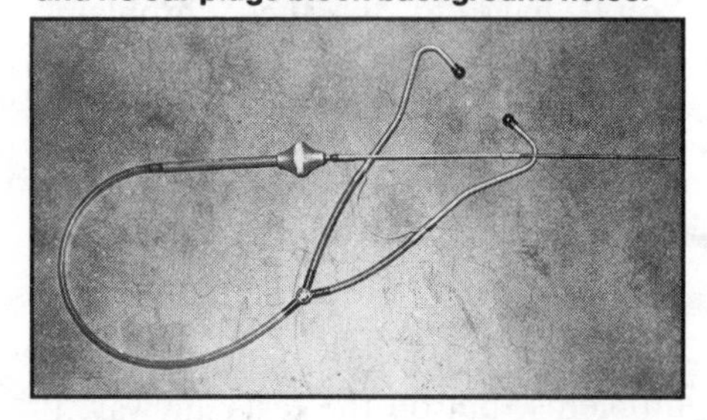

tigating an automotive sound, try different spots. Open the hood, close the hood, sit inside the car, stand to one side, stand in front, lay down in front and to the side. You won't hear all noises from each spot. And, those that you can hear will sound different from each spot.

Humans are biased toward their eyesight, so closing your eyes helps greatly. Cupping your hands around your ears may look funny, but it helps mask sounds from the sides and amplifies those in front. It's a great way to pinpoint a noise.

Finally, learn to mentally *dissect* what you are hearing. Upon first hearing a running engine, your initial impression is a big jumble of sounds. By critically identifying each sound, you can more easily block out the unimportant sounds while concentrating on those you want to hear.

A big help in locating noises is a stethoscope, length of heater hose or a wooden dowel. Unlike the stethoscope a doctor uses, an automotive stethoscope has a solid-metal probe at the business end. It works best when held against a solid part—head, block, manifold, bolt head or the like. If you suspect a noise emanating from beneath a plastic or metal cover, place the stethoscope against a nearby bolt head. For example, a noisy timing sprocket can best be heard by listening to a timing-cover bolt, not the cover itself. Cover gaskets damp sound, while a bolt's solid connection amplifies it.

As a second choice, a length of hose or dowel can be used instead of a stethoscope. With a hose, hold one end firmly against the engine and the other end to your ear. When using a dowel, position the receiving end of the dowel against your skull, just forward of the ear, so engine vibrations don't bounce the dowel into your ear.

Engine noises can be lumped into three categories: intermittent noises, those occurring each crankshaft revolution, and those occurring at every other crank revolution. First the intermittents—the oddballs. These are *external* sounds coming from loose brackets, rubbing hoses and so forth. By poking around the engine compartment, you can single out and remedy these noises.

Noises that occur at every other turn of the crank—at camshaft speed—are most likely coming from the *valve train:* valves, rocker arms, timing belt or sprockets. There is one *bottom-end* noise that can happen at every other revolution—*piston slap.* Piston-slap noise is the sound produced by the piston slamming against the cylinder as that cylinder fires at the top of the power stroke. Only when piston-to-bore clearance is excessive will piston slap be audible. And, because there's only one power stroke for every two crank revolutions, it occurs on every other crank revolution. Piston slap is easiest to detect on a cold engine, before the pistons have expanded, reducing piston-to-bore clearance.

Noises occurring at every turn of the crankshaft come from the bottom end: worn piston pins, broken rings, worn rod bearings and main bearings.

If you have trouble telling whether a noise is at one-half or at crankshaft speed, hook up a timing light and see if the noise coincides with flashes of the light. If it does, the noise is at one-half crankshaft speed—a top-end problem or piston slap. If the noise occurs twice for every flash, it's at crankshaft speed—a bottom-end problem.

Now for the hard part: What do these problems sound like? Let's start with normal engine sounds. The dominant sound should be the soft, mechanical whirr of the valve mechanism. At idle, this might sound more like soft ticking. If it is more like a clack, chances are valve clearance is excessive. Rev the engine lightly; valve-train noise should become whirr-like. If you hear one valve over the rest, it's loose. Correctly adjusted valves are quiet, so you can find the culprit with a feeler gage.

If you hear a loose valve, shut off the engine, remove the rocker cover and restart the engine. Use a 0.006, 0.008 or 0.011-in. feeler gage to slip between the rockers and valves on the idling engine. When you get to the loose valve, it will quiet considerably as the feeler gage takes up the slack. See the valve-clearance chart, page 17, and choose the appropriate-size feeler gage.

Isolating Normal Noises—The 1975—'79 1237cc engines equipped with air pumps might have a grinding noise coming from that emission-control device. Apparently, air pumps frequently shed impeller vanes. The grating noise you may hear is small plastic chunks flying around the inside of the pump. Heat and age combine to ruin air pumps. If pump internals are loose, the pump should be replaced.

Fan noise will drown out almost all other noises, especially if the hood is raised. Try to do all your testing with the fan off, if possible. Always keep in mind that the electric fan can *start at any time,* even with the key off, so keep your body and tools away from its blades. Prolonged idling is a sure way to get the fan to come on—something the engine will be doing a lot of during diagnosis.

If fan noise is a problem, try trickling water over the radiator to help it cool. You can disconnect the fan motor, but only for short periods—no more than two minutes of idling, depending on ambient temperature. Afterwards, the fan *must* be reconnected so it can cool the engine. To disconnect, disconnect the plastic connector at the fan motor.

Additionally, a certain amount of noise must be expected from the exhaust, alternator, water pump, fan belt, transaxle bearings and, in short, anything that moves. Most of these individual sounds are barely audible, but they can add up to a significant background noise.

Rocker-Arm Noise—After blanking-out normal engine sounds, there are many possible *problem* sounds. After excessive valve clearance, worn rocker arms are the next most prevalent problem noise. On high-mileage engines, rocker-arm and shaft wear can be severe enough to affect valve adjustment. The scenario goes something like this: The feeler gage feels tight during a valve-clearance check, but once valve-spring pressure begins to work on the rocker arm, more free play is found in the bushing. Excessive valve clearance causes a corresponding increase in noise.

If you suspect worn rockers, move each rocker arm by hand with the valve completely closed—easy to do as you adjust valve clearance. If the rocker arm moves in any direction other than 90° to the shaft, the rocker-arm ID *or* shaft is worn. A more thorough inspection requires disassembling the rocker-arm-and-shaft assembly and measuring the shaft and rocker-arm bores.

Piston Slap—A dull, hollow sound down in the cylinder block could be piston slap. As noted before, this is one bottom-end noise that occurs at *every other* turn of the crank. To check for piston slap, pull the sparkplug lead from that cylinder and listen again. If it was piston slap you heard, it should now be greatly diminished. By disabling the cylinder, piston slap is reduced because combustion loads no longer exist. Reconnecting the plug lead should restore the noise.

Note: If your Honda has electronic ignition, *do not* pull off a sparkplug lead and run the engine—ground that lead instead. Failure to do so may result in electronic ignition-module failure. See the nearby sidebar "Disabling the Ignition."

Piston-Pin Noise—The same plug-lead trick can be used on suspect piston-pin noises, except these will occur at crankshaft speed. If the noise is in a piston, pulling a plug wire will reduce or eliminate it. A loose piston pin makes a fairly sharp-sounding knock. Disabling the plugs one at a time is the best way to pinpoint a loose pin. Both piston slap and loose pins are hard to hear and fairly difficult to detect. So, if you don't hear ticking or knocking to begin with, don't start pulling plug leads endlessly looking for one. It's probably just not there.

Shocking News—The first time I tried pulling a plug lead from a running engine, I got a 20,000-volt jolt. There is an easier, and less painful, way! Disconnect all the leads from the plugs with the *engine off.* Then, just set the leads back on their plugs—don't push the boots on. Insulated pliers can then be used to lift the leads off to a ground—cylinder head or block—and back on the plug when you are done. Better yet, use a jumper wire to ground each plug. These are available in kit form. An adapter spring goes between each plug and wire. To *kill* a cylinder, simply touch the spring with the probe of the grounded wire.

Rod Knock—Continuing downward, the next noisemakers are rod bearings. As crankshaft-to-rod-bearing clearance increases, the high-pressure oil cushion between the connecting-rod bearing and its journal is lost. A knock is the result. Because bearings usually wear out in sets, you'll hear a castanet-like rattling with old, tired bearings. There are multiple, high-pitched knocks for each crankshaft revolution.

To test for rod knock, thoroughly warm the engine to operating temperature. With the transaxle in neutral, lightly rev the engine, say from 1000 to 2000 rpm, and *lift off the accelerator abruptly.* Engine rpm must drop sharply. As rpm drops, the rods should rattle, knock or pound, depending on how you hear it. This is because the rods *float* on their journals as they pass through the transition of being loaded, then quickly unloaded.

Main-Bearing Knock—Sounding similar to, but deeper than, worn rod bearings are bad main bearings. Main bearings knock for the same reasons as rods—excessive oil clearance—but under different conditions. To test for main-bearing knock, put the thoroughly warmed engine *under load.*

On a manual-transaxle car, load the engine by selecting second gear and letting out the clutch until the engine begins to labor. Then load the engine further by putting half of your right foot on the brake and the other half on the accelerator. Use the parking brake, too. Keep engine speed at about 1000 rpm during the test and don't let the car creep forward. With a little throttle-and-clutch juggling, the knocking main bearings will really sound off with a heavy, low-frequency pounding. Be careful. Don't do this for long or you'll burn out the clutch. You can also hear bad main bearings while going uphill, accelerating or during other periods of high engine load.

With a Hondamatic transaxle, the test for bad main bearings is easier. Place the transaxle in gear, hold the brake with your left foot and depress the accelerator slowly with your right. Don't overdo it; just load the engine so the car tries to creep. If the main bearings are going to knock, they'll do so right away. There is no need to keep the engine and transaxle

DISABLING THE IGNITION

Many diagnostic tests call for the ignition system to be disabled. With conventional ignition systems, there are several ways to short-circuit the electrical supply to the plugs; some are better than others. The best way is to remove the high-tension lead from the center of the distributor cap and ground it.

The high-tension lead is the large, heavily insulated wire running from the coil to the center of the distributor cap. To remove it, grasp the boot around the distributor-cap terminal, twist the lead slightly, and pull out the lead. The twist helps break any corrosion that resists wire removal. Now, ground the lead to the head or engine block.

On engines with *electronic ignition,* there should be *no more than* a 1/4-in. gap between the free end of the lead and ground, or *ignition-system damage may occur.* This gives the high-voltage electricity somewhere to go, instead of continuing to build voltage and trying to arc to ground inside the coil. This can destroy an expensive electronic-ignition module. Once the test is done, simply reinsert the lead into the distributor cap.

Mixed blessings: Ripped fresh-air supply hose can cause engine to breathe hot, underhood air, leading to detonation. On the other hand, it can also keep large amounts of water from bending connecting rods. Driving through deep puddles throws water up into the fenderwell/grille area where it can be drawn into engine through this hose. If enough water enters engine, a cylinder can be *hydrostatically locked*—water doesn't compress. The result is a bent connecting rod or two. Best solution is to replace torn hoses and avoid deep puddles at speed.

Manifold vacuum is an excellent indicator of overall engine condition. If you know how to use a vacuum gage, you can detect major problems within seconds.

straining—this is very tough on the torque converter. A few seconds and the test is over.

DIAGNOSTIC TESTS

Cranking Vacuum—The next diagnostic step is an engine-cranking vacuum test. This checks the pumping ability of the engine. By measuring the vacuum an engine produces while cranking, you are really testing how well-sealed it is. If all internal parts are in good shape, the engine will produce a lot of vacuum—if not, vacuum will be proportionately lower.

Begin by warming the engine to operating temperature—thermostat open. You can tell when the thermostat opens by watching the coolant-temperature gage. As the gage needle climbs with increasing coolant temperature, you'll notice it goes above normal operating temperature, then suddenly drops. The sudden drop occurs when the thermostat opens.

Another method is to hold your hand on the radiator's top tank or upper radiator hose. The top tank and hose remain cold until the thermostat opens, then warm considerably and quickly to the point where it's too hot to touch. Or, simply remove the radiator cap and watch for coolant flow. When this occurs, the thermostat has opened.

Once the engine is warm, shut it off and connect your vacuum gage to a *full manifold-vacuum* source. Any vacuum nipple on the intake manifold will do—best is the hose that supplies the brake booster. Just make sure the vacuum source you choose is *not ported vacuum*—from one of the small-diameter nipples on the carburetor. Ported vacuum exists in the carburetor primary venturi, just *above* the throttle plates. It creates a vacuum signal used for operating various emission-control switches. But, it's a vacuum signal that reads low on part-throttle applications—opposite of the high manifold-vacuum readings under the same conditions. Disable the ignition by disconnecting the primary coil wire—the small wire that's indicated by a *minus symbol* (−) on the coil.

Prop up the vacuum gage so you can see it through the windshield while you crank the engine or have a friend crank the engine. A good engine will pull a *steady vacuum* of about 10 inches of mercury (in.Hg). A worn engine with no major problems will have a steady, but lower reading. Don't be alarmed if the needle swings about 2 in.Hg—it's normal, especially on a four-cylinder engine. But if the needle drops to near zero regularly, there is a problem. Such a vacuum drop can have numerous causes: poorly adjusted valves, burned valves, worn cam lobes, blown head gasket or worn cylinders, pistons or rings. To pinpoint which cylinder is at fault, you'll have to perform more tests.

But first, a note about altitude and vacuum readings. Because atmospheric pressure *drops* as altitude *increases,* cranking vacuum will drop about 1 in.Hg for each 1000-ft increase in altitude. So, at 5000 ft—such as in Denver, Colorado—cranking vacuum should be 10 in.Hg−5 in.Hg=5 in.Hg. All other figures given previously are for sea level.

Power-Balance Test—A power-balance test shows how much each cylinder contributes to the power output of an engine. Thus, it also isolates which cylinders contribute little to manifold vacuum.

To perform a power-balance test, disable each cylinder as described earlier, page 11. The idea is to pull the lead away from the plug and ground it

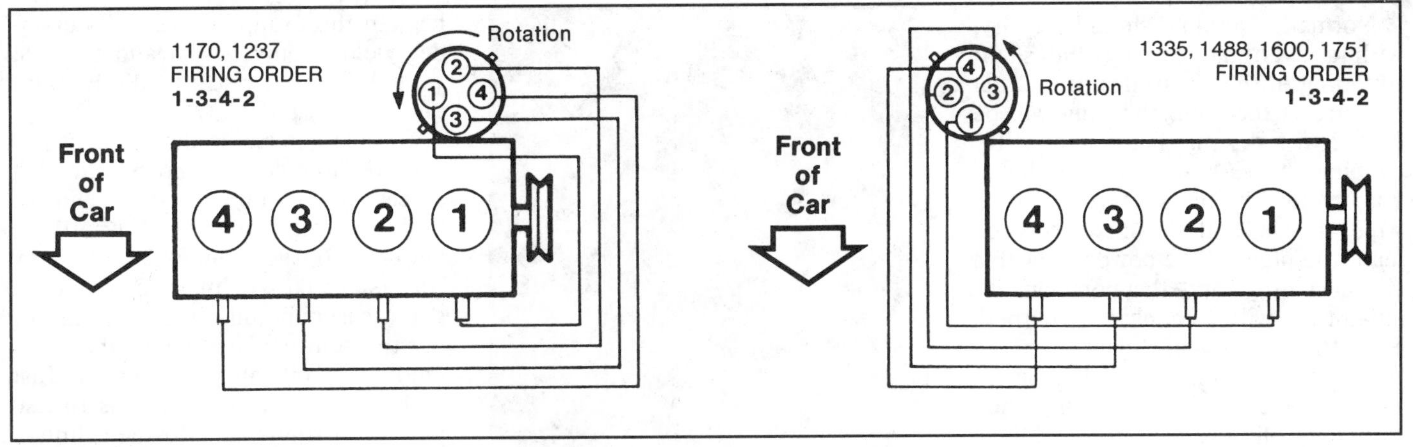

Firing order/ignition-wiring

against the block or head, stopping that plug from firing. The engine will then be running on three cylinders. By comparing rpm drop for each killed cylinder, you can determine which cylinder is at fault. Be sure to *ground* the plug lead, especially with 1980-and-later cars with *electronic ignition*. Otherwise, high voltage will arc to ground, possibly damaging the ignition system.

If you suspect a burned valve or other major problem, a quick, *ear-calibrated* power-balance test will probably tell what you want to know—which cylinder is it? Because the Honda has but four cylinders, a bad one shows right away. However, if you are looking for a more subtle problem, use a dwell/tachometer to measure rpm drop for each cylinder. The instrument-panel tachometer is not accurate enough for this test.

With the tachometer connected, disable the first cylinder and wait for engine rpm to stabilize. Now, write down the reading and reconnect the plug lead. Go to the next plug and do the same until you've done all four.

It doesn't matter so much how far rpm drops as how close the readings are to each other. Don't expect the readings to be any closer than 20 rpm. But when these readings start varying by more than 40 or 50 rpm, take notice. Remember, the cylinders with the *least* drop are the bad ones. Therefore, a really bad cylinder may not drop at all. Of course, if all cylinders are bad, none will drop very much. Good Honda cylinders usually cause a drop of about 200 rpm.

Reading Sparkplugs—If you think of a sparkplug as a removable portion of the combustion chamber, you can understand its diagnostic potential. Because the next test is the compression test, which requires sparkplug removal, let's discuss sparkplug *reading* now.

Start by removing the sparkplugs and laying them out in order. With all the electrodes facing you, compare color and texture of the four plugs.

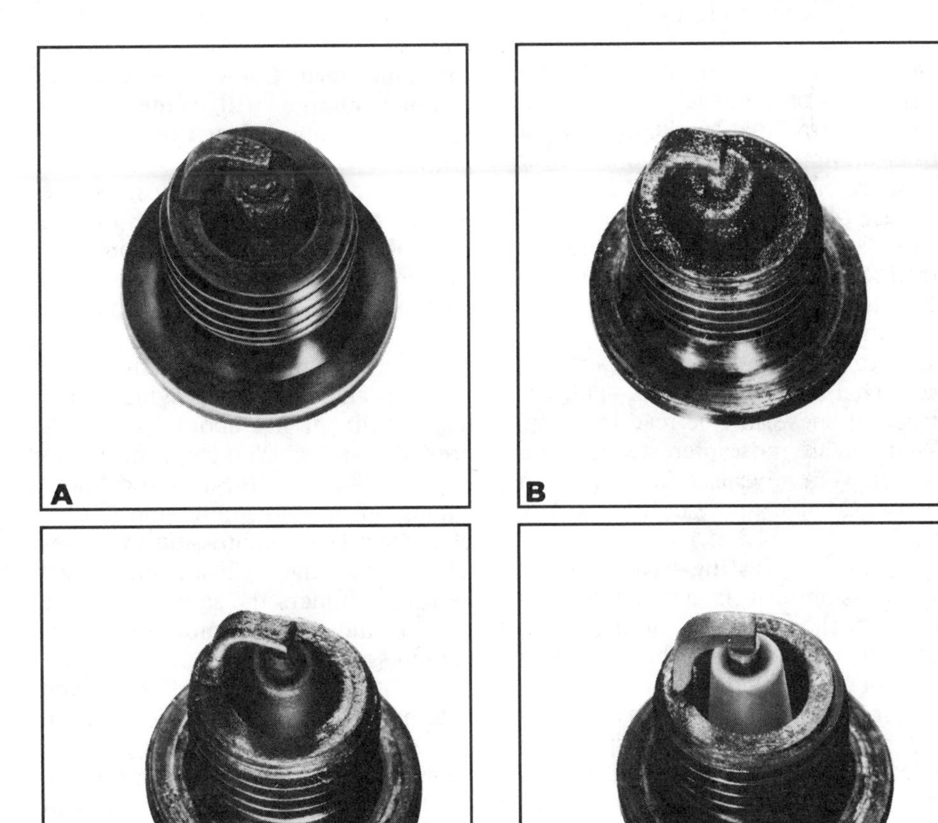

***Reading* sparkplugs can yield important troubleshooting clues. Plug A suffers from heavily rounded electrodes and pitted insulator—it's worn out. Replace such plugs and engine performance will improve. Plug B is oil-fouled. Shiny-black coating indicates excess oil consumption, possibly from worn rings and valve guides. Plug C is carbon-fouled—don't confuse it with oil-fouling. A carbon-fouled plug's dry, flat-black coating comes from excessively rich air/fuel mixture, stop-and-go driving or a too-*cold* plug heat range. Note: CVCC engines can have darker-than-normal plugs due to the rich prechamber mixture. Plug D is normal. Electrode rounding is moderate and insulator is even tan or gray, indicating all is well in the combustion chamber. Photos courtesy of Champion Spark Plug Company.**

Normally, a plug should be dry, with a even tan coating and slight rounding of the electrodes. If the fuel mixture is too rich, the plug will be coated with dry, flat-black carbon. Try rubbing the carbon into the palm of your hand. The black deposits should wipe off easily. If the mixture is too lean, the plug will be powdered with a white coating, and the porcelain insulator will appear burned. Sometimes, the porcelain turns a pastel green or yellow, which indicates *normal* operating condition when leaded gasoline is used in the Honda. Oil in the combustion chamber will leave the plug wet and shiny black. Try rubbing that into your palm and you get an oily mess that won't rub off easily.

When reading plugs, pay more attention to the porcelain insulator around the center electrode than the metal shell. It is most sensitive to *coloring,* thus is more likely to show symptoms of unusual combustion. Also be aware that sparkplugs from a street-driven engine can only show the more basic combustion conditions because of the many operating conditions a street-driven engine is subjected to. You may have heard about the ace mechanic who read the plugs, then made a one-eighth-turn adjustment to the carburetor and won the race. That's on race engines—plugs in street engines can't be read that way. What you are most interested in is the oily plug—it reveals a problem with the rings, valve guides or the auxiliary valve.

Compression Testing—The familiar compression test is a good way to measure the condition of the rings, cylinders and valves. There are two types of compression testers: a tapered rubber-cone type that is inserted into and held against the open sparkplug hole, and the screw-in type. If you use the rubber-cone type, you'll need either a remote starter switch or a helper to crank the engine.

The engine must be warmed up, ignition disabled and all sparkplugs removed. Watch out for hot parts whenever working on a warm engine, especially the 1200 exhaust manifold. The throttle and choke plate must be fully open for an accurate test—part-throttle openings result in *low* readings. So, if you are using a remote starter switch, prop the throttle linkage open with a screwdriver.

Popular diagnostic check is compression test. Screw-in tester is easiest to use, although rubber-cone type such as one shown is just as accurate.

However, if a friend helps, have him fully depress the accelerator as he cranks the engine. **Note: Avoid smoking and open flames because the air/fuel charge will come out of sparkplug holes when the engine is cranked.**

Screw-in compression testers are easier to use. Just screw it in and crank the engine yourself. If you use a rubber-cone tester, you'll be busy pushing the cone tightly against the sparkplug hole.

With either method, hold the throttle open and crank the engine so the tested cylinder has about 6—8 compression strokes. You can hear cranking speed slow as the tested cylinder comes up on its compression stroke. Note how fast compression increases and jot down the highest reading. Test all four cylinders the same way. Give each cylinder the same number of compression strokes.

Like the power-balance test, even readings are desirable. Depending on the engine, pressure can range from 75 psi to over 200 psi. So, don't be too concerned if the figures seem generally low. Trouble is much more likely if only one or two cylinders are low.

A good rule of thumb is the 75% rule: All cylinders must read within 75% of the highest cylinder. So, if the highest reading was 125 psi, simply multiply 125 by 0.75 to get 94. Therefore, if all cylinders read above 94 psi, they are acceptable. Below that, consider them faulty. Notice I said acceptable, not desirable. It is hard to set a wear limit and say anything above is good and all below are bad. In the example given, if a cylinder yielded only 97 psi and the rest were 120 or 125 psi, I would be wary of the low cylinder.

To help determine the cause of low compression, do a *wet test* by squirting a teaspoon of oil into that cylinder. SAE 30W oil is good; 40W or 50W is better. To determine how many squirts it takes to make a teaspoon, get a teaspoon and fill it while counting the squirts. Then squirt the same amount of oil into the cylinder. Just make sure your squirt can is full so you don't squirt air into the cylinder. Aim for the far side of the cylinder—it helps if the piston is down the bore somewhat.

Hand-crank the engine two revolutions or so to help spread the oil. Retest the low cylinder. If compression comes up markedly, 40 psi or more, the trouble is poor ring-to-bore sealing. A rebuild is needed to restore the lost clearances. If compression doesn't increase much—about 5 psi—then the problem is probably with the valves. It could also be a blown head gasket.

You may notice a cylinder that takes a long time to pump up. Usually, a cylinder will produce 40 psi on the first stroke, another 35 psi on the next and so on. Problem cylinders may have trouble getting to 40 psi and, instead, increase by 10 psi at a time. If you crank them enough, they'll come close to the other cylinders. Wet test such a cylinder, because this condition is usually caused by poor rings. On the other hand, a cylinder suffering from excessive oiling—from bad rings even—can yield high compression-test readings because excess oil in the cylinder seals the rings. Again, if you crank this type of cylinder enough, relatively high readings can result.

Finally, if you suspect a blown head gasket, compare the readings of adjacent cylinders. A pair of low readings on adjacent cylinders can mean the gasket between them is gone.

There are variables that affect compression testing. One is cranking speed; higher speed gives higher pressure readings and vice versa. With a small, four-cylinder engine, it isn't likely that the battery will run down during a compression test. But if it does, jump the battery to another one to maintain cranking speed.

High altitudes will lower compression readings similar to manifold-vacuum readings. A worn camshaft or jumped timing belt will do the same thing.

High-performance camshafts, with their *long-duration* profiles, will also give lower compression readings. This is because such cams sacrifice low-rpm breathing for improved high-rpm breathing. Compression testing takes place at cranking speed—well below idle speed!

Leak-Down Testing—Although it's also a measure of combustion-chamber sealing, a leak-down test is more accurate than compression testing. Accuracy is improved because variables affecting compression-test readings—those that have no bearing on the sealing capability of an engine—are eliminated. A leak-down tester uses an external air-pressure source. Testing is done with the engine stationary. Therefore, the test is not influenced by cranking speed, valve duration or altitude.

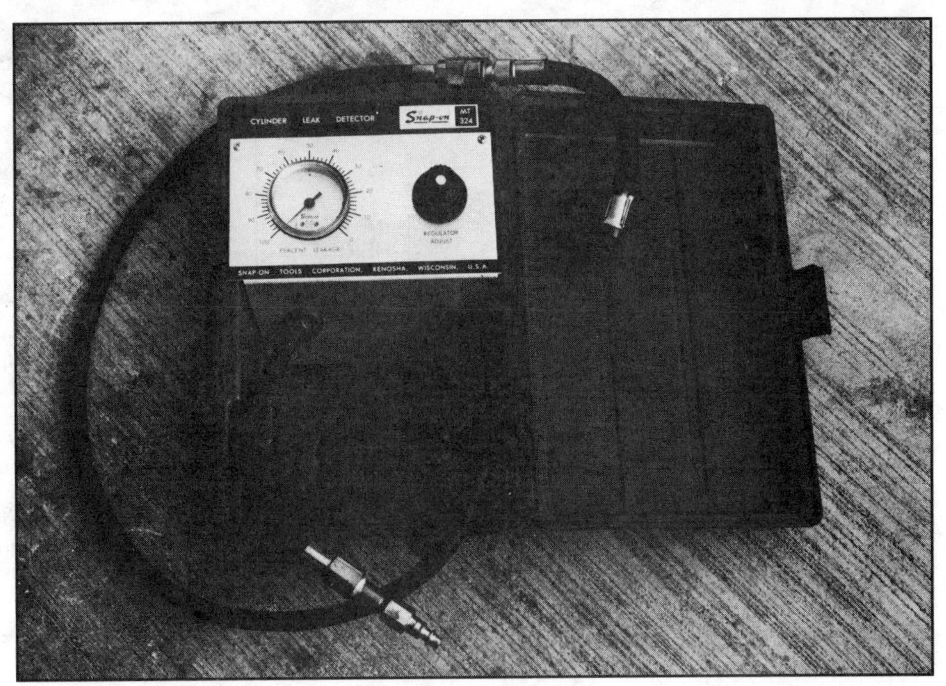

Leak-down testing can indicate more about an engine than any other single test, but cost of tester has kept it out of the do-it-yourselfer's toolbox.

Leak-down test equipment is expensive. So, unless you do a lot of engine diagnosis, this is one test to farm out. Many tuneup shops can do the test for you. The cost should be minimal.

If you insist on doing a leak-down test yourself, you'll need an air compressor—a 1/2-HP model will do—and the leak-down tester. Start by reading the instructions that came with the tester. Bring the number-1 cylinder to top dead center (TDC) of its *compression stroke.* Check the timing mark to make sure it's exactly on TDC. If it's slightly off, the engine will kick over without warning the instant the cylinder is pressurized. A good way to check for TDC is to insert a long, thin screwdriver into the combustion chamber through the sparkplug hole. With the screwdriver contacting the piston top, you can feel when the piston is at the top of its stroke.

Next, install the hose adapter in the sparkplug hole, then connect the tester to the adapter and the air compressor. Compressed air is admitted to the cylinder while the tester monitors how much air it takes to make up for cylinder leakage. The readout is in percent leakage. Remember, the cylinder *must* be at TDC of its compression stroke so both valves are closed. Otherwise, leakage will approach 100% as all the compressed air blows by an open valve, or the engine will turn over. Turn the crank 180° and check number-3 cylinder. If you have a CVCC, put a reference mark on the damper to ensure that you turn it exactly 180°. Rotate the crank back to the mark to check number 4.

Typical leakage for an engine in good shape is 10% or less. A 20%-leakage rate indicates a problem. The worse the problem is, the higher the leakage. A 90%-leakage rate indicates serious damage, such as a *holed* piston and the like. Commonly, 30% leakage is serious enough for an engine overhaul or valve job.

You can usually tell what's leaking by listening to the engine with the tester attached. If the exhaust valve is leaking, you can hear the hiss of escaping air in the tailpipe. Leakage past intake valves can be heard at the carburetor. Bad-sealing rings and cylinders can be detected at the oil-breather or dipstick holes. Use a 1-ft or so length of rubber hose with one end in your ear and the other in the carburetor, exhaust or oil-filler holes to make listening easier. Sometimes, the leakage is evenly divided and is hard to attribute to a particular valve or rings—a sure sign of multiple problems.

One way to spot a leaky exhaust valve is to hook an HC/CO meter to the tailpipe and squirt some carburetor cleaner into the cylinder. Reconnect the leak-down tester and watch the meter. It takes a minute, but if the exhaust valve is leaking, the HC portion of the meter will peg! If you go to a tuneup shop for the leak-down test, they should have an HC/CO meter available. This test doesn't require a leak-down tester, just an adapter for the sparkplug hole and a compressed-air source.

Before the leak-down test, fill the radiator to the top and leave off the cap. A rise in coolant level or stream of bubbles during the leak-down test is a sure sign of a blown head gasket. Many times, a good clue to head-gasket problems is the oily residue left around the radiator cap or overflow-tank cap. Of course, this could also be the sign of a bad cap, but high cooling-system pressures from a blown head gasket will pump coolant right by a good 12-lb cap. Try a leak-down test on a thoroughly warmed engine if you suspect the head gasket.

CHECKING VALVE LIFT

If the tests thus far point to badly worn engine internals, there is little cause to measure valve lift. You might as well get on with rebuilding the engine. On the other hand, if the engine merely seems to have one valve-related bad cylinder, check valve lift. This will give you a better idea of camshaft and valve condition.

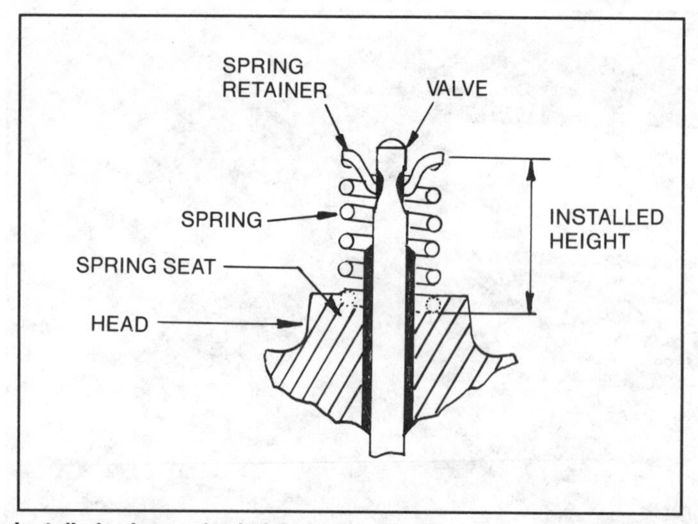

Installed valve-spring height is distance from spring seat to underside of retainer.

Best place to measure lobe lift is at spring retainer. A dial indicator is a must for this measurement; a depth gage or rule is not accurate enough.

As soon as the rocker cover is off, go straight to cylinder 4 and inspect the end cam lobe. If it is worn badly, don't bother with the valve-lift check—replace the cam. If the rest of the engine is OK, you can easily change the cam without removing the head. See Chapters 6 and 7 for details. A new cam will net renewed efficiency at minimal cost.

Are the Valves Closing?—First, you must be sure all valves are closing completely. After all, there is no way a cylinder can have any compression if a valve hangs open—even slightly. Either there's no valve clearance, causing the valve to be held open, or the valve is stuck in its guide.

To check for tight or nonexistent valve clearance, start with a 0.006-in. feeler gage and a cold engine. See page 130 for the valve-adjustment procedure for your engine. Remove the rocker cover and pass the gage between the valve-stem tips and rockers as you would during a valve adjustment—with the engine off and each cylinder at TDC. Only this time, there's no need to make precise adjustments. Just make sure all valves have *some* clearance. Pay particular attention to the valves at the cylinders that tested low during the compression or leak-down test. Look for a valve being held open by the cam and rocker arm. If you find a valve that won't accept the 0.006-in. feeler gage, try a thinner one, about 0.003 in. If that doesn't fit, try a smaller gage, if you have one.

Chances are, a valve run more than a short time with no clearance has burned. But if you caught the problem in time, adjust the tight valve to its normal clearance. See the chart on page 17. Warm the engine and give that cylinder another compression test. The reading should come up if the valve was at fault. If so, go ahead and drive the car a few thousand miles more, but keep a careful eye on valve adjustment. If a valve's lash has closed up once, it will probably do it again.

Installed Height—Suppose valve adjustment on the suspect valve was not tight, but grossly excessive. It could be stuck so that it fully opens but won't fully shut.

If this is the case, check spring *installed height.* This is the distance from the spring's bottom at the head to its top, at the spring retainer. Use a machinist rule or the depth-gage end of a vernier caliper to measure installed height. When you get your scale down around the bottom of the spring you'll notice the spring sits on a seat with a lip. You need to measure from the bottom of the spring, not the seat, so hold your scale one-third of the way up the spring-seat lip. This will allow for seat thickness. See the chart, page 102, for installed height specifications.

If installed height is close to specifications, say within 0.060 in., the valve is not stuck. But if it is noticeably short, perhaps 0.100 in., then something is keeping the valve from closing. A galled valve guide and stem, bent stem or carbon on the seat could be the problem.

Valve Lift—There's one last check for those of you with down-on-power engines, but good test results. Chances are your car has lots of miles on it, but it still gets good gas mileage, has good compression and doesn't use oil. Just the same, the engine is down on pulling power, and doesn't idle quite as smoothly as it used to.

If you've come this far in diagnosis with no problems, suspect a worn camshaft. For instance, if the power-balance test showed cylinder 4 to be questionable and the rings seem OK, you can bet cam wear is the culprit.

Honda's overhead-camshaft design makes camshaft inspection a snap—just remove the rocker cover. By bumping over the engine a little at a time with the starter, you can bring each lobe into view. Watch the way light is reflected off the cam lobes, especially the *ramp* area right before the *toe.* See the drawing on page 104 for an explanation of these and other cam-lobe terms. If a lobe is in good shape, light will reflect evenly off it with a mirror-like sheen. But if a lobe has worn, it will be striped with light and dark bands. The bands will run parallel to the shaft, and if you change your angle of view, the light will reflect differently off the multiple angles cut into the lobe. A normal lobe will have no abrupt changes in light reflection.

If this visual inspection is inconclusive, take the car to a mechanic, or measure cam-lobe lift with a dial indicator. Getting a dial indicator on Honda cam lobes is nearly impossible without removing the rocker arms. Therefore, the best way to measure Honda lobe lift is at the valve-spring retainer. Then, you'll be measuring *valve lift*—a direct product of lobe lift and *rocker-arm ratio.* Valve lift is the difference between its closed and open positions. Rocker-arm ratio is the mechanical leverage built into the rocker arm by its relationship to the shaft, cam lobe and valve tip. Rocker-arm ratio is 1.488:1. Therefore, lobe lift X 1.488 = valve lift.

Position the dial-indicator plunger square against the spring retainer with the valve completely open. The valve spring should be close to fully compressed, or its valve-open height. Zero the dial by rotating its face until the indicator needle reads 0. Then, rotate the crank pulley until the valve closes. The valve spring will be at its valve-closed, or *installed,* height. Read *valve lift* directly, then write down the finding. Because you are measuring on the "other" side of the rocker, valve clearance needs to be added to your valve-lift figure. So, measure valve clearance and add it to your dial-indicator reading.

If one or two measurements are considerably less than the rest, chances are the lobes are worn out—wiped.

Results—So, what if you do have a bad cam, but the rest of the engine is OK? Replace the camshaft, following the procedure given in Chapters 6 and 7. You'll get like-new performance from the many useful miles left in your engine. But there is no economy in running a knocking oil burner with low compression—it's time to rebuild.

VALVE CLEARANCE
in. (mm)

Year	Displacement	Intake	Exhaust	Auxiliary
73-79	1170, 1237	0.006 (0.15)	0.006 (0.15)	N/A
75-79	1488	0.006 (0.15)	0.006 (0.15)	0.006 (0.15)
76-78	1600	0.006 (0.15)	0.006 (0.15)	0.006 (0.15)
80-83	1335, 1488	0.006 (0.15)	0.008 (0.20)	0.006 (0.15)
79-83	1751	0.006 (0.15)	0.011 (0.28)	0.006 (0.15)

CVCC AND ENGINE REBUILDING

By far, the most-novel Honda engine feature is the *Compound Vortex Controlled Combustion, or* CVCC, cylinder head, fitted to 1335, 1488, 1600 and 1751cc engines. The CVCC system lowers exhaust emissions through controlled combustion. The 1170 and 1237cc engines do not have the CVCC head, but rather the more-conventional crossflow head.

Actually, Compound Vortex Controlled Combustion is somewhat of a misnomer. Vortex implies swirl or turbulence, which is currently used by many auto manufacturers to cause more complete, fast-burning combustion. The CVCC system does not aim for turbulence, however, but for stratification.

The CVCC system begins at the carburetor. CVCC carbs have three barrels: auxiliary, primary and secondary. The primary and secondary barrels operate normally, but yield a very lean mixture—up to 25:1 air/fuel ratio. The auxiliary barrel operates at all times, and dispenses a very rich mixture—about 11:1 air/fuel ratio.

This rich mixture is carried to the cylinder head by small, separate passages in the intake manifold. In the cylinder head, small, separate ports carry the mixture to the *auxiliary intake valve.* The auxiliary valve is very small, and does not open directly into the main combustion chamber. Instead, it opens into a *prechamber.* The prechamber connects with the main combustion chamber through a hole or holes. A standard sparkplug extends into the prechamber.

During the intake stroke, a very rich mixture is drawn into the prechamber. Simultaneously, a very lean mixture is drawn into the main combustion chamber via the conventional intake valve. A certain amount of blending takes place in the main combustion chamber where the prechamber empties into it. Therefore, three different mixtures exist; a rich prechamber mixture, a moderate mixture near the prechamber in the main chamber, and a lean mixture away from the prechamber in the main chamber. This is what engineers call *charge stratification.*

The sparkplug fires the rich mixture in the prechamber, establishing a *flame front.* The flame front burns the rich mixture, moves on to the moderate mixture and finally consumes the lean mixture farthest from the sparkplug. Such a lean mixture could not be ignited directly by the sparkplug. By stratifying or layering the fuel mixture, average combustion temperature is kept high, while peak combustion temperature is low. These seemingly contrary temperatures result in low hydrocarbon (HC), carbon-monoxide (CO) and oxides-of-nitrogen (NO_x) emissions.

The CVCC system adds little to the rebuilding process. On engines so equipped, the auxiliary valve and prechamber are the CVCC parts requiring attention. Most of the time disassembly, cleaning, lapping, O-ring replacement and reassembly are all that's required. And most jobs can be performed with standard hand tools and machine-shop equipment.

The only exception is the CVCC-engine three-barrel carburetor. Rebuilding this carburetor is beyond the scope of this book—and most mechanics. Even pros have trouble with this intricate piece and sometimes replace it rather than rebuild it. If you have carb problems, take them to a professional tuneup shop or Honda dealer.

Engine Removal 2

Steam cleaning or solvent blasting does great job of removing heavy, baked-on crud. Look in phone book for gas station that performs this service.

Engine removal and installation are important steps in any overhaul. Haphazardly *yanking* an engine guarantees headaches during installation. It's also dangerous. A little extra preparation and caution before and during engine removal will be handsomely rewarded when you install your rebuilt engine.

PREPARATION

Some special tools are required for engine removal, such as a floor jack, jack stands, and an engine hoist. Additionally, you should clean the engine before working on it and decide where to pull it. Containers for hardware must also be readied.

Engine Cleaning—A dirty engine is miserable to work on; wrenches slip, fasteners hide under the goo and grime gets under your fingernails. Avoid these problems by cleaning the engine before you start removing it. Three cleaning methods are generally available: steam cleaning, solvent blasting and spray degreasing.

Steam cleaning is for truly filthy cars—such as an oil leaker driven on dirt roads. It costs around $20 and takes about a half hour. Usually, a service station has the equipment. If not, check with a tractor- or heavy-equipment shop.

Solvent blasting uses compressed air and solvent to literally blow off the dirt. It works well 90% of the time. Cost is comparable to steam cleaning and practically any shop can do it. A thorough solvent blasting takes about as long as steam cleaning.

Spray degreaser can be used at home if you have a garden hose. Typically, you warm the engine, spray it with degreaser, let it sit an hour or so while the degreaser penetrates, then hose off the crud. Problem is, the stinky mess ends up on your driveway. Eliminate this problem by doing the job at a car wash. You can use high-pressure water/detergent spray and leave the mess for them to contend with. When cleaning, remember to cover the distributor, coil and carburetor with plastic bags. These parts must be kept dry or you'll have a hard time restarting the engine. With patience, this method works as well as steam cleaning or solvent blasting.

Lifting Tools—To raise and support the car during engine removal and installation, you'll need a *floor jack* and *jack stands*. The floor jack is a hydraulic jack in a wheeled frame. For your Honda, a 1-ton version is adequate, but a 1-1/2-ton jack is usually sturdier and will lift higher. Besides, if you're planning to buy a floor jack, you'll need a 1-1/2-ton version for lifting most other cars and light trucks.

Once the car is up, *you must support it with jack stands. Never* use any jack—bumper, scissors, screw or otherwise—as a stand. Jacks are for raising and lowering, not for supporting a car while you are underneath. *Jacks fail,* and if you are under the car when it does, it could be fatal. You'll need two jack stands to hold up the front of the car.

Both the floor jack and jack stands can be rented. See the telephone yellow pages under Rentals.

To lift the engine out of the chassis, you need a *cherry picker* or *chain hoist.* I prefer a cherry picker. It's basically a lifting arm on casters powered by a hydraulic jack. You merely roll it into position over the engine, hook onto and raise the engine. Once clear of the bodywork, roll the cherry picker and engine over to your workbench or engine stand. There are no long, oily, noisy chains to get tangled in engine components.

A chain hoist requires an attachment of some sort to support it and the engine. Tree limbs or garage-roof beams are commonly used—only marginally safe for lifting most engines. With either attachment, the car has to be carefully positioned under the hoist. If an adjustment is needed, you must move the car, not the hoist. But, if you do decide to use a chain hoist, the best way is to hang it from a heavy-duty metal A-frame. The A-frame provides good strength and maximum working room. You'll find a cherry picker or a chain hoist with A-frame at most equipment-rental outlets.

Get Organized—There's nothing worse than trying to install an engine someone else removed. Who knows where all those nuts and bolts go? Well, pulling an engine and throwing all the hardware in one big box amounts to the same thing. When you finally get around to installing the engine, you'll find your memory just isn't up to the task.

Save yourself considerable trouble and frustration by getting several coffee cans or boxes and labeling them. Have one for bellhousing bolts, another for engine-mount hardware and so on. Have the containers ready *before* you pull the engine. Also get a roll of wide masking tape and a permanent, *waterproof* marking pen. These are for the many vacuum-hose and electrical connections you'll disconnect.

This sounds silly, but think twice about where you plan to pull the engine. Once it's out, moving the car involves pushing or towing! The chassis will be immobile for a while,

HONDA vs INDEPENDENT

If you've been reading the Honda shop manual, you'll notice that the engine-removal instructions I give are different from Honda's.

Honda says to remove the engine and transaxle as a unit. This requires removing all engine accessories *and* transaxle connections. Drive shafts, shift linkage, tie-rod ends and steering knuckles or lower ball joints must be disconnected if you use the Honda method. Once the engine and transaxle are on the floor, they are separated and the transaxle is set aside.

In comparison, I recommend removing the *head and manifolds* as a unit, then the *block.* The transaxle is left in the chassis, with drive shafts, shift linkage, steering and suspension intact. The advantages are numerous. First, there are fewer things to disconnect, so the job goes faster. Second, lifting only the block assembly allows the use of lighter-duty equipment. Third, disconnecting the tie-rod ends usually results in two destroyed tie rods, so leaving them alone eliminates the need to replace them—a money- and knuckle-saver. Fourth, removing the head bolts is easier when the engine is still firmly bolted to the chassis. It gives something solid to pull against. Fifth, you don't have to store the transaxle because it remains in the car. Sixth, you don't have to disconnect the drive shafts—exposing the drive-shaft splines and differential to dirt and the elements. Finally, you don't have to worry if the chassis needs to be moved after engine removal—a process that would necessitate reassembling the steering and suspension. The engine-only removal requires only that the transaxle be supported with wood or wire.

The Honda system does have some advantages with manual-transaxle cars. By separating the engine and transaxle on the floor, there is a slightly better chance that the transaxle-input-shaft seal will not be damaged and the input shaft will slide more easily into the clutch-plate splines at installation time. Because replacing this seal requires transaxle disassembly, it's not an inconsequential consideration. Also, CV joints can be damaged by extreme drive-shaft angles—the kind encountered when the engine is about 1 ft off its mounts and the transaxle still hasn't separated. If the transaxle is getting overhauled along with the engine, the Honda system makes more sense. Otherwise, the benefits of the Honda system are minimal.

Therefore, I recommend leaving the transaxle in the car—unless the trans is being rebuilt also.

Still, another way to handle engine removal and installation is to pull the engine as a complete unit—with the head on the block, but less the transaxle. Advantages include: less chance of dirt and water contamination inside the engine because the cylinders aren't opened, less stoop-over work in the engine compartment and the engine can be completely assembled before installing it. Disadvantages? On some Hondas, the brake booster and master cylinder are directly above the engine. Trying to steer a fully dressed engine past the master cylinder requires the touch of Houdini. It can be done, but requires angling and tilting the engine at just the right time! Removing the head first eliminates the interference condition.

If you want to remove the engine with its head intact, you'll need to jump around this book somewhat. On teardown, start with the head and finish with the short block. That means starting in the middle of Chapter 4, going to the end of Chapter 4, then finishing with the front of Chapter 4. To assemble the engine, assemble the short block, Chapter 7, then the head, Chapters 6 and 7. You'll also be referring to Chapter 8 for head gasket installation and torque sequences. It sounds complex, but actually is a logical, step-by-step progression.

On some models, such as this Accord, hood hinges can be unbolted through grille. On other models, headlight trim rings and grille must be removed first.

Do as I say, not as I do! Steve was eager to get started, so he lifted off the hood by himself. To avoid scratching hood and fenders, use two people for hood removal.

Run the hood-hinge bolts with any shims back into their holes right after hood removal. This keeps shims in order for a proper fit later.

depending on how fast you work. Three weeks is about average. "Dead" cars attract vandals, angry landlords, even the authorities in some cities. If you don't have enough working room, try renting space at a service station. Look in the phone book for a do-it-yourself auto shop or hobby shop. Military bases often have auto-hobby shops, complete with many of the larger tools such as floor jacks and cherry pickers.

In any case, try to work on a paved surface, such as concrete or asphalt. Both are safer for jack stands than dirt or grass. You'll also stay a lot cleaner. If you must work on dirt, use plywood under the jack stands to distribute the load. Otherwise, they may sink into the ground.

ENGINE REMOVAL

Hood—Because a Honda hood opens to an almost vertical position, it doesn't necessarily have to be removed to pull the engine. Professional Honda mechanics usually leave the hood on and come in from the left fender with a cherry picker. This saves time, but increases the chance of hood damage and, with the hood off, you'll have a little extra elbowroom. The choice is yours. With an Accord, the hood is easy to remove, so why not? On Preludes and Civics, the grille and headlight trim rings must be removed to gain access to the hood-hinge bolts.

To remove the hood, unbolt the hinge bolts using a socket and extension, with the hood *closed.* This helps keep the hood from crashing down on the car's paintwork, especially when you're removing it alone. Then, release the hood latch, remove the bolts and lift off the hood. There may or may not be adjusting shims under the hinges. If so, tape each shim to its hinge so you'll know where they go. Otherwise, the hood will not fit correctly when installed.

On Civics, the windshield-washer bag should be removed, complete with hood, washer nozzles and hoses.

The best way to store the hood is back on the car once the engine is out. Or, you can tie it to a wall stud and drape an old mattress pad over it. Don't lay the hood flat where it might get stepped on, and don't stand it on end—a good way to buckle the corners.

Fender Covers—You may wish to use fender covers. However, Honda fenders are so narrow that it's hard to keep conventional fender covers from slipping off—unless you have a set of magnetic covers available from Mack Tools and other tool suppliers. Otherwise, you'll fill half the engine compartment with the fender cover so it won't fall off, covering most of what you are trying to work on. If you don't have magnetic fender covers, use racer's—duct—tape, doubled back on itself to hold each fender cover to the inner fender. If you go without fender covers, remove your belt and keep tools and keys out of your pockets. Otherwise, you'll scratch the paint.

Battery—At the battery, disconnect the negative, then positive battery cables. You'll need a 10mm wrench. If a cable is stuck on a post, don't yank or pry it off. You may break a battery post. Invest in an inexpensive battery-cable puller and use it.

Leaving the battery in the car won't get in the way of pulling the engine, but you'll have to keep tools from arcing across the posts. If you remove the battery, wash it with water and baking soda. Then store it off the floor on a pair of 2x4s. Concrete floors magically discharge batteries.

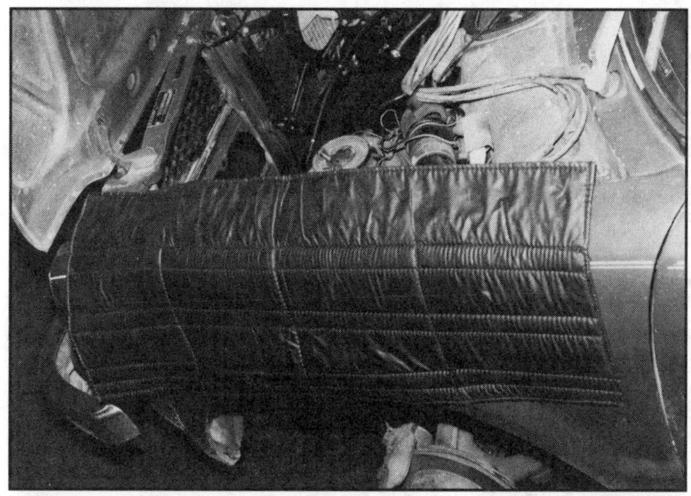

Fender covers protect paint from scratches. Magnetic covers that don't fall off were developed by Terry Haney of Haney Brothers Honda, Tucson, AZ, and are now sold by Mack Tools. Photo by Ron Sessions.

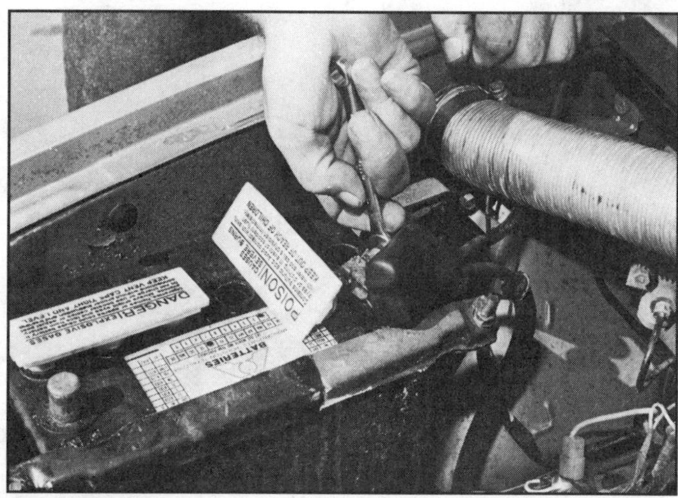

Disconnect battery cables, negative first. Then, remove battery, clean it with baking-soda solution and store it out of the way in a cool corner off of a concrete floor.

Drain Oil—Slide a drain pan under the engine and remove the pan plug. It takes a 17mm wrench. You can let the oil drain while you move on to other jobs, but replace the plug when the oil is out. The longer you let it drain, the less oily mess you'll have at teardown.

Drain Coolant—If the coolant is clean and less than a year old, save it by draining it into a clean container. Honda made the draining job easy by providing a nipple on the radiator drain cock to attach a hose. If you don't use a hose, the coolant drains through a hole in the sheet metal. Draining goes faster if you remove the radiator cap and open the bleeder nipple at the distributor housing. Old antifreeze jugs or plastic milk cartons are great for storing coolant. Be sure to store the coolant out of children's reach—ethylene glycol is poisonous.

On late-model Hondas, there is also a coolant drain plug for the block. It's above and slightly forward of the oil filter. Remove it to drain coolant trapped in the block. Do not remove the similar-looking hex plug under number-4 sparkplug.

Oil Cooler—Check the left side of the radiator for an engine-oil cooler. All 1979—'81 Preludes and Accords are factory-fitted with a cooler. Beginning in 1982, only automatic-transaxle 1751cc models use an engine-oil cooler. For now, unclamp and drain the two oil lines. Lay them back over the engine. Mark the lines; inlet is at the top, outlet at the bottom.

Air Cleaner—Unclamp the fresh-air duct at the fender and air cleaner and

If coolant is new, drain into clean pan and store in old antifreeze container. Coolant drains faster with radiator cap removed.

Unclamp and remove fresh-air and hot-air ducts. This fresh-air duct didn't have to be removed. It cracked and fell off earlier.

remove the duct. This will get it out of your way. Under the air-cleaner snout you'll find another large-diameter hose. This one routes heated air from the exhaust manifold. Unclamp it from the air cleaner, but leave it attached to the exhaust. Atop the air-cleaner snout is a small vacuum motor. Remove the vacuum hose from the motor nipple. Use masking tape and *permanent* marking pen to *flag* the hose.

With a 10mm wrench, remove two bolts attaching the air-cleaner bracket to the rocker cover. Remove the wing nut atop the air cleaner and raise the housing, but don't remove it. There are more hoses underneath. On 1980-and-later models, remove the four bolts and nuts inside the air cleaner to release the housing from the carburetor and vacuum-hose bundle. Mark

Begin air-cleaner removal by unbolting any brackets. This bracket requires removal of bottom two bolts only.

and disconnect all hoses, then remove the air cleaner.

Hoses—Because there are so many engine variations, all with different

Check around air cleaner, removing any clipped-on wires or parts. Remove wing nut.

Wires and hoses *must* be marked when disconnected. Even on '80-and-later models with underhood hose-routing diagrams, marking hoses eases installation.

Cruise-control mechanism disconnects at three points. Mark and remove vacuum hose, then unscrew nut.

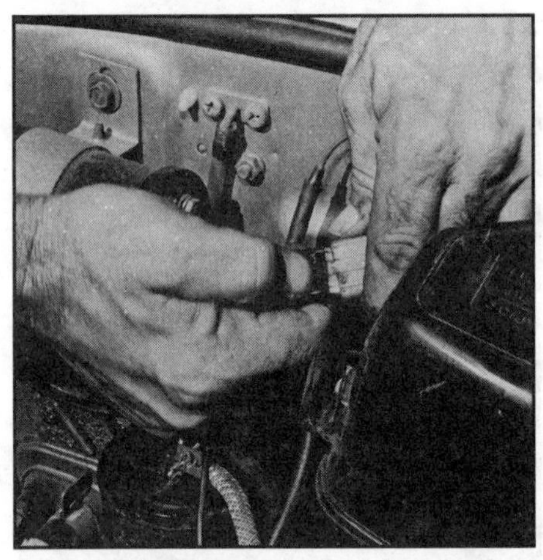

Label and disconnect electrical lead at emission-control box.

At other end, remove cotter key and washer. Slip off arm and remove cruise-control diaphragm.

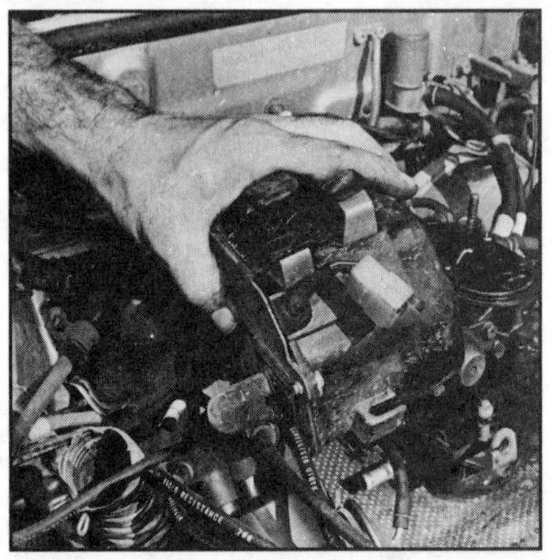

Slide box out of mount and prop next to carburetor. Some engines have two or three "black boxes."

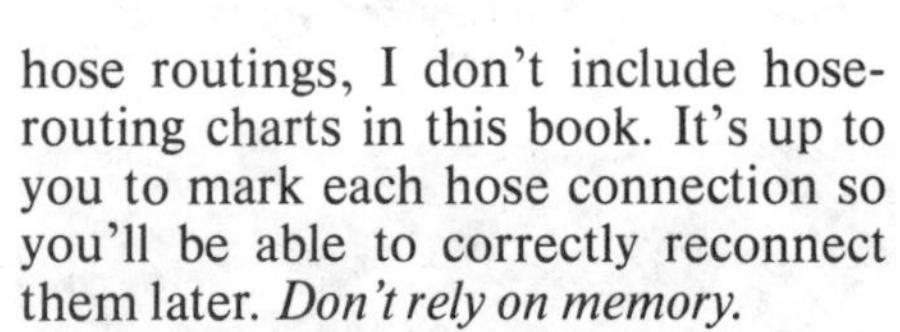

hose routings, I don't include hose-routing charts in this book. It's up to you to mark each hose connection so you'll be able to correctly reconnect them later. *Don't rely on memory.*

An excellent way to mark hoses is with letters or numbers. Wrap one piece of tape around the hose, another on its connection. Mark both with the same letter or number—A or 1 in our present example. You don't have to know the function of each hose—just that the 12 hose goes on the 12 connection. Don't use pencil or ball-point pen unless you'll have the engine back in within a short time, and keep the car out of direct sunlight and the rain. If the marks fade or wash off, you'll be lost. Permanent marker works best.

Black Box—You've just uncovered a whole bunch of vacuum hoses. Again, *mark* and remove them as necessary, after you find the *emission-control box*—black box. Late-model Hondas may have two or three black boxes.

Each box sits high on the fire wall and has numerous vacuum hoses sprouting from it. It functions as a manifold for vacuum controls and solenoids, eliminating clutter from the engine compartment. You can easily pull the black box up and out of its mount, disconnect any electrical connectors and lay the box across the intake-manifold heat shield. In one step you've disconnected most of the vacuum hoses. The control box, hoses, carburetor and manifolding all come out with the head. You'll remove them later during engine disassembly.

More Hoses—Trace the power-brake-booster vacuum line from the booster to the intake manifold. Mark and disconnect the hose at the manifold. Loop the hose out of the way towards the brake booster. Depending on the model, there are more vacuum hoses going to valves, solenoids and canisters. Mark and disconnect them at the carburetor.

Some disconnections are not vacuum lines. On CVCC engines, the fuel line T's off before entering the carburetor. Unclamp and disconnect it at the T-fitting. On 1200s, disconnect the inlet and outlet fuel lines at the fuel pump.

On high-altitude and California models, mark and disconnect the three air-jet-controller hoses. Look just forward of the left shock tower of the inner fender for the air-jet controller.

Choke & Throttle Cables—Both cables use two jam nuts against a bracket to anchor the cable housing. You'll need a 12mm wrench for the throttle nuts and 10mm for the choke nuts. Completely unthread the nut toward the carburetor end of the

On CVCC engines, disconnect fuel line at T-fitting near carburetor. On 1200s, undo inlet line at fuel pump.

Throttle and choke cables are removed as follows: Loosen nut closest rubber boot, run nut off its threads, pull cable-housing end out of bracket and slip cable end out of actuating arm.

cable. Now pull back the cable housing until the cable will pass through the bracket slot. Partially rethread the nut back onto the cable housing. The cable end has a cylindrical end. Take some slack from the cable and rotate the cylindrical end in the slot cut in the choke or throttle-actuating arm. Feed the cable through the slot; the cylindrical end will pass out the open side of the arm. Repeat for the other cable. Lay the cables aside.

More Wires—On CVCC engines, mark and disconnect the four electrical connectors at the carburetor.

Flag and disconnect the distributor-ground and high-tension wires. On '80-and-later models with electronic ignition, disconnect the plastic distributor connector.

Rear Torque Rod—Next, undo the torque rod—a rod running from the fire wall to the rear of the cylinder head or rear engine-mount bracket. You can unbolt it from the engine and work around the rod, or remove the bolt at the fire wall, then remove the torque rod.

Exhaust—There are several types of exhaust connections, depending on the year and model. In all but 1200s, the exhaust manifold connects to the exhaust pipe at the back side of the engine. At this connection there is also a bracket from the block that braces the exhaust system. The exhaust manifold must be disconnected from the exhaust pipe and the bracket removed from the block. Leave the bracket attached to the exhaust manifold so it helps brace the manifold away from the block. This makes a

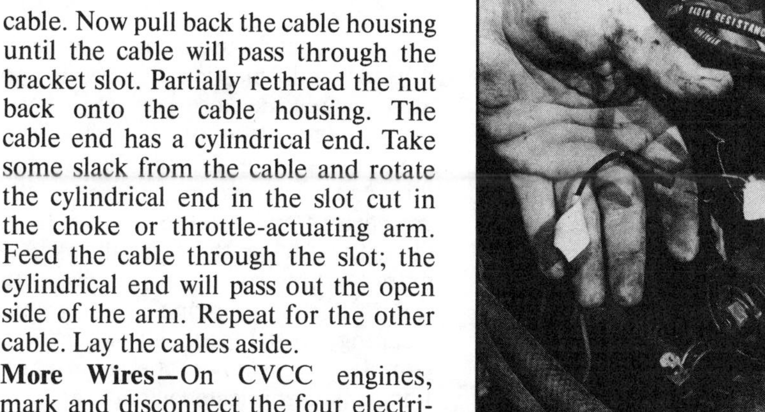

Remove coil negative and high-tension wires at distributor. With electronic ignition, disconnect plastic connector instead of coil negative lead.

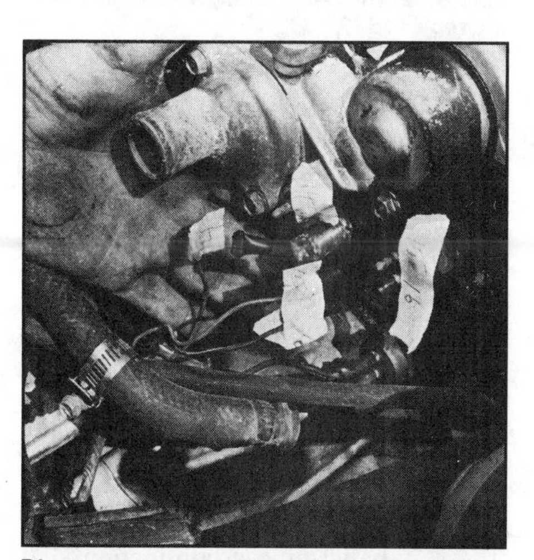

Disconnect coolant-temperature and thermosensor leads. On some CVCC engines, coolant-temperature sensor mounts vertically under thermostat. On 1200s, sensor is on intake manifold.

tripod out of the head so it won't tilt off the back of the block. The tripod is formed by the head bolts and exhaust bracket.

Some engines use three studs between manifold and pipe, others only two. The bracketry differs also; some are U-shaped channels, others are simple 90° brackets. Furthermore, Civics can have a heat shield—two small nuts retain it—that must come off before you can reach the manifold/pipe connection. Lastly, there are differences concerning where the air filter gets its hot air. Early Accords, like the one pictured, take their heat from around the exhaust manifold, well above the manifold/pipe connection. Later Accords pipe heat up from below the connection, through two hoses. You'll have to remove the hoses, in that case.

Things are simpler with 1200s. Their exhaust manifold runs down the front of the engine and connects to the pipe. Unbolt the manifold from the pipe. There will be some sheet metal around the manifold, but it won't be in your way. The 1200 manifolds have a bracket that's mounted to the oil-pan rail. Remove the bracket at the block.

No matter which style exhaust is used, disconnecting it means getting under the car. On Preludes, first remove the engine and transaxle shields. This will give you and the jack more room. You can reach the three bolts securing each shield without

Although there are several different rear-torque-rod designs, all remove like this one. Remove through bolt and swing rod to one side. Rubber bushing in rod needs to be replaced.

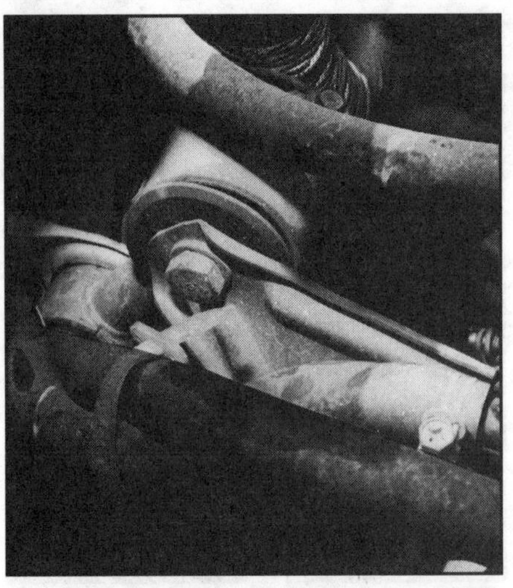

Rear torque rod for all Preludes and '80-and-later Civics: It's shorter and goes from fire wall to bellhousing bracket. Earlier torque rod attached to head.

When working under car, *use jack stands.* Jack under crossmember, as shown. Place jack stands under raised, reinforced sections of rocker-panel sill behind front wheels. Photo by Ron Sessions.

jacking the car. Remove the shields, then place the floor jack under the forward attaching point of the fore-and-aft crossmember. This is the piece that runs between the engine and transaxle. The transaxle mounts at the crossmember's center. Raise the car high enough to get jack stands under the *front supports*—reinforced sections inboard of the rocker panels just behind the front wheels. Set the parking brake and block the rear wheels.

Slide under the car and look up at the exhaust-pipe connection. If there's a shield around the connection, use a 1/4-in. drive, 10mm flex socket to remove the two shield nuts. If you encounter the U-shaped bracket, take out all nuts and bolts except the two that hold the bracket to the exhaust-pipe clamp. Don't remove the two clamp bolts. With 90° brackets, just undo the two bolts holding the bracket to the block. Undo the manifold-to-pipe connection and any hot-air ducts.

You might have better luck reaching some of these nuts and bolts from up top after you see where they are from below. Although you'll be working *blind* from the top, at least the grit won't fall into your eyes.

Radiator Hoses—It's time to stand up again. Immediately below the distributor on CVCC models, you'll find two radiator and two heater hoses; disconnect all four. Hiding under the thermostat housing on early CVCC engines is the water-pump-to-cylinder-head bypass hose. This short hose can be difficult to slide off. If it gives you any trouble, slit it lengthwise and pull it off. You can easily replace it later. Later CVCC engines have two bypass hoses.

On 1200 models, disconnect the radiator and heater hoses. Also, remove the *two* formed bypass hoses, one from the bottom of the intake manifold, the other from the water-pump connecting pipe. These join the heater hoses at an H-joint.

Besides the hoses, there are electrical leads to disconnect. On CVCC engines, disconnect the coolant-temperature sending unit—one wire—and two thermosensors—two wires, each next to the thermostat housing. The 1200 engines have no thermosensors, but disconnect the sending-unit wire at the intake manifold.

Also, if your engine has a mechanical-tachometer drive, unscrew the tach cable at the end of the cylinder head.

Alternator—At the other end of the engine is the alternator. On 1200 engines, the alternator is mounted on the fire-wall side of the engine; CVCC powerplants mount it in on the radiator side. Earlier air-conditioned (A/C) cars have the compressor installed where the alternator usually goes and the alternator installed above the left engine mount, facing the timing belt. Later air-conditioned cars keep the alternator in its usual spot, low on the side of the block, and mount the A/C compressor above the left engine mount.

All alternators use a combination of nutted and plastic connectors to attach wires. Because you don't have to completely remove the alternator from the chassis to pull the engine, you can just disconnect the plastic connectors and set the unit aside as far as the slack in the wires allows. For extra working room, or if you want to do a thorough engine-compartment cleaning, disconnect all wires and completely remove the alternator. Label all disconnected wires.

Depending on the model, you may find only one nutted connection. This leads to the oil-pressure sending unit. If the wire does not pass through the wire loom, you can remove the lead from the sending unit and leave the alternator wire connected.

With the electrics out of the way, remove the alternator from its mount. Cars with the alternator alongside the block use a simple mount, with an adjusting bolt on top and a through bolt on the bottom. Remove the adjusting bolt, swing the unit aside and slip off the drive belt. Remove the through bolt and lift off the alternator.

Air-conditioned cars with the alternator mounted facing the timing belt differ only in that the adjusting bolt also holds down a stamped-steel belt guard. Remove the adjusting bolt and guard, the through bolt, then pull off the alternator.

Alternator Bracket—Most cars use a simple alternator bracket mounted to the side of the block. Removal is simple. Unbolt the upper arm at the head, and the lower bracket at the

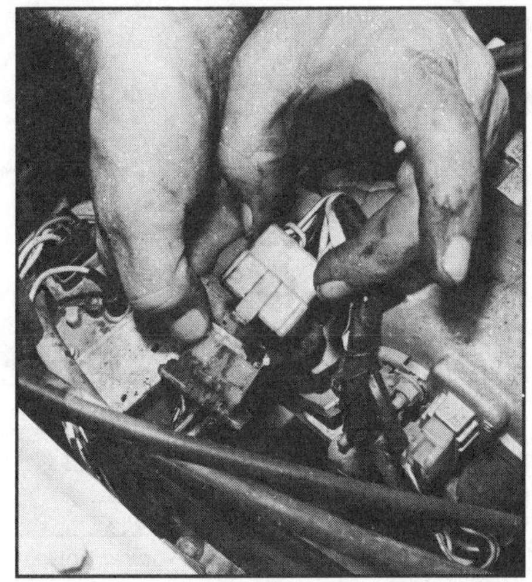

Disconnect alternator electrical leads here. On non-air-conditioned cars, remove oil-pressure-sending-unit wire as well.

This style mount is used for both alternators and A/C compressors. After removing adjusting bolt, remove through bolt.

block—two bolts. Alternators mounted by the timing belt are another story. These brackets attach to both sides of the head and the engine mount with a total of six bolts. Two bolts run horizontally into the head, parallel with the crankshaft, on the front side of the engine. Two more bolts run vertically into the engine mount in front of the timing belt. The last two run into the head horizontally, on the manifold side. Remove all six and lift off the bracket.

Left Engine Mount—To remove the alternator drive belt, the left engine mount must be disconnected. Using a 12mm wrench, remove the two bolts attaching the mount to the engine. Then use a 14mm wrench to remove the bolt that clamps the mount to the left inner fender. With this third bolt out, lever the engine mount into the fender with a screwdriver. *On 1200s, reconnect the left engine mount after removing the belt; otherwise, the engine may fall out when the crossmember is removed later.*

Air Conditioning—This section is for A/C compressors mounted so they face the timing belt. If the A/C compressor is mounted alongside the block, read the A/C warning below, then skip ahead to the next section. You will remove your compressor after pulling the head. On non-A/C cars, go to the rocker-cover section that follows.

A/C Warning—Before doing anything with the A/C system, you must be aware of the dangers involved. All A/C systems use high-pressure freon

There's no need to completely remove alternator, unless you want to replace bushings, diodes and brushes. Set it aside with its mounting bolts and guard. If you didn't disconnect battery, sparks will fly if alternator terminals contact metal.

With accessories out of way, bracket bolts can be removed. These two are part of left engine mount.

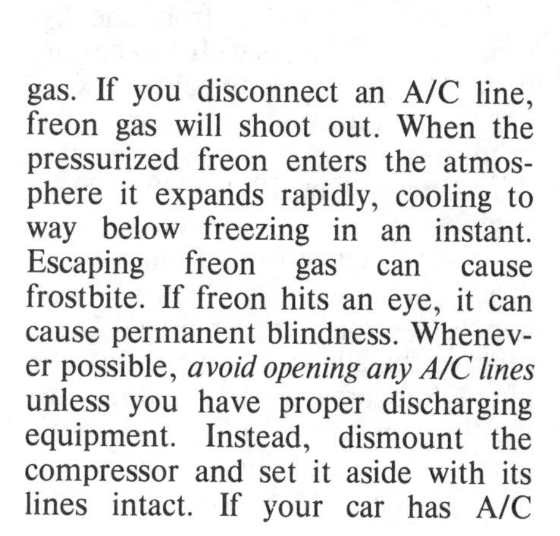

gas. If you disconnect an A/C line, freon gas will shoot out. When the pressurized freon enters the atmosphere it expands rapidly, cooling to way below freezing in an instant. Escaping freon gas can cause frostbite. If freon hits an eye, it can cause permanent blindness. Whenever possible, *avoid opening any A/C lines* unless you have proper discharging equipment. Instead, dismount the compressor and set it aside with its lines intact. If your car has A/C

On each side of head are two bracket bolts. These two pass into side of head. Other bolts pass into end of head.

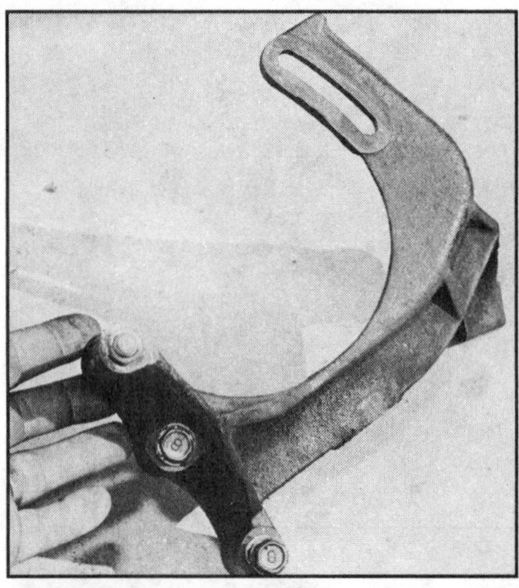

With bracket removed, two end mounting bolts are visible. Top nut is on alternator through bolt.

Unscrew both acorn nuts and lift off rocker cover.

Remove two timing-belt upper-cover bolts.

troubles, go to an A/C specialist after the engine is back in and running.

As you lean over the left fender, the A/C-compressor adjusting mechanism is to your left. Remove the adjuster from both compressor and bracket. Remove the compressor-pivot bolt—on the right side of the compressor. Slip off the drive belt. The compressor will now lift out of its bracket. Set it aside as far as the A/C lines will allow.

Unbolt the bracket from the head and set it aside. To remove the drive belt, the left engine mount must be disconnected as previously discussed.

Rocker Cover—Use a 10mm wrench to remove the two acorn nuts atop the rocker cover. There is a ground strap under the driver's-side nut. Fold the strap back, out of the way.

Timing-Belt Cover—Use the same 10mm wrench to remove the two timing-belt-upper-cover bolts. Lift off the cover and carefully inspect the timing belt. If you think you might reuse the belt, mark its direction of rotation with an arrow. As you face the belt from the left fender, the belt rotates counterclockwise. Your arrow should point to the front of the car if you put it above the cam sprocket. If the belt is obviously cracked, or you plan to replace it, don't bother. Just grasp the belt with both hands and pull it off the camshaft sprocket.

Oil-Pump-Drive Cover—Honda engines drive the oil pump off the camshaft. This requires a long drive shaft between the cam and oil pump, which is at the bottom of the block.

The oil-pump drive gear is at the middle of the camshaft. The drive gear turns a driven gear that's mounted on the oil-pump drive shaft. The gear is out of sight under an aluminum cover. You need to remove this cover, both to remove the oil-pump drive shaft and one of the head bolts. Unbolt the cover bolts with a 10mm wrench, then lift off the cover. The cover will fight you slightly because it's doweled around each bolt. Don't lose the dowels; they tend to fall out. Set the cover and bolts aside, then grasp the drive-shaft gear and pull out the cover and drive shaft. The gear may hang up in the valve train, but should come out with a little fiddling.

Fuel Pump—On 1200 engines, unbolt and remove the mechanical fuel pump and spacer from the flywheel end of the head. It's necessary to remove the pump to gain access to one of the four nuts for the cylinder-head studs.

Air Pump—On 1975–'79 1237cc engines, an air-injection pump is used to control emissions. Remove the three hoses—supply, bypass and manifold—from the pump. Then, remove the adjuster bolt beneath the pump. Support the pump and remove the through bolt up top. Set aside the pump, belt and hoses. The air-injection manifold stays on the engine.

Head Bolts/Nuts—It's now time to loosen the head bolts and remove the head. Honda says engine temperature must be below 100F before the head can be removed. This is to prevent the head from warping from uneven loading while it's expanded by heat. So, make sure the engine is cold before unbolting the head. Considering all the work you've done so far, the engine has probably cooled enough. As an additional precaution to avoid warpage, Honda specifies a *cylinder-head-bolt removal sequence,* opposite of the torque sequence. Also, they recommend loosening the bolts in 30° increments until finger tight. Honda has a history of blown head gaskets. Although many mechanics feel the head-removal recommendations are over-conservative, better safe than sorry.

On most Hondas, special *12-point* 10mm bolts secure the head. This group includes the '76-and-later 1488, and all 1335s, 1600s and 1751s. A smaller group of engines use familiar hex-head head bolts with 10mm shanks and 14mm heads, including the 1975 1488, and all 1170s and 1237s. In addition to these head bolts, 1170, 1237 and 1335 engines have nuts on the cylinder-head-to-block studs to remove. On 1200s, one of these nuts is hidden inside the fuel-pump mount. See the nearby illustration.

Of course, removing and installing

If reusing timing belt, mark direction of rotation with arrow—counterclockwise—then pull it off. This belt was obviously cracked, so I didn't bother marking it.

One head bolt is hidden under oil-pump drive gear. Start by unbolting and removing drive-gear cover. Watch for falling dowels.

the 12-point, 10mm head bolts requires a special socket. Chances are you may have to buy a 3/8-in. drive, 12-point 10mm socket and adapt it to a 1/2-in.-drive ratchet. Finding this socket in 1/2-in. drive is pretty tough, but if you can, all the better. Whichever way you do it, be extra sure the socket is not worn out. A 10mm hex is relatively small. Dividing it into 12 points makes for narrow flats. Your socket must be top quality and like-new, with sharp corners, or you'll round off the head-bolt heads. Then you're in big trouble!

Follow the unbolting sequence shown nearby. When all bolts and nuts are loose, pull them out of the head, making sure you remove the hardened washer under each bolt and nut. Store the bolts in order by punching them into a piece of cardboard. There are several different lengths, so each must go back into its original hole.

Remove the Head—Double-check around the head, making sure all the disconnections have been made. Have a place ready to set the cylinder-head assembly. An empty drain pan or the edge of a workbench work well.

Now, get a firm grip on the head-and-manifold assembly and lift. Although one person can do this job, the weight of the assembly and the potential harm to your back call for two people. The head may also be stubborn to remove. Both the head gasket and dowels can cause backaches here. A light pry where shown, will break the head loose. But, before you do any prying, make doubly sure all the bolts and nuts are out! Check for the bolt next to the oil-pump drive gear. It can hide under a puddle of oil. And don't forget about disconnecting the exhaust-pipe flange and thermostat bypass hose. Both are stronger than you in a tug of war.

Find a spot to pry where the head overhangs the block. On 1200s, you can *lightly* pry with round bar stock or a stout screwdriver shaft inserted into the fuel-pump and distributor-mounting holes. *Never* pry between the block and head. You could scratch the mating surfaces, which can cause a head gasket to blow. Any prying you do should be with moderate force. If

Simply pull drive gear and shaft out of engine.

Loosen and remove all head bolts and nuts. Magnet is big help with washers.

The 1200 head is secured with combination of bolts and studs. It's easy to overlook studs in three corners. This one is visible through fuel-pump mounting hole.

Head gasket and dowels can make head seem immovable. A light pry, such as between these two accessory brackets, should break loose head. If not, stop! Check for overlooked bolts or nuts.

Get someone to help lift the head. Each person should lean over a fender and lift straight up. Set head aside where it won't get damaged. Be particularly careful with the head-gasket surface.

you find you are exerting yourself, something is wrong. Stop, take a look around, and find out why the head is not coming free.

Once you have the head off, carry it to your predesignated spot and set it down. Beware of the oily mess if you stand the head on end.

Air Conditioning—If the A/C compressor is mounted to the front of the block, remove it after reading the A/C warning on page 25. Don't disconnect any A/C lines. For now, all you need to do is unbolt the compressor from its mount and set it aside.

The A/C compressor is held on by four bolts. Two are above the compressor in plain view; the other two are directly below it. Remove the bottom two bolts. These adjusting bolts are extra long bolts with 14mm-hex heads. Don't lose them. With the bottom bolts out, loosen the top pair. The compressor will fall toward the block, allowing you to remove the belt. Support the compressor with one hand and remove the bolts with the other. Now you can set the compressor out of the way. Handle the compressor lines with care. Bend them as little as possible. Above all, don't hang the compressor by the lines.

Although it can be left on, you'll gain extra clearance if you remove the A/C-compressor bracket. Remove the two bolts to release the bracket.

Oil-Pressure Sending Unit—If you haven't already done so, remove the oil-pressure sending-unit lead. Don't forget to mark it. Double-check that the A and B thermosensor wires above the bellhousing have been disconnected.

Without disturbing freon lines, remove both A/C-compressor upper mounting bolts. Opening a pressurized freon line without proper discharging equipment can result in bodily injury.

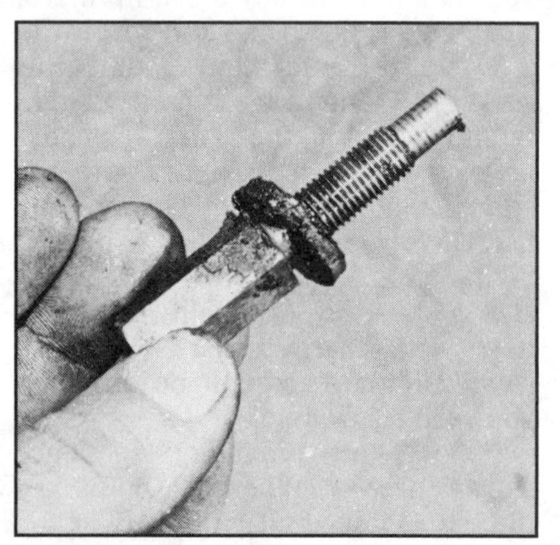

Removing this lower A/C-compressor mounting bolt and its mate requires working blind under compressor.

Power Steering—As with air conditioning, it's best to leave the power-steering (P/S) lines connected when you remove the pump. It isn't that P/S fluid is dangerous. Instead, dirt can enter the system if the lines are disconnected. Plus, leaking P/S fluid makes a mess.

The P/S pump uses a simple, two-bolt mount. Remove the adjusting bolt, slip off the belt, pull out the through bolt and set the pump aside. Take care not to overstress the hoses.

Crossmember—The car should still be on jack stands. If it isn't, raise it back up because you need to unbolt the crossmember. Don't jack the car too high because the cherry picker has limited travel. It's not much fun to lift the engine as high as it will go, only to discover it won't clear the front sheet metal. Raise the car just enough so you can comfortably reach the rear crossmember bolts—no higher. Don't forget to place jack stands underneath before going underneath yourself.

Note: On 1200s, support the transaxle with a floor jack or it may fall

With lower bolts removed, set compressor aside. There is no need to remove it from engine compartment.

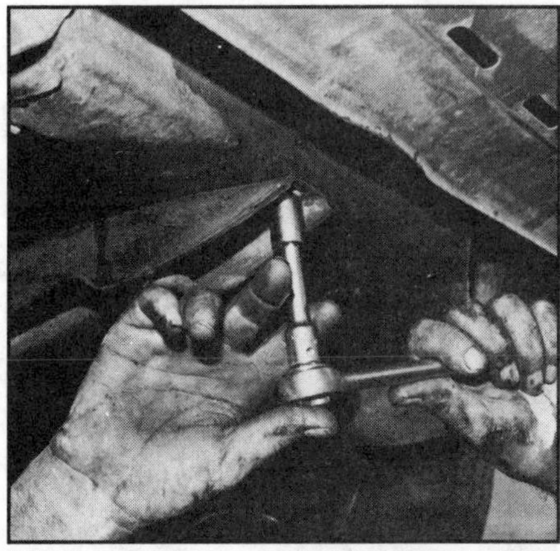

With transaxle supported by floor jack or engine hanging from chain hoist, remove crossmember.

out when the crossmember is removed!

On 1979-and-earlier models, you'll need a 12mm or 14mm socket to remove the bolts retaining the crossmember between the engine and transaxle. On these early models, there are three bolts at each end of the crossmember. Remove them, but leave the two at the center. The two center bolts hold the transaxle mount to the crossmember. Watch out! With the six attaching bolts removed, the crossmember will drop free of the transaxle mount, unless you have a 1980-or-later model. Set it aside for the moment.

I know what you're thinking: What's holding the transaxle in place when the crossmember is removed? Don't worry. As long as the engine is bolted to the transaxle, the engine mounts keep the transaxle from moving longitudinally, laterally or vertically.

On 1980-and-later Civics and all Preludes, you need to first remove the torque rod that's between the bellhousing and radiator support before the crossmember. Undo the two torque-rod bolts under the front engine mount and two others at the bellhousing. This lets the torque rod swing out of the way. If you need more room, completely remove the rod at the radiator support.

The 1980-and-later crossmember has two bolts at either end and also bolts to a center engine mount. Remove the two nuts inside the crossmember's *U*-shaped section, then the end bolts to free the crossmember.

Dust Cover—Remove the three dust-cover bolts and two pan bolts to remove the flywheel dust cover, or splash shield. This is the sheet-metal piece covering the lower bellhousing area. You'll now have access to the clutch or torque converter.

Support Transaxle—Before separating the engine from the transaxle, the transaxle must be supported. If you haven't done so already, place the floor jack under the transaxle with a 2x4 between the jack pad and transaxle to protect the aluminum case. Also, make sure the jack is squarely under the transaxle case, not the shift linkage, a drive shaft or other fragile part. Raise the jack until it *just* contacts the transaxle—no more.

Torque Converter—On automatic-transaxle cars, disconnect the torque converter from the *driveplate,* or flexplate. The driveplate is used in place of a flywheel on automatic-transaxle-equipped cars. Use a 10mm wrench to remove the eight bolts holding the torque converter to the driveplate. On most Honda automatics with the bolt-together torque-converter housing, you might think there are 16 driveplate bolts. Actually, every other one is part of the torque converter and should be skipped. Many late-model Hondas have welded torque-converter housings, so you'll see only eight bolts.

Rotate the crankshaft to expose all eight bolts. Do this with a 17 or 19mm socket on the crank-pulley bolt. This bolt is accessible through a hole in the left wheelwell.

Attach Hoist—Once you are satisfied the car is at the proper height, go ahead and hook up the engine hoist.

With the head-off method of pulling the engine, you can attach a strap or chain to the block with two head bolts and large washers. The large washers ensure that the chain can't slip over the bolt heads. Place the bolts at diagonally opposite ends of the block/head mating surface. Use a wrench to run the bolts in until they bottom, no more.

Although it may look shaky with the chain at the heads of the bolts, it's OK to lift the block using the long head bolts because the block is light. Worse comes to worse, you'll bend two head bolts—replacements are available from your local Honda dealer. Limit the chances of this happening by leaving as much slack as possible in the chain. Even better would be to use two shorter bolts—10mm thread diameter X 1.25mm thread pitch X 50mm long. When snugged down, bolts this length will clamp the chain links between the bolt head and block deck. It eliminates the bending load on the bolts.

With the head-on method of pulling the engine, attach the chain to the sides of the head. On 1200s, use the

When removing engine with head, bolt lifting chain to head. On 1200s, good lift points for chain are distributor hold-down bolt and fire-wall-to-engine strut-rod-bracket bolt. Photos by Ron Sessions.

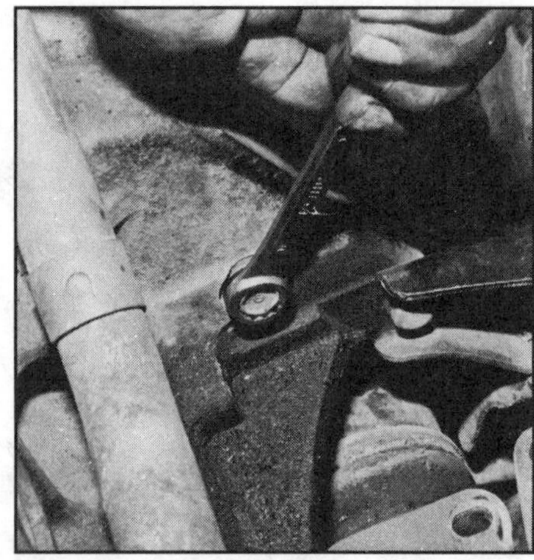
This *backward*-facing starter/bellhousing bolt must be completely removed. Second starter bolt is at top on starter side of bellhousing.

Left engine mount uses three bolts. Remove all three . . . then push chassis half of mount into fender. If you don't, engine half won't clear chassis.

distributor hold-down bolt at the front and the torque-rod-bracket bolt at the rear, as shown.

Raise the hoist just enough to take all slack out of the lifting chain, no more. Above all, don't put a load on the engine mounts.

Starter & Bellhousing Bolts—Six bolts attach the engine to the transaxle. Two of the six double as starter bolts.

Begin with the starter. Disconnect the nutted and quick-disconnect electrical leads. If your starter has a bare wire between the solenoid and starter, leave it connected. Some engines have two starter bolts coming in from the transaxle side of the starter. Others, however, have one bolt entering from the transaxle side and another coming from the engine side. Furthermore, one bolt may have a nut at the end instead of a threaded blind hole. In any case, support the starter, remove both bolts and pull the starter from the bellhousing.

Remove the four additional bellhousing bolts. All can be reached from above. Surprisingly, there are no bellhousing bolts at the bottom because of the crossmember and transaxle mount.

Engine Mounts—The left engine mount may already have been removed. If not, remove the bolts, then slide the mount toward the fender. It will slide completely out of the way. A *slight* pry with a bar doesn't hurt. Both the front and rear engine mounts must also be disconnected. The rear mount may have a heat shield. Remove it with a 10mm wrench. Under the shield you'll find

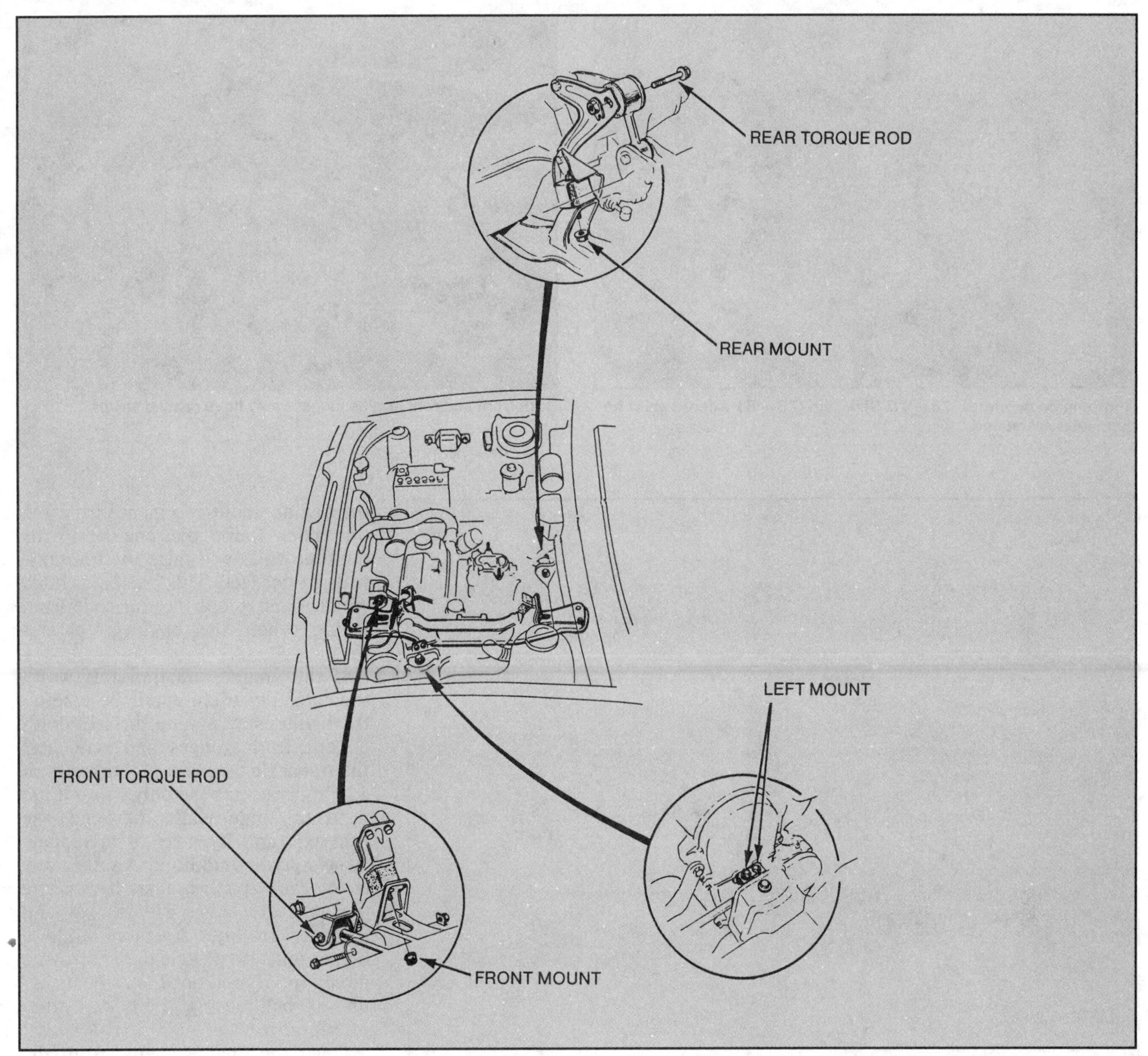

Typical engine-mount/torque-rod layout for Prelude and '80—'83 Civic. 1982—'83 Accord layout is similar, but with no front torque rod and longer rear torque rod. Courtesy of American Honda Motor Co., Inc.

two nuts; remove them. One doubles as a ground strap attachment. Also, remove the two bolts that go into the transaxle and the other going into the block.

At the front mount, remove only the two smaller nuts. Again, there's a ground strap under one of the nuts.

Lift the Engine—With everything disconnected, the engine should be ready to lift. Lift gently. The engine should separate easily from an automatic transaxle. Manual transaxles are somewhat more difficult. If the engine has not separated from the transaxle after lifting the engine an inch or so, stop. Give the engine a tug toward the left fender. Have a helper hold the transaxle or it will just slide along with the engine. Remember, the drive shafts are still connected—don't allow the transaxle to pull against them. Same goes for the transaxle input shaft. This shaft passes from the transaxle, through the bellhousing and into the clutch. To disengage the input shaft, move

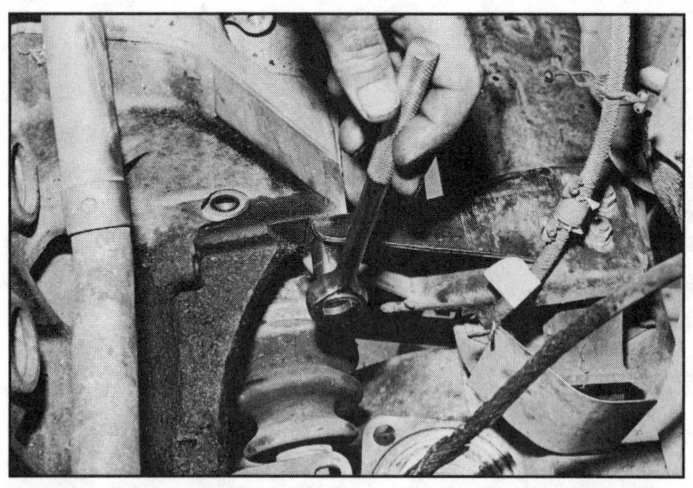

Rear engine mount of '73—'79 Civic and '76—'81 Accord must be completely removed.

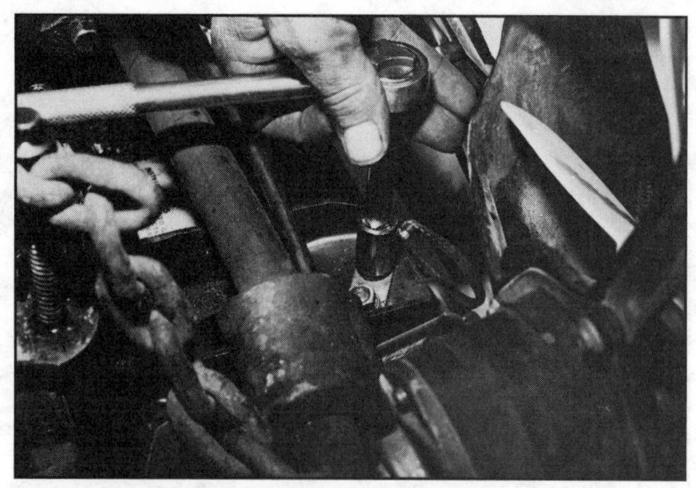

Both front and rear engine mounts may have ground straps.

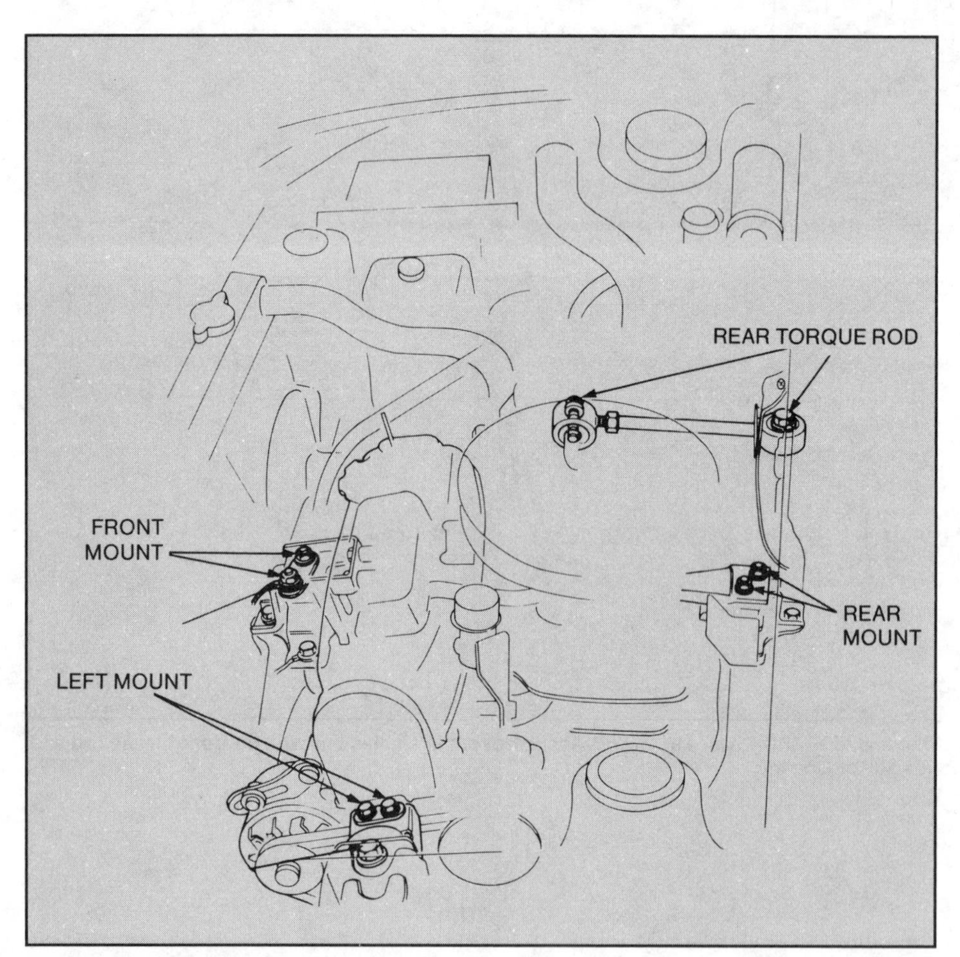

Typical engine-mount/torque-rod layout for 1973—'79 Civic and 1976—'81 Accord. Courtesy of American Honda Motor Co., Inc.

the engine about two inches from the transaxle. Raise the engine off its mounts while you raise the transaxle with a floor jack. This prevents clutch damage and keeps the transaxle from falling when the engine separates from it.

If the engine and transaxle won't separate, pry them apart. Be careful, the bellhousing mating surfaces don't benefit from gouges and scratches; the transaxle case is cast aluminum. If a light pry seems the only way out, do it from underneath, between the transaxle and flywheel or driveplate. A *light* pry should do it. Another way is to *shock* the two apart by striking the transaxle case with a 2x4 and hammer in one direction while a helper pulls the engine in the opposite direction. If not, look for the hang-up—a bellhousing bolt or other interference.

Once the engine is free from the transaxle, check around it for cables, wires and the like. In some chassis, the speedo cable will hook on the front pulleys, which can be fatal to the cable if not discovered in time. As you are checking, keep your hands out from under the engine. It's better to get down on your hands and knees to look under the engine than to lose a hand while feeling around.

As you lift the engine, increase its exit angle so it clears the left inner fender. I prefer to raise the pulley end and push down on the flywheel end while a friend works the hoist. It takes a fairly steep attitude. Go gently and steer around obstacles as you pass them. Don't be afraid to stop and look around the engine compartment. Lift-

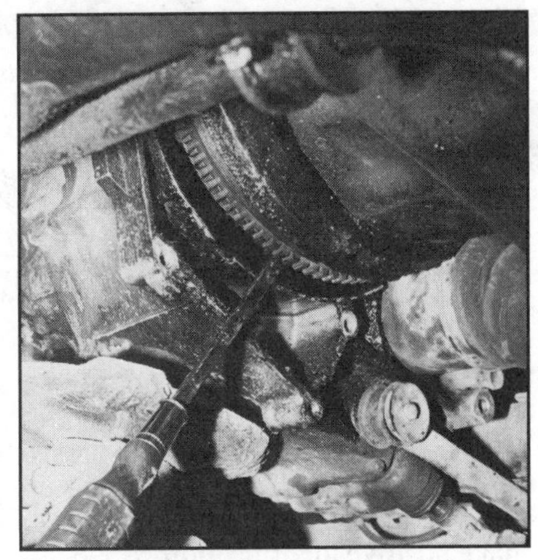

To separate engine and transaxle, pry lightly against flywheel or driveplate. Make sure all bellhousing bolts are removed first.

As engine is raised, position A/C compressor to one side.

While lifting engine, tilt block slightly to clear master cylinder. Photo by Ron Sessions.

Two people or more are best during engine removal. There's less chance of damage to engine-compartment parts.

ing the engine is not a *clean and jerk,* it's a slow, gentle process.

Once the engine is clear of the bodywork, roll the cherry picker away from the car and lower the engine. If you're using a chain hoist, you'll have to roll the chassis away, then lower the engine. Don't forget to support the transaxle from the fire wall with coat hanger or baling wire.

Clean-up—Now's the time to put everything away before it gets scattered. All the hardware can be stored, but don't bury it too deep because you'll be cleaning it shortly. The best place for the hood is back on the car. Rags, fender covers and tools can all be scooped up.

Before returning the cherry picker and floor jack, turn to Chapter 4 and start engine teardown. There are several parts that can be removed while the engine hangs from the hook.

Parts Identification & Interchange 3

Engine number is on machined pad on block bellhousing flange. Engine-number chart lists all applications. EF1 prefix identifies block from 1600 Accord. The 1 after dash denotes first model year, '76. This is the 14,360th Accord engine built that year.

Besides being interesting, knowing the differences between Honda parts and how they fit together can be valuable information. For example, if your engine's crankshaft is bad, you can save money by finding a used one in a junkyard. But first, you must know exactly which crankshaft you have and which other cranks interchange with it.

IDENTIFICATION

Engine identification can be accomplished by using number codes and a physical description. In this chapter, physical descriptions are given for each part under its heading—blocks, crankshafts and so on. The numbers are discussed here.

There are a lot of numbers in a Honda engine compartment: *Vehicle-identification number (VIN), chassis number, transaxle number* and *engine number.*

Of these, the engine number is the most important for engine rebuilding. It specifies the displacement and *destination* of the engine. Destination refers to *California, High Altitude,* and *except California and High Altitude,* or *49 states.*

Most of the time, destination is a moot point. Differences are largely external—emission controls, fuel- and ignition-system settings. A California car will have some extra vacuum hoses, switches and relays compared to a 49-state version. But inside, the two engines are identical. So, if you are looking for a cylinder head, crankshaft, connecting rod or the like, engine destination is not important. On the other hand, if you're looking for a carburetor or complete engine assembly, sticking to the original destination is wise. That way, you'll have the correct vacuum-line fittings and small parts that can be so bothersome to replace piecemeal.

Additionally, in the chart on page 36, the year, model and body style are given for each engine number.

Another often-referred-to number is the chassis number or VIN. It's used for registration and insurance purposes to determine the vehicle specifications and model year. In a junkyard, it can help identify engine-related parts on an engineless chassis. Junkyard cars can get so chopped up that the chassis number is needed to make positive identification.

As for the transaxle number, it's on the transaxle housing and refers only to the gearbox. I had to stop somewhere, so transaxle numbers are not included in this book.

The identification (ID) plate is found inside the engine compartment on the radiator support. The ID plate is stamped with the chassis and engine numbers. These numbers duplicate those found on the original engine and chassis, respectively.

Number Location—Just where are the engine and chassis numbers? The engine number is on a machined pad on the left side of the bellhousing mounting flange, near the starter motor. Lean over the right fender—the numerals are visible with the engine in the chassis. You may have to move around a little because the rear torque rod on late models can block your view.

The chassis number is stamped both into the top center of the fire wall and onto the VIN plate. The VIN plate is visible through the windshield atop the driver's side of the instrument panel.

ENGINE FAMILIES

There are three distinct Honda car-engine families. By engine family, I mean a group of different engines that are basically the same. Within an engine family you can expect to find all the major components in the same relative positions, although bore, stroke or valve sizes may be different.

1200 Family—The first Honda engine family is casually referred to as the *1200.* It consists of the 1170cc and 1237cc powerplants, both with *cast-aluminum cylinder blocks and heads.* The 1170, designated *EB1,* is fairly rare. It was available in 1973 only, a

Vehicle identification number—VIN—is visible through windshield on driver's side of instrument panel.

VIN and engine number also appear on ID plate. Plate is in engine compartment, near hood hinges.

short model year for the Honda Civic. In 1974, the 1237 superceded the smaller engine—it's called the *EB2.* The only difference between the two is the 1237's 2mm (0.079-in.) larger bore. Consequently, these two engines are practically identical, hence the common, 1200 name.

In 1975, the 1237 needed an air-injection-reactor (AIR) pump to meet emissions regulations. Note: These are not CVCC engines. From then on, the base Civic with its air-pumped 1237cc engine became known as the *AIR car* in Honda circles.

One important change was made to the 1237cc engine for its last two years of production, 1978—'79. The cylinder head was changed to the Japanese home-market design, with larger-diameter valves. Also, slightly domed pistons were used. The changes were significant enough for Honda to redesignate the engine the *EB3.*

Large water pump at left is from 1200; center pump is from 1975-'79 1488 or 1600. '80-and-later pump is at right. The 1200 pumps have larger pulleys and turn slower than pumps for cast-iron blocks. Later 1488 pump has smaller vanes for less horsepower loss—increased gas mileage.

CVCC Family—The second engine family is the largest. It includes the 1488, 1600 and 1751cc engines. Common design features include a *cast-iron* block and aluminum CVCC cylinder head. There are numerous bore, stroke, valve-size, flywheel, distributor, carburetor, valve-order and other changes in this family, so parts don't necessarily interchange within the family.

The 1488cc engine was introduced in 1975, debuting the CVCC emission-control system. Bore and stroke are 74 X 86.5mm (2.91 X 3.41 in.). The '75 1488 is chock-full of detail differences compared with later 1488s. Some of the major ones are valve order, intake-manifold design, head-bolt spacing, auxiliary-valve design and head-gasket material. Therefore, almost nothing from the '75 1488 will interchange with other Honda CVCC engines.

Keep in mind that 1975 was the first year for the 1488 and the CVCC system. The improvements made for 1976 saw the 1488 all the way through to 1979. 1976—'79 1488s are plentiful and many parts interchange between them.

In 1980, the Honda line received an engineering face-lift. Besides updated bodywork, improvements were made to all engines, including the 1488. A new cylinder head and related changes distinguish these engines, so keep 1980 in mind as a dividing line.

Although the 1488 continues beyond 1983, it shares little more than a name with previous 1488s. Significant changes include a new oil pump, head, and manifolds. Therefore, 1984-and-later models are not covered in this book.

Perhaps the easiest engine to chronicle is the 1600cc engine. It was introduced in the 1976 Accord and was replaced by the 1751cc engine in 1979. Thus, the 1600 was produced only three years—1976—'78.

The best way to describe the 1600 is as a 1488 with a 6.5mm-longer stroke.

ENGINE NUMBERS

Year	Engine Number	Model	Destination	Displacement
CIVIC				
73	EB1-1000001 to 1069000	1200	All	1170cc (71 CID)
74	EB2-1000001 to 1076643	1200	All	1237cc (75 CID)
75	EB2-2000001 to 2025148	1200	All	1237cc (75 CID)
	ED1-1000001 to 2000000	CVCC exc. S.W.	All	1488cc (91 CID)
	ED2-1000001 to 2000000	CVCC S.W.	All	1488cc (91 CID)
76	EB2-2025149 to 3000000	1200	All	1237cc (75 CID)
	ED3-2000001 to 2500000	CVCC exc. S.W.	All	1488cc (91 CID)
	ED3-2500001 to 3000000	CVCC exc. S.W.	exc. Calif.	1488cc (91 CID)
	ED4-2000001 to 3000000	CVCC S.W.	All	1488cc (91 CID)
77	EB2-3000001 to 4000000	1200	All	1237cc (75 CID)
	ED3-3000001 to 3500000	CVCC exc. S.W.	Calif.	1488cc (91 CID)
	ED3-3500001 to 3900000	CVCC exc. S.W.	exc. Calif. & Hi Alt'd	1488cc (91 CID)
	ED3-3900001 to 4000000	CVCC exc. S.W.	Hi Alt'd	1488cc (91 CID)
	ED4-3000001 to 3500000	CVCC S.W.	Calif.	1488cc (91 CID)
	ED4-3500001 to 3900000	CVCC S.W.	exc. Calif. & Hi Alt'd	1488cc (91 CID)
	ED4-3900001 to 4000000	CVCC S.W.	Hi Alt'd	1488cc (91 CID)
78	EB3-1000001 to 1500000	1200	exc. Calif.	1237cc (75 CID)
	ED3-4000001 to 4500000	CVCC exc. S.W.	Calif.	1488cc (91 CID)
	ED3-4500001 to 4900000	CVCC exc. S.W.	exc. Calif. & Hi Alt'd	1488cc (91 CID)
	ED3-4900001 to 5000000	CVCC exc. S.W.	Hi Alt'd	1488cc (91 CID)
	ED4-4000001 to 4500000	CVCC S.W.	Calif.	1488cc (91 CID)
	ED4-4500001 to 4900000	CVCC S.W.	exc. Calif. & Hi Alt'd	1488cc (91 CID)
	ED4-4900001 to 5000000	CVCC S.W.	Hi Alt'd	1488cc (91 CID)
79	EB3-2000001 to 2500000	1200	exc. Calif.	1237cc (75 CID)
	ED3-5000001 to 5500000	CVCC exc. S.W.	Calif.	1488cc (91 CID)
	ED3-5500000 to 5900000	CVCC exc. S.W.	exc. Calif. & Hi Alt'd	1488cc (91 CID)
	ED3-5900001 to 6000000	CVCC exc. S.W.	Hi Alt'd	1488cc (91 CID)
	ED4-5000001 to 5500000	CVCC S.W.	Calif.	1488cc (91 CID)
	ED4-5500001 to 5900000	CVCC S.W.	exc. Calif. & Hi Alt'd	1488cc (91 CID)
	ED4-5900001 to 6000000	CVCC S.W.	Hi Alt'd	1488cc (91 CID)
80	EJ1-1000001 to 2000000	1300	exc. Calif.	1335cc (81 CID)
	EM1-1000001 to 1500000	1500	Calif.	1488cc (91 CID)
	EM1-1500001 to 1900000	1500	exc. Calif. & Hi Alt'd	1488cc (91 CID)
	EM1-1900001 to 2000000	1500	& Hi Alt'd	1488cc (91 CID)
81	EJ1-2000001 & Up	1300	Calif.	1335cc (81 CID)
	EJ1-2500001 & Up	1300	exc. Calif. & Hi Alt'd	1335cc (81 CID)
	EJ1-2900001 & Up	1300	& Hi Alt'd	1335cc (81 CID)
	EM1-2000001 & Up	1500	Calif.	1488cc (91 CID)
	EM1-2500001 & Up	1500	exc. Calif. & Hi Alt'd	1488cc (91 CID)
	EM1-2900001 & Up	1500	& Hi Alt'd	1488cc (91 CID)
82	EJ1-3100001 & Up	1300	Calif.	1335cc (81 CID)
82	EJ1-3500001 & Up	1300	exc. Calif. & Hi Alt'd	1335cc (81 CID)
82	EJ1-3900001 & Up	1300	& Hi Alt'd	1335cc (81 CID)
82	EM1-3100001 & Up	1500	Calif.	1488cc (91 CID)
82	EM1-3500001 & Up	1500	exc. Calif. & Hi Alt'd	1488cc (91 CID)
82	EM1-3900001 & Up	1500	& Hi Alt'd	1488cc (91 CID)
83	EJ1-2400001 & Up	1300	Calif.	1335cc (81 CID)
83	EJ1-3000001 & Up	1300	exc. Calif. & Hi Alt'd	1335cc (81 CID)
83	EJ1-3500001 & Up	1300	& Hi Alt'd	1335cc (81 CID)
83	EJ1-4000001 & Up	1500	Calif.	1488cc (91 CID)
83	EM1-4500001 & Up	1500	exc. Calif. & Hi Alt'd	1488cc (91 CID)
83	EM1-4900001 & Up	1500	& Hi Alt'd	1488cc (91 CID)
ACCORD				
76	EF1-1000001 to 2000000	1600	All	1600cc (98 CID)
77	EF1-2000001 to 2499999	1600	Calif.	1600cc (98 CID)
	EF1-2500001 to 2899999	1600	exc. Calif. & Hi Alt'd	1600cc (98 CID)
	EF1-2900001 to 3000000	1600	& Hi Alt'd	1600cc (98 CID)
78	EF1-3000001 to 3499999	1600	Calif.	1600cc (98 CID)
	EF1-3500001 to 3899999	1600	exc. Calif. & Hi Alt'd	1600cc (98 CID)
	EF1-3900001 to 4000000	1600	& Hi Alt'd	1600cc (98 CID)
79	EK1-1000001 to 1499999	1800	Calif.	1751cc (107 CID)
	EK1-1500001 to 1899999	1800	exc. Calif.	1751cc (107 CID)
	EK1-1900001 to 2000000	1800	& Hi Alt'd	1751cc (107 CID)
80	EK1-2000001 to 2300000	1800	Calif.(5-spd)	1751cc (107 CID)
	EK1-2300001 to 2500000	1800	Calif.(Auto)	1751cc (107 CID)
	EK1-2500001 to 2900000	1800	exc. Calif. & Hi Alt'd	1751cc (107 CID)
	EK1-2900001 to 3000000	1800	& Hi Alt'd	1751cc (107 CID)
81	EK1-3000001 & Up	1800	Calif.	1751cc (107 CID)
	EK1-3500001 & Up	1800	exc. Calif. & Hi Alt'd	1751cc (107 CID)
	EK1-3900001 & Up	1800	& Hi Alt'd	1751cc (107 CID)

ENGINE NUMBERS

Year	Engine Number	Model	Destination	Displacement
ACCORD				
82	EK1-4000001 & Up	1800	Calif.	1751cc (107 CID)
82	EK1-4500001 & Up	1800	exc. Calif. & Hi Alt'd	1751cc (107 CID)
82	EK1-4900001 & Up	1800	& Hi Alt'd	1751cc (107 CID)
83	EK1-5000001 & Up	1800	Calif.	1751cc (107 CID)
83	EK1-5500001 & Up	1800	exc. Calif. & Hi Alt'd	1751cc (107 CID)
83	EK1-5700001 & Up	1800	exc. Calif. & Hi Alt'd	1751cc (107 CID)
83	EK1-5900001 & Up	1800	& Hi Alt'd	1751cc (107 CID)
PRELUDE				
79	EK1-1000001 to 1499999		Calif.	1751cc (107 CID)
	EK1-1500001 to 1899999		exc. Calif. & Hi Alt'd	1751cc (107 CID)
	EK1-1900001 to 2000000		& Hi Alt'd	1751cc (107 CID)
80	EK1-2000001 to 2300000		Calif. (Std Trans)	1751cc (107 CID)
	EK1-2300001 to 2500000		Calif. (Auto Trans)	1751cc (107 CID)
	EK1-2500001 to 2900000		exc. Calif. & Hi Alt'd	1751cc (107 CID)
	EK1-2900001 to 3000000		& Hi Alt'd	1751cc (107 CID)
81	EK1-3000001 & Up		Calif.	1751cc (107 CID)
	EK1-3500001 & Up		exc. Calif. & Hi Alt'd	1751cc (107 CID)
	EK1-3900001 & Up		& Hi Alt'd	1751cc (107 CID)
82	EK1-4000001 & Up		Calif.	1751cc (107 CID)
	EK1-4500001 & Up		exc. Calif. & Hi Alt'd	1751cc (107 CID)
	EK1-4900001 & Up		& Hi Alt'd	1751cc (107 CID)

CHASSIS NUMBERS

Year	Chassis Number	Model	Body Type	Displacement	Transaxle
CIVIC					
73	SBA-1007001 to 1100000	1200	2 Dr. exc. Hatchback	1170cc (71 CID)	4 Spd. Man
	SBB-1007001 to 1100000	1200	2 Dr. exc. Hatchback	1170cc (71 CID)	2 Spd Auto
	SBC-1007001 to 1100000	1200	2 Dr.Hatchback	1170cc (71 CID)	4 Spd Man
	SBD-1007001 to 1100000	1200	2 Dr.Hatchback	1170cc (71 CID)	2 Spd Auto
74	SBA-2100001 to 2300000	1200	2 Dr. exc. Hatchback	1237cc (75 CID)	4 Spd Man
	SBB-2100001 to 2300000	1200	2 Dr. exc. Hatchback	1237cc (75 CID)	2 Spd Auto
	SBC-2100001 to 2300000	1200	2 Dr.Hatchback	1237cc (75 CID)	4 Spd Man
	SBD-2100001 to 2300000	1200	2 Dr.Hatchback	1237cc (75 CID)	2 Spd Auto
75	SBA-3300001 to 4000000	1200	2 Dr. exc. Hatchback	1237cc (75 CID)	4 Spd Man
	SBB-3300001 to 4000000	1200	2 Dr. exc. Hatchback	1237cc (75 CID)	2 Spd Auto
	SBC-3300001 to 4000000	1200	2 Dr.Hatchback	1237cc (75 CID)	4 Spd Man
	SBD-3300001 to 4000000	1200	2 Dr.Hatchback	1237cc (75 CID)	2 Spd Auto
	SG-A1000001 to 2000000	CVCC	2 Dr. exc. Hatchback	1488cc (91 CID)	4 Spd Man
	SG-B1000001 to 2000000	CVCC	2 Dr. exc. Hatchback	1488cc (91 CID)	2 Spd Auto
	SG-C1000001 to 2000000	CVCC	2 Dr.Hatchback	1488cc (91 CID)	4 Spd Man
	SG-D1000001 to 2000000	CVCC	2 Dr.Hatchback	1488cc (91 CID)	2 Spd Auto
	SG-E1000001 to 2000000	CVCC	2 Dr.Hatchback	1488cc (91 CID)	5 Spd Man
	WB-A1000001 to 2000000	CVCC	4 Dr.S.W.	1488cc (91 CID)	4 Spd Man
76	SBA-4000001 to 5000000	1200	2 Dr. exc. Hatchback	1237cc (75 CID)	4 Spd Man
	SBB-4000001 to 5000000	1200	2 Dr. exc. Hatchback	1237cc (75 CID)	2 Spd Auto
	SBC-4000001 to 5000000	1200	2 Dr.Hatchback	1237cc (75 CID)	4 Spd Man
	SBD-4000001 to 5000000	1200	2 Dr.Hatchback	1237cc (75 CID)	2 Spd Auto
	SG-A2000001 to 3000000	CVCC	2 Dr. exc. Hatchback	1488cc (91 CID)	4 Spd Man
	SG-B2000001 to 3000000	CVCC	2 Dr. exc. Hatchback	1488cc (91 CID)	2 Spd Auto
	SG-C2000001 to 3000000	CVCC	2 Dr.Hatchback	1488cc (91 CID)	4 Spd Man
	SG-D2000001 to 3000000	CVCC	2 Dr.Hatchback	1488cc (91 CID)	2 Spd Auto
	SG-E2000001 to 3000000	CVCC	2 Dr.Hatchback	1488cc (91 CID)	5 Spd Man
	WB-A2000001 to 3000000	CVCC	4 Dr.S.W.	1488cc (91 CID)	4 Spd Man
	WB-B2000001 to 3000000	CVCC	4 Dr.S.W.	1488cc (91 CID)	2 Spd Auto
77	SBA-5000001 to 6000000	1200	2 Dr. exc. Hatchback	1237cc (75 CID)	4 Spd Man
	SBC-5000001 to 6000000	1200	2 Dr.Hatchback	1237cc (75 CID)	4 Spd Man
	SBD-5000001 to 6000000	1200	2 Dr.Hatchback	1237cc (75 CID)	2 Spd Auto
	SG-A3000001 to 4000000	CVCC	2 Dr. exc. Hatchback	1488cc (91 CID)	4 Spd Man
	SG-C3000001 to 4000000	CVCC	2 Dr.Hatchback	1488cc (91 CID)	4 Spd Man
	SG-D3000001 to 4000000	CVCC	2 Dr.Hatchback	1488cc (91 CID)	2 Spd Auto
	SG-E3000001 to 4000000	CVCC	2 Dr.Hatchback	1488cc (91 CID)	5 Spd Man
	WB-A3000001 to 4000000	CVCC	4 Dr.S.W.	1488cc (91 CID)	4 Spd Man
	WB-B3000001 to 4000000	CVCC	4 Dr.S.W.	1488cc (91 CID)	2 Spd Auto
78	SBA-6000001 to 7000000	1200	2 Dr. exc. Hatchback	1237cc (75 CID)	4 Spd Man
	SBC-6000001 to 7000000	1200	2 Dr.Hatchback	1237cc (75 CID)	4 Spd Man
	SBD-6000001 to 7000000	1200	2 Dr.Hatchback	1237cc (75 CID)	2 Spd Auto
	SG-A4000001 to 5000000	CVCC	2 Dr. exc. Hatchback	1488cc (91 CID)	4 Spd Man

CHASSIS NUMBERS

Year	Chassis Number	Model	Body Type	Displacement	Transaxle
	SG-C4000001 to 5000000	CVCC	2 Dr. Hatchback	1488cc (91 CID)	4 Spd Man
	SG-D4000001 to 5000000	CVCC	2 Dr. Hatchback	1488cc (91 CID)	2 Spd Auto
	SG-E4000001 to 5000000	CVCC	2 Dr. Hatchback	1488cc (91 CID)	5 Spd Man
	WB-A4000001 to 5000000	CVCC	4 Dr. S.W.	1488cc (91 CID)	4 Spd Man
	WB-B4000001 to 5000000	CVCC	4 Dr. S.W.	1488cc (91 CID)	2 Spd Auto
79	SBA-7000001 to 8000000	1200	2 Dr. exc. Hatchback	1237cc (75 CID)	4 Spd Man
	SBC-7000001 to 8000000	1200	2 Dr. Hatchback	1237cc (75 CID)	4 Spd Man
	SBD-7000001 to 8000000	1200	2 Dr. Hatchback	1237cc (75 CID)	2 Spd Auto
	SG-A5000001 to 6000000	CVCC	2 Dr. exc. Hatchback	1488cc (91 CID)	4 Spd Man
	SG-C5000001 to 6000000	CVCC	2 Dr. Hatchback	1488cc (91 CID)	4 Spd Man
	SG-D5000001 to 6000000	CVCC	2 Dr. Hatchback	1488cc (91 CID)	2 Spd Auto
	SG-E5000001 to 6000000	CVCC	2 Dr. Hatchback	1488cc (91 CID)	5 Spd Man
	WB-A5000001 to 6000000	CVCC	4 Dr. S.W.	1488cc (91 CID)	4 Spd Man
	WB-B5000001 to 6000000	CVCC	4 Dr. S.W.	1488cc (91 CID)	2 Spd Auto
80	SL-A1000001 to 2000000	1300	2 Dr. Hatchback	1335cc (81 CID)	4 Spd Auto
	SL-C1000001 to 2000000	1300	2 Dr. Hatchback	1335cc (81 CID)	5 Spd Man
	SR-A1000001 to 2000000	1500	2 Dr. Hatchback	1488cc (91 CID)	4 Spd Man
	SR-C1000001 to 2000000	1500DX	2 Dr. Hatchback	1488cc (91 CID)	5 Spd Man
	SR-D1000001 to 2000000	1500	2 Dr. Hatchback	1488cc (91 CID)	2 Spd Auto
	SR-E1000001 to 2000000	1500GL	2 Dr. Hatchback	1488cc (91 CID)	5 Spd Man
	WD-A1000001 to 2000000	1500	4 Dr. S.W.	1488cc (91 CID)	5 Spd Man
81	SL-332 BS000001 & Up	1300	2 Dr. Hatchback	1335cc (81 CID)	3 Spd Auto
	SL-431 BS000001 & Up	1300	2 Dr. Hatchback	1335cc (81 CID)	4 Spd Man
	SK-532 BS000001 & Up	1300	2 Dr. Hatchback	1335cc (81 CID)	5 Spd Man
	SR-332 BS000001 & Up	1500	2 Dr. Hatchback	1488cc (91 CID)	3 Spd Auto
	SR-532 BS000001 & Up	1500	2 Dr. Hatchback	1488cc (91 CID)	5 Spd Man
	SR-533 BS000001 & Up	1500GL	2 Dr. Hatchback	1488cc (91 CID)	5 Spd Man
	WD-352 BS000001 & Up	1500	4 Dr. S.W.	1488cc (91 CID)	3 Spd Auto
	WD-552 BS000001 & Up	1500	4 Dr. S.W.	1488cc (91 CID)	5 Spd Man
	ST-343 BS000001 & Up	1500	4 Dr. Sedan	1488cc (91 CID)	3 Spd Auto
	ST-543 BS000001 & Up	15000	4 Dr. Sedan	1488cc (91 CID)	5 Spd Man
82	SL-431 CS000001 & Up	1300	2 Dr. Hatchback	1335cc (81 CID)	4 Spd Man
	SL-532 CS000001 & Up	1300	2 Dr. Hatchback	1335cc (81 CID)	5 Spd Man
	SR-332 CS000001 & Up	1500	2 Dr. Hatchback	1488cc (91 CID)	3 Spd Auto
	SR-532 CS000001 & Up	1500	2 Dr. Hatchback	1488cc (91 CID)	5 Spd Man
	SR-533 CS000001 & Up	1500GL	2 Dr. Hatchback	1488cc (91 CID)	5 Spd Man
	ST-343 CS000001 & Up	1500	4 Dr. Sedan	1488cc (91 CID)	3 Spd Auto
	ST-543 CS000001 & Up	1500	4 Dr. Sedan	1488cc (91 CID)	5 Spd Man
	WD-352 CS000001 & Up	1500	4 Dr. S.W.	1488cc (91 CID)	3 Spd Auto
	WD-552 CS000001 & Up	15000	4 Dr. S.W.	1488cc (91 CID)	5 Spd Man
83	SL-431 DS000001 & Up	1300	2 Dr. Hatchback	1335cc (81 CID)	4 Spd Man
	SL-532 DS000001 & Up	1300FE	2 Dr. Hatchback	1335cc (81 CID)	5 Spd Man
	SR-332 DS000001 & Up	1500	2 Dr. Hatchback	1488cc (91 CID)	3 Spd Auto
	SR-532 DS000001 & Up	1500DX	2 Dr. Hatchback	1488cc (91 CID)	5 Spd Man
	SR-533 DS000001 & Up	1500S	2 Dr. Hatchback	1488cc (91 CID)	5 Spd Man
	ST-343 DS000001 & Up	1500	4 Dr. Sedan	1488cc (91 CID)	3 Spd Auto
	ST-543 DS000001 & Up	1500	4 Dr. Sedan	1488cc (91 CID)	5 Spd Man
	WD-352-DS000001 & Up	1500	4 Dr. S.W.	1488cc (91 CID)	3 Spd Auto
	WD-552 DS000001 & Up	1500	4 Dr. S.W.	1488cc (91 CID)	5 Spd Man
ACCORD					
76	SJ-D1000001 to 2000000	1600	2 Dr. Hatchback	1600cc (98 CID)	2 Spd Auto
	SJ-E1000001 to 2000000	1600	2 Dr. Hatchback	1600cc (98 CID)	5 Spd Man
77	SJ-D1000001 to 2000000	1600	2 Dr. Hatchback	1600cc (98 CID)	5 Spd Man
	SJ-E2000001 to 3000000	1600	2 Dr. Hatchback	1600cc (98 CID)	5 Spd Man
78	SJ-D3000001 to 4000000	1600	2 Dr. Hatchback	1600cc (98 CID)	2 Spd Auto
	SJ-E3000001 to 4000000	1600	2 Dr. Hatchback	1600cc (98 CID)	5 Spd Man
	SJ-G3000001 to 4000000	1600LX	2 Dr. Hatchback	1600cc (98 CID)	2 Spd Auto
	SJ-H3000001 to 4000000	1600LX	2 Dr. Hatchback	1600cc (98 CID)	5 Spd Man
79	SM-D1000001 to 2000000	1800	2 Dr. Hatchback	1751cc (107 CID)	2 Spd Auto
	SM-E1000001 to 2000000	1800	2 Dr. Hatchback	1751cc (107 CID)	5 Spd Man
	SM-H1000001 to 2000000	1800LX	2 Dr. Hatchback	1751cc (107 CID)	5 Spd Man
	SM-J1000001 to 2000000	1800	4 Dr. Sedan	1751cc (107 CID)	2 Spd Auto
	SM-K1000001 to 2000000	1800	4 Dr. Sedan	1751cc (107 CID)	5 Spd Man
80	SM-D2000001 to 3000000	1800	2 Dr. Hatchback	1751cc (107 CID)	3 Spd Auto
	SM-E2000001 to 3000000	1800	2 Dr. Hatchback	1751cc (107 CID)	5 Spd Man
	SM-G2000001 to 3000000	1800LX	2 Dr. Hatchback	1751cc (107 CID)	3 Spd Auto
	SM-H2000001 to 3000000	1800LX	2 Dr. Hatchback	1751cc (107 CID)	5 Spd Man
	SM-J2000001 to 3000000	1800	4 Dr. Sedan	1751cc (107 CID)	3 Spd Auto
	SM-K2000001 to 3000000	1800	4 Dr. Sedan	1751cc (107 CID)	5 Spd Man
81	SM-332 BC000001 & Up	1800	2 Dr. Hatchback	1751cc (107 CID)	3 Spd Auto
	SM-333 BC000001 & Up	1800LX	2 Dr. Hatchback	1751cc (107 CID)	3 Spd Auto
	SM-532 BC000001 & Up	1800	2 Dr. Hatchback	1751cc (107 CID)	5 Spd Man

CHASSIS NUMBERS

Year	Chassis Number	Model	Body Type	Displacement	Transaxle
	SM-533 BC000001 & Up	1800LX	2 Dr Hatchback	1751cc (107 CID)	5 Spd Man
	SM-342 BC000001 & Up	1800	4 Dr Sedan	1751cc (107 CID)	3 Spd Auto
	SM-542 BC000001 & Up	1800	4 Dr Sedan	1751cc (107 CID)	5 Spd Man
82	SZ-332 CC000001 & Up	1800	2 Dr. Hatchback	1751cc (107 CID)	3 Spd Auto
	SZ-333 CC000001 & Up	1800LX	2 Dr. Hatchback	1751cc (107 CID)	3 Spd Auto
	SZ-342 CC000001 & Up	1800	4 Dr. Sedan	1751cc (107 CID)	3 Spd Auto
	SZ-532 CC000001 & Up	1800	2 Dr. Hatchback	1751cc (107 CID)	5 Spd Man
	SZ-533 CC000001 & Up	1800LX	2 Dr. Hatchback	1751cc (107 CID)	5 Spd Man
	SZ-542 CC000001 & Up	1800	4 Dr. Sedan	1751cc (107 CID)	5 Spd Man
83	SZ-342 DA000001 & Up	1800	4 Dr. Sedan	1751cc (107 CID)	4 Spd Auto
	SZ-532 DC000001 & Up	1800	2 Dr. Hatchback	1751cc (107 CID)	5 Spd Man
	SZ-533 DC000001 & Up	1800LX	2 Dr. Hatchback	1751cc (107 CID)	5 Spd Man
	SZ-542 DC000001 & Up	1800	4 Dr. Sedan	1751cc (107 CID)	5 Spd Man
	SZ-732 DC000001 & Up	1800	2 Dr. Hatchback	1751cc (107 CID)	4 Spd Auto
	SZ-733 DC000001 & Up	1800LX	2 Dr. Hatchback	1751cc (107 CID)	4 Spd Auto
	SZ-742 DC000001 & Up	1800	4 Dr. Sedan	1751cc (107 CID)	4 Spd Auto
PRELUDE					
79	SN-B1000001 to 2000000		2 Dr. Sedan	1751cc (107 CID)	2 Spd Auto
	SN-F1000001 to 2000000		2 Dr. Sedan	1751cc (107 CID)	5 Spd Auto
80	SN-B2000001 to 2000000		2 Dr. Sedan	1751cc (107 CID)	3 Spd Man
	SN-F2000001 to 2000000		2 Dr. Sedan	1751cc (107 CID)	5 Spd Auto
81	SN-322 BC000001 & Up		2 Dr. Sedan	1751cc (107 CID)	3 Spd Auto
	SN-522 BC000001 & Up		2 Dr. Sedan	1751cc (107 CID)	5 Spd Man
82	SN-322 CC000001 & Up		2 Dr. Sedan	1751cc (107 CID)	3 Spd Auto
	SN-522 CC000001 & Up		2 Dr. Sedan	1751cc (107 CID)	5 Spd Man

That makes the bore and stroke 74 X 93mm (2.91 X 3.66 in.); an *undersquare* engine—bore is significantly smaller than stroke. Typical undersquare engines have better low- and mid-range power and torque than do oversquare engines of the same displacement. Good low-range torque is welcome in the heavier Accord.

The 1600 is identical to the 1488 in many respects, providing numerous interchange possibilities.

Last of the cast-iron-block CVCC family is the 1751cc engine. The increased displacement was needed for the sporty Prelude, introduced in 1979. A new bore-and-stroke combination was used to get the extra displacement—77 X 94mm (3.031 X 3.701 in.). Besides powering the Prelude, the 1751 also replaced the 1600 in 1979-and-later Accords. In 1983, the Prelude was totally redesigned and fitted with a new 1830cc engine. The 1830 is substantially different than the engines in this book, so it isn't covered. In the Accord, though, the 1751 finished out the 1983 model year.

Civic 1300—The last engine family includes only one engine, the 1335. The 1335cc engine was used from 1980 through 1983 as the base-level Civic powerplant. Even though it is in a class of its own, the 1335 doesn't hold many surprises because it's a combination of the other two engine families. The block is similar to the 1170/1237cc design, but is a different casting. The major visual difference is the water-pump casting on the opposite side of the block.

On the other hand, the 1335's CVCC cylinder head looks like that from a 1488. Like the 1600, the 1335 was manufactured for only a few years, 1980—'83. Changes during that period were fairly minimal. In 1984, the 1335 was replaced by the very different 1342cc engine, which is not covered in this book.

CYLINDER BLOCK

Because the 1200—*1170* and *1237*—blocks are so similar, there is no sense in discussing them separately. Although the two blocks are different castings, the only meaningful difference between them is bore size. The 1170cc engine uses a 70mm (2.76-in.) bore; the 1237, a 72mm (2.83-in.) bore. Both blocks are aluminum with cast-in steel *liners.* The liners provide the working surface of the bore—the area the piston rings contact.

Because this liner is cast integrally with the block, it cannot be replaced. Some people have tried boring out the liner completely and pressing in a new one, but cost and liner, or *sleeve,* availability make this impractical. Also, 1200 blocks are an *open-deck* design. With the head removed, you can look down into the water jacket. Except for cylinders 1 and 2, and cylinders 3 and 4 being *siamesed*—joined full length at their closest points—the tops of the cylinders are not joined together—or at the sides of the block. The cylinders are *free-standing.* This open-deck design, combined with the block's aluminum construction, doesn't provide sufficient strength to withstand the rigors of pressing in new liners. If cylinder wear on your 1200 engine is past the limit, replace the block.

If you have a worn-out 1170 block, I recommend replacing it with a 1237 block. All you'll need to convert are the 1237 pistons and rings to match the larger bores. You were going to have to bore anyway, so you'd be buying pistons regardless. Starting life as a 1237, your engine will have more power with comparable fuel economy.

A word about liners: If you look at the steel liners from the deck, they appear very thick. This apparent thickness gives the impression that these blocks could be bored considerably larger than stock. Unfortunately, it's not so. If you could saw a 1200 block in half, you would see that the outside of the liners is machined into a series of barbs, page 64. The barbs help anchor the steel liner to the aluminum that's cast around them. They also limit boring to the limits specified by Honda: up to 1mm (0.04 in.) in *four* steps for 1170s, and 0.5mm (0.02 in.) in *two* steps for 1237s.

If you are combing the junkyards for a 1200 block, take a vernier caliper along to measure the cylinders. This will alert you to blocks that are already bored to the limit. It will also settle any identity questions between 1170 and 1237 blocks. If you buy a block, be sure to get the main-bearing caps with it. On 1200 blocks, the caps are joined together into a *cage,* or *cradle.* If you don't get the original bearing cage, the block must be *align-bored,* page 67.

1335—At first glance, it's easy to mistake a 1335cc engine block for one of the 1200s. Both are aluminum with steel cylinder liners and caged main bearings. The 1335 block casting is quite different, however. Use the water-pump casting for identification. The 1200 water pump is on the radiator side of the block; the 1335 pump is on the fire-wall side.

The 1335 block will not interchange with the 1200s for several reasons. The cylinder-head stud and bolt pattern looks the same, but has one less stud. The 1335 block has equal bore spacing. The 1200 crossflow head and 1335 CVCC head won't even pass over the other's block studs. Even if you could get the 1200 head on the 1335 block, the oil filter would interfere with the exhaust manifold. This is because the 1335 oil filter is at the center of the block rather than at the timing-belt end.

Keep in mind that outside of the U.S., Honda has sold other engine variations. Unlike the CVCC-like version of the 1335 sold in the U.S. from 1980 through 1983, Honda exported to Canada a 1335cc version of the 1237 engine during those years. Therefore, the Canadian, or more properly, the export 1335 block, will interchange with the U.S. 1237 block. They are actually the same part.

1488 & 1600—As noted earlier, the 1488 uses a cast-iron block. It is a completely different design than the 1170, 1237 and 1335 blocks, and will not interchange with any of them.

Features shared by all cast-iron blocks are: separate main-bearing caps, water pump on fire-wall side and *closed decks*—end of water jacket is closed, except for block-to-head water passages. The tops of the cylinders are joined together, forming a rigid cylinder-head mating surface, or deck.

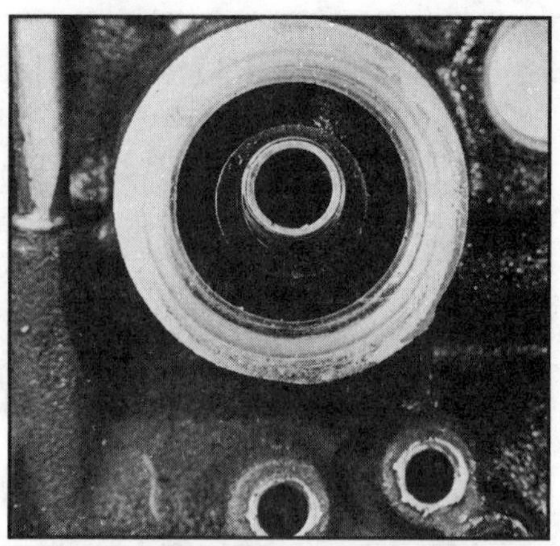

Oil-cooler-fitting holes at 5 and 6 o'clock under oil-filter pad identify this as 1751 block. Holes may have hose nipples or are blocked off, but all 1751 blocks have them.

The 1488 and 1600 blocks are the only CVCC-family blocks that will interchange. They share a 74mm (2.91-in.) bore and use different crankshafts and connecting rods for different displacements. Therefore, to make a 1600 from a 1488, you'll need a 1600 crankshaft and connecting rods. The opposite is needed to make a 1488 from a 1600. Pistons are the same for both engines.

There is one exception to the 1488/1600-block swap—the '75 1488. Because of its unique cylinder head, this block has a different head-bolt pattern. That makes the '75 1488 block an oddball; it won't interchange with any other 1488.

1751—To make a larger engine for the Prelude, Honda started with a new, 1751 block casting. Its 77mm (3.031-in.) bore is 3mm larger than the 1488/1600 block.

The 1751 block will not interchange with any other family. And, some 1751 blocks won't interchange with each other! The difference lies in the cylinder-head-bolt pattern. Starting in 1982, valve order was changed in the 1751 cylinder head. To accommodate this change, the bolt pattern was also changed. So, 1979—'81 blocks have the early bolt pattern; 1982—'83 blocks have the latter. A block must be replaced with one from the same group—early or late. To double-check a block, hold a head gasket from the original head/block combination against it. If the block is from the same group, the head-bolt holes will line up. Check the bolt holes along the water-pump (left) side of the block. That is where the difference between early and late types is.

Differentiating 1751 blocks from 1488 or 1600 blocks is easy because of the oil-cooler fittings. All 1751 blocks are drilled and tapped with two extra oil passages, next to the oil filter. Usually, you'll find two hose nipples there. On 1982—'83 manual-transaxle models, you'll find two plugs instead—an oil cooler wasn't used on these 1751s. If you buy a 1751 block, either keep the old hose fittings or get two new ones with the new block.

CRANKSHAFTS

All Honda crankshafts use select-fit, color-coded bearings, as described in Chapter 5, page 70. Furthermore, all Honda engine families use the same-size main bearings. They measure 20mm (0.787-in.) long and have a *nominal* wall thickness of 2mm (0.0791 in.). *Actual* wall thickness varies with the color of the bearing. All crankshafts have 50mm (1.969-in.) main-bearing journals. All blocks have 54mm (2.126-in.) main-bearing bores.

There are two different connecting-rod-bearing sizes. The 1170, 1237 and 1335 aluminum blocks use crankshafts with 40mm (1.575-in.) rod-bearing journals and rods with 43mm (1.693-in.) big ends. The cast-iron blocks—1488, 1600 and 1751—use 42mm (1.654-in.) rod journals and rods with 45mm (1.772-in.) big ends. All connecting-rod bearings are 17.5mm (0.689-in.) long and nominally 1.5mm (0.0591-in.) thick.

1170 & 1237—Both engines use 76mm- (2.99-in.-) stroke crankshafts. Additionally, the 1335 crankshaft with its 82mm (3.20-in.) stroke will interchange with the 1170/1237 crankshaft. Because the 1335 engine has the same bore as the 1200s, fitting its crankshaft to a 1237 block will yield 1335cc. You also need the shorter, 1335 rods to complete the swap. Of course, you could go the other way, too. If, for some strange reason, you wanted to make a 1237 from a 1335 block, all that's needed is the shorter-stroke crank and rods.

1488, 1600 & 1751—All three of these crankshafts have different strokes. The 1488 stroke is 86.5mm (3.41 in.), the 1600, 93mm (3.66 in.) and the 1751, 94mm (3.70 in.). As noted under the cylinder-block description, 1488 and 1600 crankshaft-and-rod combinations will interchange.

All Honda engines except 1751 use 17mm pin at left. Larger-diameter 18mm 1751 pin is at right.

Honda 1600 crank (top) is fully counterweighted; 1488 crank (bottom) is not. Additional counterweights were necessary due to 1600's longer stroke. All 1200 and 1335 cranks look like a 1488, but are smaller. The 1751 crank is counterweighted like a 1600.

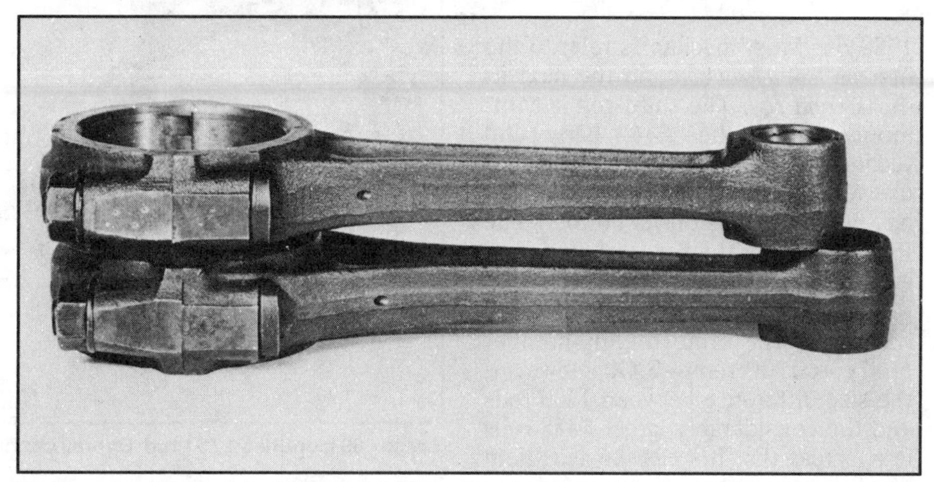

Best way to differentiate 1200 and 1488 rods is by length. Shorter 1200 rod is on top. Big ends are also different diameter, but harder to see.

The 1751 crank can *only* be used in 1751 blocks. First, the longer throws will not clear the smaller blocks. Second, there is no piston available that has the 1751 piston-pin diameter and the 1488/1600 bore size. Third, 1751 blocks have equal *bore spacing*—distance between bore centers. The 1488 and 1600 cylinders 1 and 2, 3 and 4 have equal bore spacing, but cylinder 2-to-3 spacing is greater. The front and rear pair of cylinders are siamesed. The non-siamesed bores of the 1751 provide for better cooling efficiency—a desirable feature as displacement increases.

With 1488 cranks, there is less counterweighting than with 1600 or 1751 cranks. The 1488 has no counterweights between the 1—2 and 3—4 throws. The 1600 and 1751 cranks are fully counterweighted. Comparing the 1600 and 1751 cranks side-by-side, you'll notice that the 1751 part has thicker counterweights.

Neither the 1170/1237 nor 1335 crankshaft has counterweighting between the 1—2 and 3—4 throws, so that visual clue is no help with them.

When swapping 1751 cranks, watch out for different flywheel-bolt sizes. Any 1751 crankshaft will bolt into any 1751 block, but a different flywheel may be required. See page 44 for details.

CONNECTING RODS

All Honda rods use *pressed-in* piston pins. A 0.0006—0.0016-in. (0.014—0.04mm) interference fit retains the pins in the small end of the rods. A 17mm piston pin is used on all engines except the 1751. It has an 18mm pin. The small end does not use a bushing.

As noted earlier, the 1170, 1237 and 1335cc engines share a common big-end diameter—43mm. The iron-block engines, 1488, 1600 and 1751cc, all use rods with 45mm big ends.

Differences among the rods are *strength* and *center-to-center length.* As displacement and horsepower increases, the rods get beefier, with more material. Center-to-center length is the distance between the big- and small-end centers—in other words, from the center of the rod-bearing hole to the center of the piston-pin bore. While this distance is difficult to measure without special equipment, the relative lengths of the different rods help identify them.

An even better identifier is the number forged into the rod *beam*—the section between the big and small ends. These *forging numbers* can be found centered on the beam of all Honda rods.

1170 & 1237—These engines use the same rod. Look for a forging number prefixed with an S, such as SXX or SXXX. The number changes, but the S is always there.

1335—Due to its longer stroke, the 1335cc engine uses a shorter connecting rod than the 1170/1237cc engines. If stacked one atop the other, the 6mm (0.210-in.) difference will show. The 1335 rod forging number is PAO.

1488—There are three 1488 rods, two used before 1980 and one used from

Three 1488 rods differ mainly in their caps. Bottom rod is early style and features smooth cap. Middle rod is same, but with different cap. Look for two ribs at cap's outer edges. Upper rod is totally different PA6 piece. Its cap looks like cross between two earlier versions, two ribs with center smoothed over. All rods interchange if used in sets, but engine may have to be rebalanced.

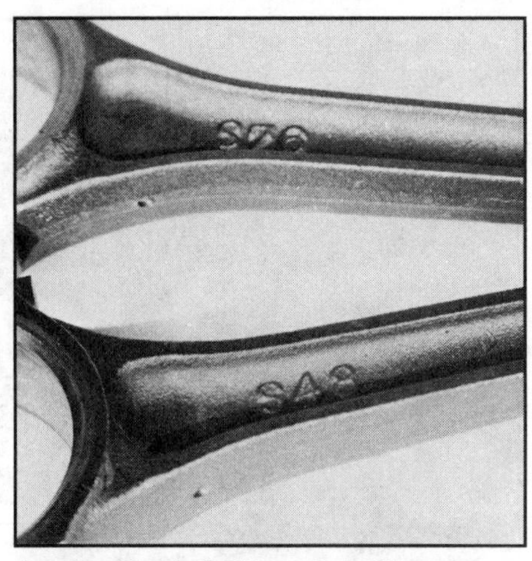

Because both 1200 and 1488 rods use SXX number, identification is done by physical comparison. The 1200 rod (top) has larger *bell* where beam meets big end.

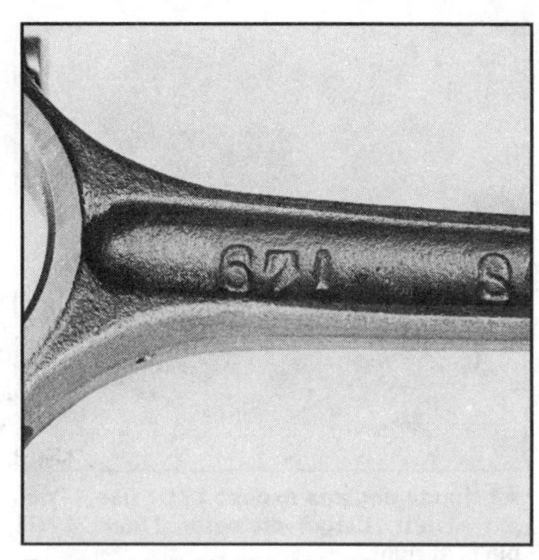

Forging number 671 identifies this as a 1600 rod. Compared to 1488 rod, 1600 rod is slightly shorter.

1980 on. Most mechanics refer to the first as the *early rod* and the next as the *second rod.* The third rod is commonly called the *PA6.* Early and second rods are really the same, but use a different cap. The second cap has two extra ribs on its bottom. The PA6 features thicker webbing and generally looks beefier.

Both early- and second-type 1488 rods have the same forging-number prefix as 1200 rods—SXX. However, the size difference between 1200 rods and the considerably larger 1488 rods is so great that it's easy to tell them apart. Nevertheless, it's a good idea to take a 1488 rod along for comparison. No such identification problems exist for the PA6 rod because it has PA6 cast into it.

All three 1488 rods have the same center-to-center length and will interchange without any modification. Because of slight weight differences, replace the rods in *sets* only. Engine rebalancing will be required when such a change is made.

1600—Look for the numerals 671 on 1600 rod beams. The 1600 rod is stronger than the first two 1488 rods because more material is used throughout. One area where this shows is the channel formed by the sides, or *flanges,* of the beam. The bottom of this channel widens slightly where the beam meets the big end. The widening is more pronounced on the 1488 than the 1600 rod.

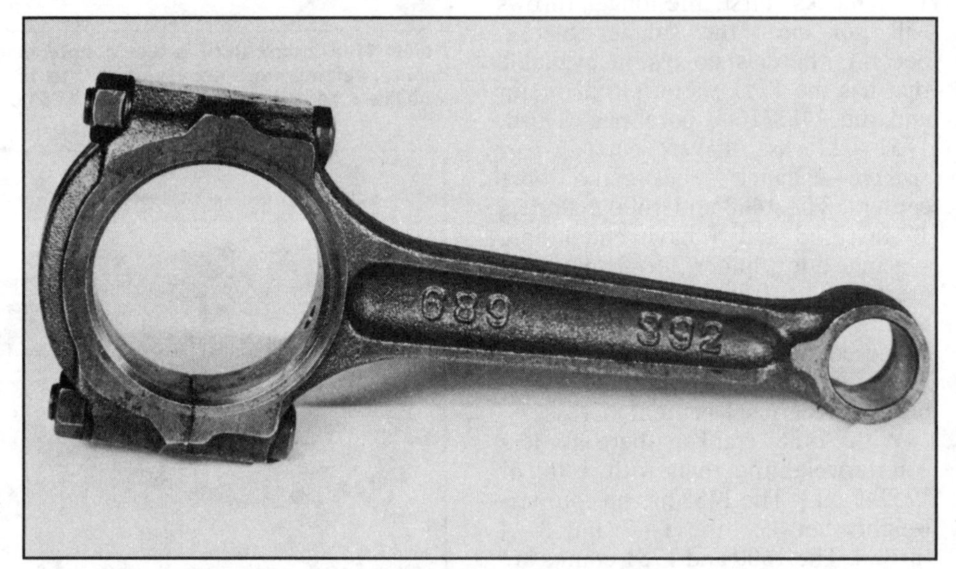

Large 689 identifies 1751 rod. Overall beefiness and 18mm pin bore are other clues.

Like the 1335 versus 1237 rod, the 1600 rod is 6.5mm shorter than its 1488 cousin to accommodate a longer stroke.

1751—For the more powerful Prelude engine, Honda naturally used a stronger connecting rod. There is no widening of the channel flange on the 1751 rod. Instead, the bottom section of the flange neatly curves around in a semi-circle. This leaves a lot of material where the beam meets the big end. When viewed from the side, the beam tapers in a straight line from the big end to the piston-pin bore; 1488 and 1600 rods are slightly convex here when viewed from the side. The 1751 rod also has a much larger machined area around the big-end bore. Look for a 689 casting number on 1751 rods.

PISTONS

From an identification and interchange viewpoint, there isn't much to say about pistons. After all, you aren't going to buy pistons in a junkyard—I hope. The only kind of piston you want is a new one. Only then will you get a quiet-running, long-lasting, oil-tight engine. However, if you are swapping other parts, such as rods, crankshafts and heads, pistons can make a difference. So, let's take a look at some piston-design features.

Compression Height—Compression height is the distance from the center of the pin bore to the top of the piston. Raising or lowering compression height is one way engineers can juggle different-length strokes in the same cylinder block. The other way is to change rod length. As we've seen

EB2 piston (left) and EB3 (right). Slightly higher compression height, raised dome and wider oil ring of EB3 piston shows here.

Dished EB2 and domed EB3 pistons work with different heads to arrive at same compression ratio.

in the previous section, Honda engineers would rather change rod length. Consequently, there are few changes in piston compression height between Honda engines.

Another consideration linked to compression height is compression ratio. Several factors influence compression ratio: cylinder volume, combustion-chamber volume, head-gasket thickness, and piston compression height and shape. With Honda engines, these factors seldom vary. The end result is little change in compression ratio between engines. When combined with different head-bolt patterns, very little swapping is possible.

One major difference occurs between 1237cc engines. The 1974—'77 EB2 1237 engine uses a dished piston—piston top is recessed. The 1978—'79 EB3 1237 engine uses a *domed* piston—piston top is raised. To keep compression ratio the same between the two engines, Honda made the EB3 combustion chamber larger. In physical terms, the change to the cylinder head isn't much; about 9cc. However, compression-ratio change from swapping heads is significant.

Mounting the EB2 head on the EB3 short block causes an excess compression-ratio increase. A typical compression-test reading for that combination is 220 psi. That means the compression ratio is about 12:1—too high for today's best pump gasoline.

Going the other way, putting an EB3 head on the dished-piston EB2 short block drops compression ratio to about 7.5:1. This engine-block-and-cylinder-head combination

This 1751 piston shows how larger pin cuts into groove just below oil ring. Pre-'82 pistons like this one are dished slightly.

will run without pinging or detonating on regular fuel, but will be down on power.

If you want or need to swap these two cylinder heads, use the piston originally used with each head. That will keep compression at 8.0:1—a good compromise that provides decent power without detonation.

Ring Thickness—For those of you searching for the last ounce of reduced engine drag for fuel economy or rpm, there isn't too much choice when it comes to ring thickness. Most Hondas use a 1.5mm (0.059-in.) compression ring. On 1982-and-later 1335 and 1488cc engines, compression-ring thickness is reduced—to 1mm for the top ring and 1.2mm for the second ring.

Oil-ring thickness varies also. The 1170 uses 2.75mm- (0.108-in.-) thick

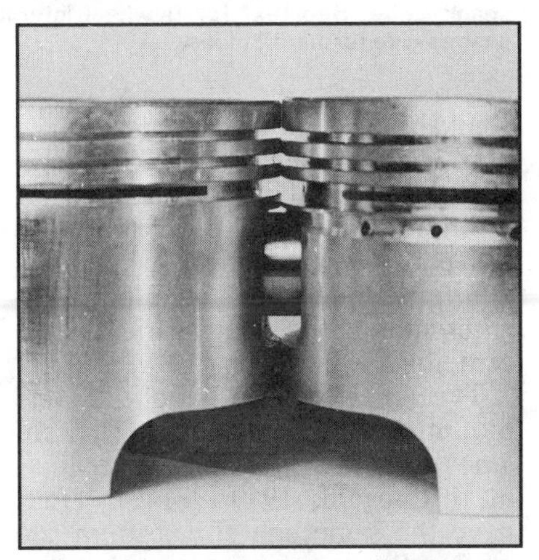

1975—'79 and 1980—'81 1488 pistons have several differences. Earlier piston (left) is perfectly flat on top and has slightly less compression height. Later piston is dished and has oil-drainback holes below oil rings.

oil rings. The 1975—'77 1237 uses 2.5mm- (0.0975-in.-) thick oil rings. All 1981-and-earlier 1335, 1488, 1600 and 1751cc engines, as well as '78—'79 1237s, use 4.0mm- (0.157-in.-) thick oil rings. Beginning in 1982, 1335 and 1488cc engines use 2.8mm- (0.110-in.-) thick oil rings.

FLYWHEELS & DRIVEPLATES

With a few exceptions, most flywheels and driveplates will bolt onto any Honda crankshaft. But, timing marks vary considerably. If the marks are different, you can repaint or mark the flywheel as necessary. This is not necessary with aluminum-block engines because timing marks are at the crankshaft pulley.

Flywheels used with 1200s have no *dowel hole*. Therefore, a 1200 flywheel won't work on a doweled, non-1200

Two 1751 cranks: At left is 1981-1/2-and-earlier type. Flywheel mounting flange is 74mm and uses 10mm-shank bolts. 1980-1/2-and-later crank at right has a 76mm flange and 12mm-shank bolts. See text for flywheel interchange. Counterweight shapes give further ID clues.

Three Honda oil-pump-rotor types: At left is 1973—'80 rotor; slightly taller 1981—'83 rotor is in middle. At right are two rotors used in two-stage pumps.

crankshaft. But you could bolt a non-1200 flywheel on a 1200 crankshaft if you had the engine balanced.

Another major difference involves the pilot bearing. The 1751 engines don't use one; all other Honda engines mount the pilot bearing in the center of the flywheel.

The last exception is bolt size and mounting-flange diameter. The engines except the 1751 were identical in this regard. 1979—'81-1/2 1751s have the same bolt size, pattern and mating-flange diameter as earlier engines. About midway through 1981, bolt size was increased and mating-flange diameter was increased about 0.050 in. A wider rear-main seal was used on large-bolt cranks as well. Earlier flywheels can be used on the later crankshafts by drilling out the bolt holes and machining about 0.050 in. off the mounting flange of the flywheel. These machining operations make the swap more costly, so use the matching flywheel, if possible.

The later, large-bolt flywheel cannot be used on earlier crankshafts. The smaller bolts don't pilot in the later flywheel's larger bolt holes. And, drilling and tapping the bolt holes in the crankshaft flange is not practical.

Because the change took place mid-year, take a flywheel bolt along when looking for a flywheel. If the bolt fits the holes, you can use the flywheel. If the bolt is too large, the flywheel can be used after it's modified, but it would be cheaper to find the correct flywheel. If the bolt is much smaller than the flywheel holes, the flywheel

Canted valves and deeply recessed sparkplug holes are indicative of crossflow, hemi-head, as with this EB2.

Intake side of EB2 head: Distributor mounts in hole at left (arrow), fuel pump uses studs and hole at right; holes between intake ports are manifold water passages.

cannot be used. And, if you swap crankshafts, use the correct rear-main seal for that crank.

OIL PUMP

Except for the oil-cooler-equipped 1751cc engine, all Honda oil pumps use the straightforward *trochoid* design. This is the common inner-and-outer-rotor pump. See accompanying photo. All 1979—'80 1751 pumps are of the same design, but are *two-stage.* Instead of having one set of inner and outer rotors, there are two sets, each in its own chamber. The larger set supplies oil under pressure through the engine via the filter. The smaller set sends oil to the oil cooler, which returns it to the pan. Beginning in 1982, the oil cooler and two-stage pump is restricted to 1751s with an automatic transaxle. Manual-transaxle Preludes and Accords of this period have single-stage pumps and no oil cooler.

Simply stated, single-stage and two-stage oil pumps cannot be interchanged. On an oil-cooler-equipped 1751, you need the two-stage pump. If you have an engine other than the 1751, the two-stage pump will not bolt on, nor would it do any good if it did. On a non-oil-cooled 1751, the number-3 main-bearing cap is not drilled with the two additional oil passages the two-stage pump requires. So, if you want to add an oil cooler to a 1751, you must drill these two extra passages, and use the two-stage pump. Or, install a screw-on, sandwich-type adaptor with oil-cooler fittings between the block and oil filter, and use the single-stage pump.

Visual differences between EB2 and EB3 cylinder heads are subtle. It's difficult to see 2mm valve-size difference. Side-by-side, EB2 combustion chamber is shallower than . . .

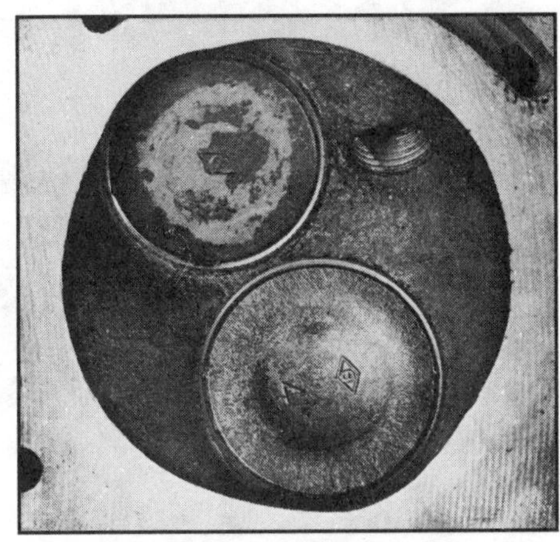

. . . EB3 combustion chamber. Valve size and chamber shape necessitate small relief to unshroud EB3 exhaust valve.

Valve springs from 1170, '74 1237 (left) and 1975—'79 1237 (right). Earlier springs use *plain* retainers; later, tall exhaust valves use *two-dot* retainers and thrust washers. Intakes are same for all years except '79.

CYLINDER HEAD

Most changes made to Honda engines have been confined to cylinder heads, and most of these to CVCC heads. Many cylinder-head changes also affected other parts of the engine. Think of the cylinder head as the entire upper half of the engine: head, camshaft, manifolds, carburetor, distributor and so on. By treating these parts as a unit, you'll be less likely to stumble over some of the changes.

1170 & 1237—The 1200 uses a *cross-flow head*—intake ports enter from one side of the head and exhaust ports exit from the other. The intake manifold is at the rear—fire-wall side—and the exhaust manifold at the front. The camshaft operates the valves through the two shaft-mounted rocker-arm assemblies; one for the intake and one for exhaust valves. This arrangement allows canting the valves toward the outside of the head for better breathing. It also allows the use of a *hemispherical,* or *hemi,* combustion-chamber shape. Hemi heads are known for good breathing, which helps the small 1200s keep up with traffic.

Also operated by the camshaft are the distributor and fuel pump. The distributor mounts at the cam-sprocket end; the fuel pump at the flywheel end.

Distinguishing the 1200 head from CVCC units is easy because of the 1200's smaller size and crossflow design.

The EB1 1170 and 1974 EB2 1237 cylinder heads are identical. The 1975—'77 EB2 head differs only in that it is drilled and tapped for an air-pump mount. The 1978—'79 EB3 head has 2mm (0.078-in.) larger intake and exhaust valves than the EB2 head. The later head also has straighter ports for better breathing. Additionally, the combustion chamber is 9cc larger. The increased size makes the chamber more of a true hemi and angles the sparkplug closer to the chamber's center. These changes make the EB3 head the one to have for horsepower. If you are interested in changing to the EB3 head, read about the swap on page 43.

CVCC—The non-crossflow design places both intake and exhaust manifolds on the backside of the head, toward the fire wall. The camshaft is centrally located, as on the 1200, and operates the valves through rocker arms. As with the 1200 rocker-arm assembly, CVCC heads have two rocker shafts, but one is for main-intake and exhaust valves, the other for auxiliary-intake valves. The main-intake valves are positioned closest to

Three-dot retainer (right) is used with '79 1237 intake spring. This late spring has larger OD than those used with plain retainers; it requires a different spring seat. All 1200 engines should be updated to taller, two-dot exhaust spring/retainer combination. Three-dot, large intake-spring combination is preferable, but not absolutely necessary on earlier 1200 heads.

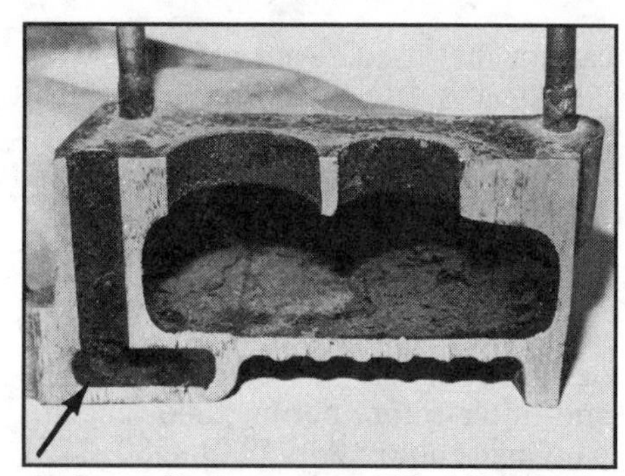

Jackson Racing cut a '75 1488 intake manifold in half to show source of clogging problem. Small passage at left is auxiliary-valve intake runner. Runner 90° bend (arrow) is where carbon-clogging problem begins.

First CVCC head was '75 1488 eight-port version. Intake ports are in upper row, exhausts in lower. CVCC auxiliary-intake ports look like bolt holes at 10 and 2 o'clock to main intake ports.

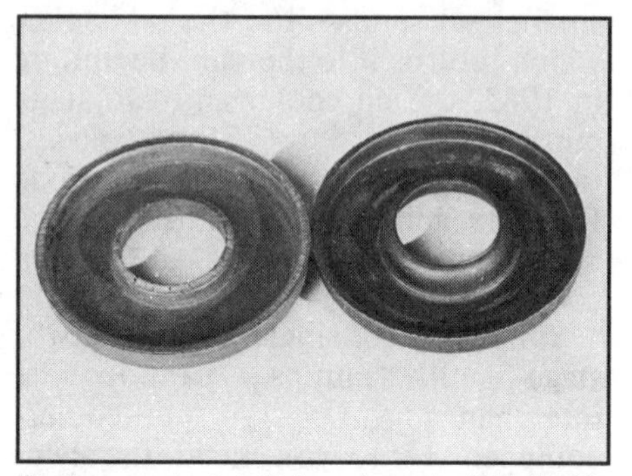

Spring seat at left is unique to '75 1488. All other 1488 and 1600 engines use seat with rolled inner ring at right. Seats do not interchange.

the manifolds; auxiliary-intake valves are on the far, or front side. The distributor is driven off the flywheel end of the cam. The distributor drive is covered and sealed by the distributor housing.

The auxiliary intake-valve assembly is nothing more than a very small valve with its spring and associated parts. The valve and spring are assembled into a valve holder, which fits into the cylinder-head precombustion chamber from the top. The precombustion-chamber port enters the main combustion chamber through a hole or series of holes in its roof. The main combustion chamber is round with smooth radii, especially near the auxiliary-valve hole(s).

1975 1488—The 1975 1488 head features eight ports—one each for all intake and exhaust valves. Starting from the cam-sprocket end, the main valves are arranged: EIIEEIIE. Other differences with later heads include: no valve-stem seals on the exhaust valve, unique rocker arms for intake and exhaust, a thrust washer under the exhaust-valve spring seat, a locknut/spring-seat combination on the auxiliary valve and a unique head-bolt pattern. Metal-clad exhaust-valve-stem seals are now available from Honda, and can be retrofitted to '75 1488 heads.

The major complaint with the 1975 1488 head is that it runs too cool. Its normal operating temperature would be fine for a non-CVCC head, but it runs too cool to keep the CVCC passages from building up and clogging with carbon. The problem is especially bad in cold climates when the car is used mostly for short trips. The rich CVCC auxiliary mixture coats its passages and auxiliary valves until they require disassembly and cleaning.

On the other hand, the 1975 1488cc engine isn't all bad. Most of the carbon-clogging problem is cured by an updated intake manifold from Honda. The original auxiliary passages drop straight down from the carburetor, then turn 90° before entering the head. The sharp 90° bend is where clogging begins. The updated manifold has an enlarged radius at this junction, which reduces flow restriction. As a result, flow velocity is increased and carbon buildup reduced. So, aside from the clogging, the eight-port 1975 1488 head gives good power without overheating problems associated with later heads. That means *no head-gasket problems,* unlike the hot-running 1976—'79 1488 and 1600cc engines. In fact, the 1975 1488 originally used a *sticky* head gasket, unlike the slippery, moly-coated gaskets of later engines. However, *only moly-coated* head gaskets are presently available as service replacements.

1976—'79 1488 & All 1600s—For the next model year, Honda redesigned the CVCC head to run hotter. The eight-port head was scrapped and a six-port design—four intake and two exhaust—substituted. By siamesing the exhaust into two ports, the ex-

All 1600 and 1976—'79 1488 heads are six-port designs, such as this one from junkyard. Changes in a head's port layout always mean different manifolds. Photo by Ron Sessions.

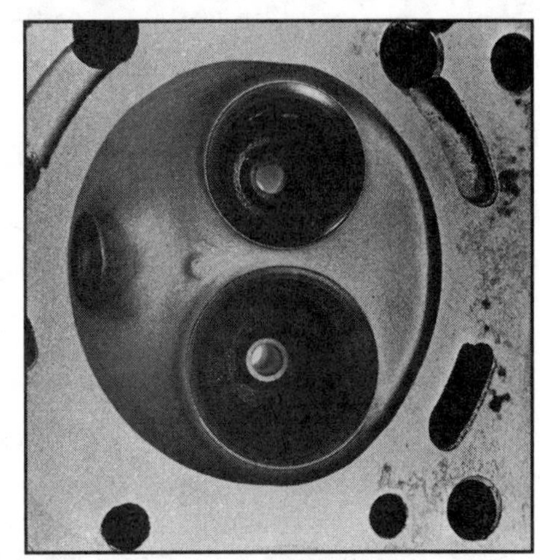

Typical 1488, 1600 combustion chamber is smooth, with prechamber hole connected through side of main combustion chamber.

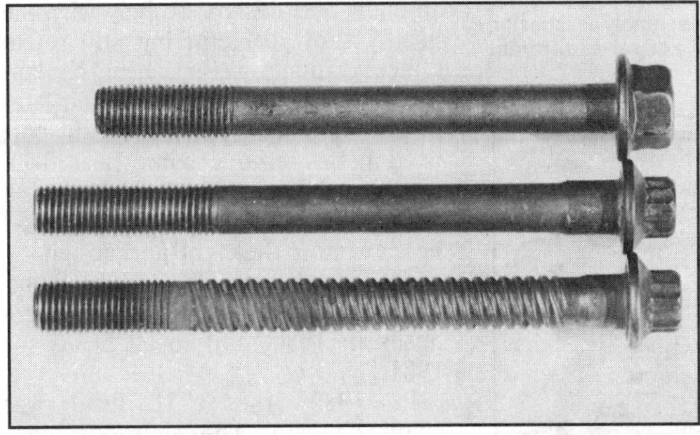

Three 10mm-shank head bolts. Top bolt is a 1200 part and has conventional 14mm hex head. Middle bolt is early, non-knurled CVCC bolt used in '75 1488 engines. Bottom bolt is for all later CVCC engines. Spiral-shank bolt was retrofitted to 1976—'77 CVCC engines and factory installed in all later engines. CVCC head-bolt heads are 10mm, 12-point.

Three prechamber collars may look identical. They go with first three auxiliary-valve assemblies pictured on page 48. 1975 1488 collar (left) has smaller OD than 1976—'77 1488/1600 and '78 1600 collar (middle). 1978—'79 1488 collar (right) is taller than other two. 1975 1488 collar has gasket against its lip in this photo. All lips are same thickness.

haust manifold maintained more concentrated heat in the head.

To get the exhaust into two ports, the valve order had to be changed to: IEEIIEEI. It also meant a different intake manifold, exhaust manifold and camshaft. Furthermore, the new port design necessitated a different head-bolt pattern, so the block was changed also.

The 1975 1488 may look like all other 1488s, but it is unique. Therefore, while the 1976—'79 1488 can use any 1976—'79 1488 or 1600 head, the 1975 1488 accepts only one.

1980—'83 1335—You'd think the 1335 heads would look just like 1488 parts, but they actually mimic the 1751 heads. All 1980 heads and 1981 heads on engines with manual transaxles are six-port designs and use 6.5mm diameter auxiliary-intake valves. But 1981 heads on engines with automatic transaxles and all 1982—'83 heads are eight-port types. As with 1488 and 1751 heads, the '82 1335 head got smaller combustion chambers and different valve angles. These changes and new pistons raised the compression ratio. Valve/piston interference is a problem with these engines if the timing belt breaks or camshaft is incorrectly timed.

1980—'83 1488—Many changes were made to the 1488cc engine for 1980. The head was changed to a better-breathing 1975-style eight-port design. All 1980—'83 heads will interchange with each other. And they'll bolt onto 1976—'79 blocks. You must also use the later manifolds.

Combustion-chamber volume was reduced in 1982, along with different valve angles and more-deeply dished pistons. The net effect was compression ratio being raised from 8.8:1 to 9.3:1. The higher compression gives improved power and fuel efficiency. If you want to use the 1982—'83 1488 heads on an earlier 1488, you'll also need the 1982-and-later deep-dish pistons with valve cutouts. Otherwise, compression will be too high for pump gas, and the valves and pistons may touch due to reduced clearance. In fact, valve-to-piston clearance is so close that even with the correct pistons, 1982—'83 1488s are very sensitive to valve timing. If timing is off more than a few degrees, valve and piston damage may result.

1979—'83 1751—The 1979 and 1980 49-state 1751 cylinder head looks just

Lineup shows why many auxiliary-valve parts don't interchange. From left: '75 1488; 1976—'77 1488, 1600 and '78 1600; 1978—'79 1488; 1979—'81 1751. Two center 1488 assemblies look similar, but have different ODs. 1975 1488 assembly is shorter, uses different holder and valve. The 1751 valve assembly has taller, thicker valve, different spring, retainer and locating pin.

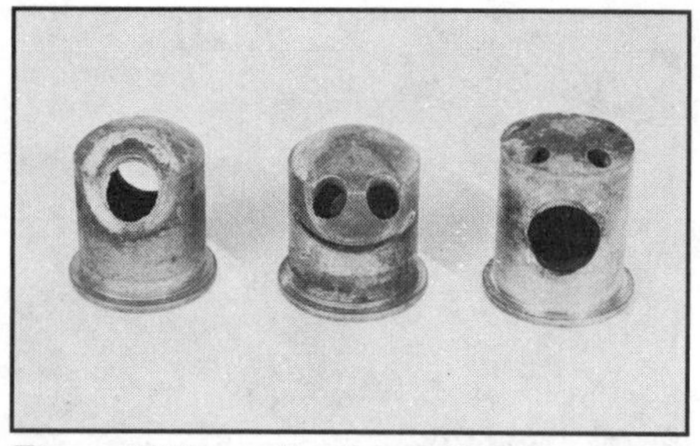

Three 1751 prechamber collars: At left is 1979 collar; one at center is 1979-1/2 update and replacement part; five-hole collar at right is used in '81 California and 1982—'83 engines. Single-hole collar at left can be updated with collar at center.

1982 1751 combustion chamber uses roof-mounted prechamber collar typical of '80-and-later CVCC heads. Main valve order is reverse of earlier 1488 layout.

Auxiliary-valve springs don't change much, except between engines. At left is 1488/1600 spring; 1751 spring is at right. Besides obvious height difference, 1488/1600 spring has larger-diameter wire. Note progressive design of both springs. End with closer-spaced coils goes against head.

Comparing eight-port layout of this '82 1751 head to '75 1488 head on page 46 shows how exhaust valves were isolated. CVCC passages are small holes at 10 and 2 o'clock to the main intake ports. Oblong hole at bottom center is integral EGR passage.

like the 1976—'79 1488/1600 head, but will not interchange with it. Because of different head-bolt spacing, 1751 heads will interchange only with other 1751 heads.

As with the 1488 heads, the 1980 California 1751 head was revised to an eight-port design. To help with the head-gasket problem, but still retain effective intake-passage heat, coolant was routed through 1980-and-later intake manifolds. Thus, the eight-port head helps remove some heat from the head, but the intake charge is still coolant heated. In 1981, all 1751 heads went to the eight-port design.

An exhaust-gas recirculation (EGR) port was added to California heads in 1980, and to all heads for 1981.

In 1982, the 1751 head was changed again. The eight-port configuration was retained, but valve order was changed to separate the exhaust valves. This last 1751 valve order is EIEIIEIE. With this arrangement, cylinder-head temperatures are further reduced, and the almost equal-length intake runners give better breathing. As with all valve-order changes, the 1982—'83 head uses different manifolds and camshaft.

And, as also seen on the 1488 head, the 1982—'83 1751 cylinder head has smaller combustion chambers and *must* be used with notched pistons from those years. Unlike the 1335 and 1488 heads, the 1751 head cannot be swapped onto earlier blocks because the bolt pattern was changed to accommodate the different valve order. Therefore, the 1979 and 1980 49-state 1751 head is unique. 1980 California and all '81 heads will interchange and 1982—'83 heads will interchange, but no head will interchange with another outside of its group.

Teardown 4

Engine with head before teardown: Carb, crank pulley and accessories are missing because I bought this 1237 as is. Start with carburetor and work down when disassembling a complete engine.

Tearing down a Honda engine goes quickly, but don't go too quickly. Take time to look for signs of wear as you go. Too many mechanics scatter parts in record time, losing valuable clues that could help with parts inspection and reconditioning. Don't fall into this trap. Study internal-wear patterns as you remove each part. Then you'll be able to spot the problem areas in your engine and estimate the work and parts required to fix them. Teardown is an opportunity to observe the effects of maintenance—or the lack of it.

As when you pulled the engine, separate hardware according to function. Store parts in labled containers. Pay special attention to the location of internal parts and their positions relative to each other. This is your last chance to see them in their factory-installed order—if this is the engine's first rebuild. Sketches or photographs can be helpful—make them before you remove the parts.

EXTERNALS

If you pulled the engine as a complete assembly, start engine disassembly by removing the head, page 27. Then, come back here to finish disassembling the short block.

If you removed the cylinder head and block separately, start stripping the block while it hangs from the cherry picker or hoist. Lightening the engine makes it easier to manhandle later.

Oil Filter—Start with the oil filter. If it won't unscrew by hand, try an oil-filter strap wrench. Or, you can hammer a large screwdriver through the filter. Then lever the screwdriver counterclockwise—very messy, but it gives enough leverage to break loose the filter.

Right above the oil filter is the oil-pressure sending unit. Use an adjustable wrench to remove it.

Mounts—Both engine and accessory mounts can be unbolted now. With the engine shown, this meant the A/C-compressor and rear engine mounts. Each is attached with two bolts.

Water Pump & Pipe—One of the more unusual features of Honda engines is the external water pipe that's routed alongside the block. To remove the water pipe, use a 10mm socket to remove the clamp bolt, grasp the pipe and pull it out of the

Other side of 1237 shows air-injection manifold. Air-pump hose had been cut off and pump discarded; emission-control-equipment tampering is illegal in some states. Photo by Ron Sessions.

While engine hangs from cherry picker, small parts can be easily removed. Engine-mount bracket, A/C-compressor bracket, oil filter and oil-pressure sending unit can be removed from front of block.

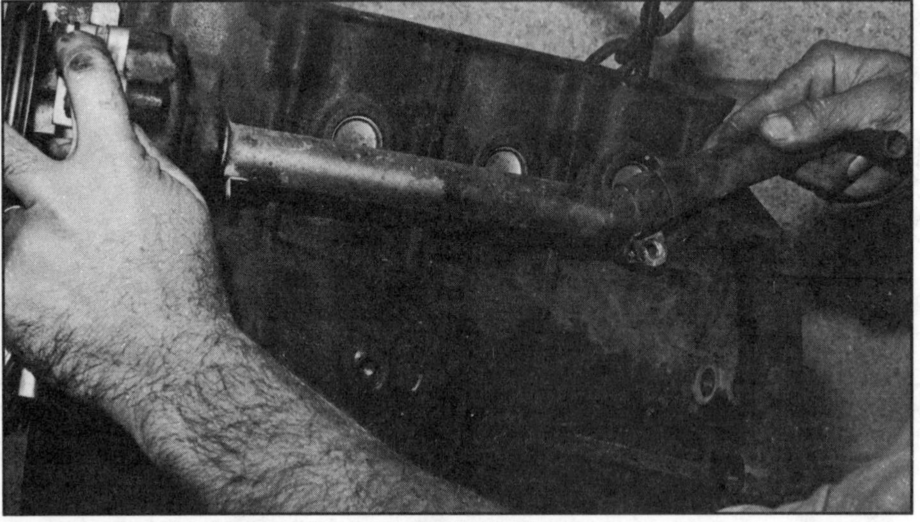

Remove water pipe. Once bracket is unbolted, pull out pipe.

Remove water pump and pulley. Water pump is high-failure-rate part on Hondas; most never make it to 70,000 miles. Your rebuilt engine should get a new one.

water-pump housing. An O-ring at the water-pump end provides slight resistance.

Set aside the pipe and unbolt the water-pump pulley, if possible. Some pumps have press-fit pulleys. On these, slip a 10mm open-end wrench behind the pressed-on pulley and unscrew the three or four bolts. Hold the pump away from the block while removing some of the bolts or they'll butt against the back of the pulley. When removing the pump, make sure you get the D-shaped O-ring.

On 1975—'79 1237cc engines, removing the water pump will also remove the air-injection-pump mounting bracket—the two share common bolts.

Clutch—On manual-transaxle cars, the clutch is nestled inside the flywheel. The pressure plate is indexed to the flywheel with two dowels. Therefore, its unnecessary to matchmark the pressure plate and flywheel. Remove the six bolts, and the pressure plate and disc will fall out. If you are reusing the clutch disc, mark which side faces the flywheel to avoid confusion when installing the disc.

Flywheel—By removing the clutch, you uncovered six flywheel bolts. These can be tough customers. It isn't that they are so tight—30 to 50 ft-lb—it's their half-high hexes. The sides of the hex are made short so they won't interfere with the clutch.

Because pressure plate is doweled to flywheel, matchmarking their relationship is not needed.

This decreases the bearing area between the 14mm-socket and hex flats, which translates into slipping tools and rounded bolt heads. For this reason, use a six-point instead of a

Use box-end wrench to smack loose pressure-plate bolts—open-end wrench might slip off and damage flats. Remove bolts one turn at a time to avoid warping pressure plate.

Lower pressure plate with clutch disc. Try not to touch friction material with greasy fingers. Goo on clutch causes disc chatter.

With clutch out of way, remove flywheel bolts. Flywheel weighs over 20 lb, so have a good grip on it before pulling it off crank.

12-point socket. You can *try* to remove these bolts with a breaker bar and socket. Hold the crank stationary with one hand, the tool in the other, and with your "third," keep the engine from moving. A helper should steady the engine while you pull on the breaker bar. If possible, use a *flywheel lock plate*—a toothed tool that bolts to one bellhousing-flange bolt hole and engages two flywheel teeth—to free one hand.

A much better way to remove these bolts is with an impact wrench. Impact torque is applied directly to the bolts because the inertia of the engine resists rotation. If you don't have an impact wrench, take the engine to a service station and have them zap off the flywheel bolts. While you're at it, have them remove the crank-pulley bolt. This requires a 17- or 19mm-hex socket.

Best way to remove crank pulley is with impact wrench. If you don't have one, take short block to service station and have pulley bolt zapped off.

Driveplate—Engines mated to automatic transaxles have a driveplate in place of the flywheel. Because there's no clutch to worry about, all you need to do is remove the six driveplate bolts and pull the sheet-metal plate off the crank. Except on 1200s, driveplates and flywheels are positioned by a dowel jutting from the crank flange. Therefore, on these engines, you don't have to mark the driveplate's relationship to the crank. The dowel does it for you. On 1200s, center punch the crank and flywheel as you did for the pressure plate.

Engine Stand—Once the engine exterior has been stripped, I recommend mounting the short-block assembly on an engine stand. A stand is not absolutely necessary, but it makes things a lot easier. With a stand bolted to the bellhousing mating flange, you can rotate the engine about the crankshaft centerline. When it's time to install the crank, or rods and pistons, it takes but a second to roll the engine over. On the bench, it means moving a bunch of wooden blocks or supports, wrestling the block over and refitting the blocks. You can rent engine stands from rental yards and some auto-parts outlets. If you do occasional engine rebuilding, buy a stand.

If you can't find three or four 10mm bolts long enough to extend through the engine-stand mounting pedestals and thread into the block, use cylinder-head bolts. Depending on which engine you have, these 85—105mm-long bolts will work.

Crank Pulley—It's time to work on the other end of the engine. Start by loosening the bolt in the center of the crank pulley. If you didn't already do this with an impact wrench, you'll have to keep the crank from turning. As previously mentioned, you can do this with a flywheel lock plate. If you don't have one of these, take two flywheel bolts and run them into their holes in the flywheel flange. Now position a hammer handle, drift or other stout *handle* through the bolts. You can now hold onto the crank at one end and break loose the pulley bolt with a wrench at the other end. With the bolt removed, slide the pulley off the crank *snout*. This is a slip fit and

Four 10mm head bolts are perfect for mounting engine to stand. Bellhousing-to-block bolts are not long enough to fit through pedestals of most engine stands. Photo by Ron Sessions.

When stripping complete engine, upper belt cover must come off before bottom cover. CVCC bolts are on cover sides; 1200 timing-belt-cover bolts are in cover face.

Be sure to catch woodruff key when removing crank pulley. In photo, key has fallen inside pulley. More often, key can be plucked from crankshaft groove with needle-nose pliers.

should not require a puller. Be very careful when removing the pulley; the *key* that positions it to the crank will probably fall out. Unlike a *woodruff key,* which is half-round, the Honda key is *flat* on all sides. It is like a section of small bar stock.

To store the key, tape it to the pulley, or put it in a small box. Small parts are easy to lose. If you do lose the key, it means a special trip to the Honda dealer—special part, special trip.

Lower Belt Cover—Four bolts hold on the timing-belt lower cover.

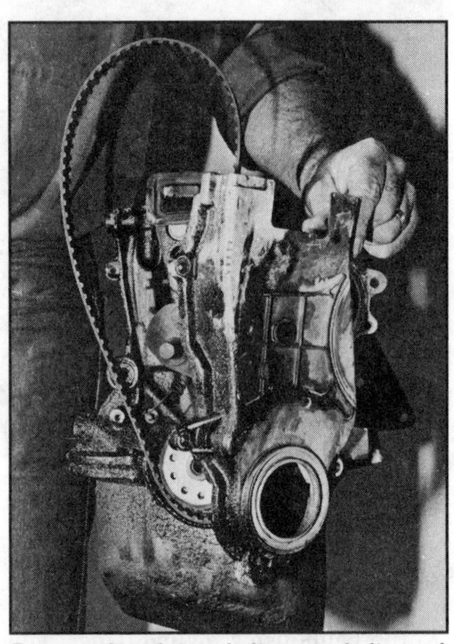

Remove four lower belt-cover bolts and pull off lower cover.

Unbolt the plastic cover, and pry it away from the block. The gasket should not resist much. Remove the timing belt by disengaging it from the tensioner and crank sprocket. At the top of the cover is a square rubber grommet that fits around the engine-mount bracket. Pull the grommet off over the bracket, then unbolt the bracket. It attaches with two bolts.

Belt, Sprocket & Tensioner—The crankshaft timing-belt sprocket is next. Actually, three separate parts make up this lower sprocket: the sprocket and its two end *fences.* The fences, which look like discs with upturned edges, keep the belt from running off the ends of the sprocket. All three pieces merely slip off the crank.

Next to come off is the timing-belt tensioner. It's immediately above the crank snout. Remove the two bolts and the tensioner is off. Removing the tensioner frees a small spring that hooks onto a block-mounted pedestal below the tensioner. Don't lose the spring. Make sure you collect all of the spring washers and rubber grommets, too.

Oil Pan—It's bottoms-up time. Roll the engine over so you have clear access to the oil pan's 14 bolts and four nuts. This is no problem if the block is on an engine stand. Just have a drip pan underneath. If you are working on a bench, you can turn the engine completely upside down. Another way is to roll the engine over on one side. Use a short 2x4 under the block to prop up the engine so you can reach the lower row of pan bolts. Turning the engine on its side will help keep trapped oil from draining out the top of the block and all over the bench—until you remove the pan.

If you haven't done so already, remove the sheet-metal dust cover for the flywheel or driveplate. Then, finish removing the oil-pan bolts. You'll need a 10mm socket on a long

Slip off crank sprocket *fence* and unthread timing belt.

Turning engine upside down is easy with engine stand. Loosen pan bolts from *center out*.

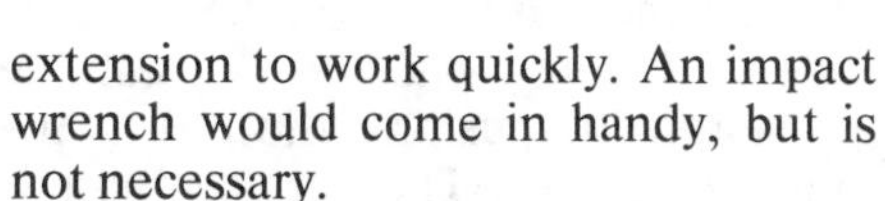

extension to work quickly. An impact wrench would come in handy, but is not necessary.

Just because all bolts are out doesn't mean the oil pan will fall off. Usually, a good straight pull works. Avoid driving screwdrivers between the pan and block. This can distort the pan and cause leaks later. Instead, take the screwdriver and hold it at a shallow angle to the oil-pan lip. Rap the screwdriver with a hammer. Or, just rap the pan a few times with a rubber hammer. When the gasket seal breaks, the pan emits a hollow sound. Give the pan a tap in the four corners where the crankshaft cutouts are. Then lift off the pan.

Oil Pump—Welcome to the *bottom end* of the engine. The first job is to remove the oil pump. On CVCC engines, three bolts attach the oil pump to the block. Two are in the open, a third hides inside the pickup screen. Pull the screen off the pump pickup with a small screwdriver. Remove all three bolts and pull the pump from the block.

On 1170, 1237 and 1335cc engines, the main-bearing caps are incorporated into a cage—the oil pump mounts to this cage. There are three bolts, but all are exposed. You don't have to get under the screen. Additionally, there are two bolts in the *pump-bypass block*. This is the tubular extension that also attaches to the main-bearing cage. Remove these two bolts, then remove the pump.

Connecting Rods—All that's left in the engine are the connecting-rod/piston assemblies and crankshaft.

Slide off crank sprocket and rear fence.

Knock pan loose with rubber or rawhide hammer. Once pan is loose, lightly pry up pan with screwdriver, then lift off corner studs. *Don't drive screwdriver between pan and block*, especially on aluminum-block 1200s.

Loosen both tensioner bolts, then remove tensioner-pivot bolt. This allows tensioner to rotate so spring will unclip. Now remove remaining bolt and tensioner.

Hondas absolutely require frequent oil changes. Engine smelled like a burned oil refinery and was full of gritty oil. Crank was nearly ruined.

1200 oil screen was half clogged by carbon particles. Lost oil pressure caused spun rod bearing, ruining crank and rod. Frequent oil changes or cleaning screen would have saved engine.

Pry screen from iron-block-engine oil pump to get at third oil-pump retaining bolt. If examination with magnet reveals metal particles, check crank and rods for damage.

Don't forget to remove third bolt inside CVCC oil pump.

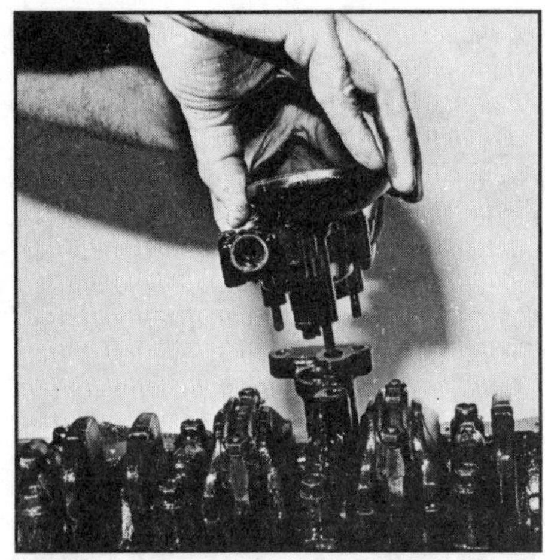

A little wiggling and straight pull removes CVCC pump.

Numerals on Honda rods indicate bearing sizes. You *must* mark rods for position.

Oil pump on aluminum-block engines uses four bolts. Two pass through pump body and two through outlet block. Remove all four and lift pump and outlet section as unit. Screen fits over outside lip.

Remove the rod/piston assemblies first.

Before removing the connecting rods, mark them. Honda does not mark the connecting rods according to their position in the engine. You *must* do it. While inspecting the rods, you'll see numbers that *look* like position numbers, but aren't. Upon close inspection, you may see two 3s, or no 1s or some other unusual combination. These numerals are for bearing selection only and have nothing to do with connecting-rod bearing-journal or bore position.

If you have a numbering set to mark the rods, great. If not, use a hammer and *sharp* center punch. Rotate the crankshaft so the big end of rod 1 is where you can get the numbering set or punch on it. Don't use the side that's already factory-marked. Stamp both rod and cap at their parting line.

You must stamp both because each rod and cap are a matched pair. Your markings are the only way to ensure the pair go together later. With a numbering set, mark the rods 1—4, with 1 at the pulley end and 4 at the flywheel end. With a punch, make one indentation on the first rod, two on the second, and so on. Hammer only hard enough to make a mark. Excessive force will damage the rod.

Checking Rod Side Clearance— Another pre-removal necessity is checking connecting-rod *side clearance.* This is the distance between the side of the rod big end and crankshaft-bearing-journal end. Measure it by pushing the rod in one direction until it butts against the crankshaft. Then, find the feeler gage that fits between the rod and crankshaft on the opposite side of the rod big end. Repeat the procedure by pushing the rod in the other direction. The service limit is 0.016 in. (0.4mm).

If wear is beyond the service limit, you have two choices; replace the rod and/or crankshaft, or find a shop that will weld up the crank. You are better off replacing the rod or crank. Subsequent checks will determine whether the rod or crank (or both) are worn past their respective limits.

Remove Rods— After checking them, remove each rod-cap nut using a thin-wall 12mm socket and extension. Grasp the rod cap with your thumb and forefinger and wiggle the cap off the rod bolts. These bolts are designed to accurately position the cap, so it won't just slide off. With the cap off, slide a piece of hose over each rod

Check rod side clearance. Maximum is 0.016 in. (0.4mm) for all iron blocks, 0.012 in. (0.3mm) for aluminum blocks.

After numbering, remove rod nuts. On aluminum blocks, use extension to clear main-bearing cage, as shown.

Wiggle rod cap between thumb and forefinger to loosen it. Once loose, pull straight up to remove cap.

bolt. This protects the crank journal from being scratched by bolt threads.

Push on the rod/piston assembly to move it out of its bore. Don't let the rod bang against the cylinder. Unlike most engines, Hondas rarely develop a ridge at the top of the cylinders. Many Honda-engine builders attribute this to the *low-tension* ring design—the rings don't press hard against the bores. Therefore, dressing the cylinder tops with a ridge reamer is seldom needed. Be ready to catch the rod/piston assembly when it pops out of the block.

With the crankshaft in this position, you should be able to get rods 1 and 4 out. Then turn the crank 180° and remove rods 2 and 3. After removing the rod/piston assemblies, slip the caps back on their rods and start the nuts on the bolts.

Spin Crankshaft—Turn the crankshaft over a couple of times. It should spin freely. You are checking for a bent crank. A bent five-main-bearing crank is not common, but this check is too easy to skip. You should be able to turn the crank by pushing lightly on a throw. If the crank won't rotate easily through 360°, and has a *sticky* spot, it may be bent. If you suspect a bent crank, pay close attention to *reading* bearings, page 69.

Checking Crankshaft End Play—Before you remove the crankshaft, check its *end play.* Use a lead mallet to knock or a screwdriver to lever the crankshaft forward or rearward as far as it will go. Then, measure between the thrust washer on main-cap 4 and the crankshaft fillet. Or, set up a dial indicator with its plunger square against either end of the crank. Lever

Guide rod-and-piston assembly from bore with both hands. Use journal protectors over rod bolts to avoid nicking crankshaft.

the crank toward one end, zero the indicator and lever the crank in the opposite direction. Read end play directly.

End-play service limit is 0.018 in. (0.45mm). If end play exceeds the service limit, mike the crankshaft thrust washers after removing them. New thrust washers are 0.097-in. (2.5mm) thick. What you must determine is if new thrust washers will restore end play to specifications. The desired end-play range is 0.004—0.014 in. (0.1—0.36mm). If not, the crankshaft may be unusable.

As an example, let's say end play is 0.025 in. Subtracting the service limit of 0.018 in. nets 0.007 in. that must be restored by using new thrust washers. Let's also suppose that both old thrust washers mike 0.091-in. thick. Subtracting the old thrust-washer thickness—0.091 in.—from the new thrust-washer thickness—0.097 in.—gives 0.006 in. Therefore, using new

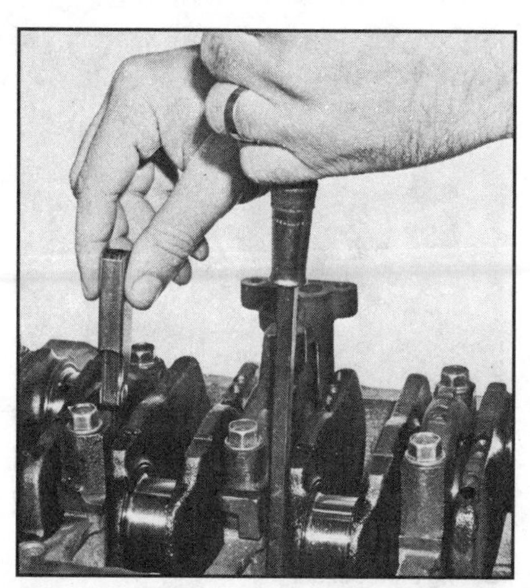

Check crankshaft end play before removing crank. Lever crank fore and aft in block with large screwdriver. Measure with feeler gages between crank and thrust washers on number-4 main cap. Service limit on end play is 0.018 in. (0.45mm). If end play is excessive, measure thrust-washer thickness and compare to new thrust washer. New thrust washers are 0.097-in. (2.5mm) thick. Oversize thrust washers are not available. If end play cannot be corrected with new thrust washers, replace crankshaft.

thrust washers will add 0.006 in. X 2 = 0.012 in. Reducing the 0.025-in. end play by 0.012 in. will bring it to 0.013 in., within the desired 0.004—0.014-in. end-play range. The crankshaft can be reused.

Now, if new thrust washers will not reduce end play to acceptable limits, the crankshaft must be welded and turned to specs, or replaced. Usually, a quick visual check can determine whether the crank or thrust washers are the culprits.

Breaking loose main-bearing bolts is easy with 1/2-in.-drive ratchet. With a breaker bar, it's much easier.

Loosening main-bearing caps with lead mallet works great. Light taps are sufficient.

Unlike rod caps, iron-block main-bearing caps are numbered. Cage design of aluminum-engine main caps makes numbering unnecessary.

If you don't have a lead mallet, try this trick. Place two open-end wrenches or bolts in main-cap bolt holes. Grasp wrenches or bolts and rock cap out of its register.

Caged main caps of aluminum engine can be popped loose with length of wood. *Pry lightly* and alternate from side to side, or you'll crack cage.

Main-Bearing Caps—On all iron-block engines, the main-bearing caps are numbered at the factory. The number is cast into the bottom of each main-bearing cap. With the engine upside down, the numbers will face up, except for number 3; it is cast into the side of the cap because of the oil-pump mounting.

Aluminum-block engines differ because their main-bearing caps are cast into one integral unit—a cage. Remove one cap and you remove them all. Because you can't install these caps backward, there is no need for numbers.

With a ratchet and 14mm socket, remove the main-bearing-cap bolts. On aluminum-block engines, lift off the cap assembly, or cage. Rock the cage off its dowels, first one end and then the other.

Engines with separate main-bearing caps are similar. Rock the caps toward the sides of the block, not parallel to the crank centerline. To get more leverage on the caps, partially insert a main-bearing bolt into each cap hole. Wrap both your hands around them and rock the cap up and off. Remove all five caps.

Oil Seals—By removing the main-bearing caps, you have exposed two oil seals, one at each end of the crankcase. Pull them out with your fingers and set them aside. When you get new seals, compare them to the old ones to make sure you got the right seals.

Remove Crankshaft—All that's left to do in the bottom end is lift the crankshaft out of the block. Lift straight up so you don't bang the crank against the block and nick a journal. Now's a good time to read bearings as described on page 69.

STRIPPING THE HEAD

Don't take the bare block off the engine stand, yet. It makes a great hold-down for the head as you strip it. On CVCC engines, find two head bolts, set the head assembly on the block and run in the two bolts so the head won't fall off. On aluminum-block engines, slide the head down over the cylinder-head studs. Use the old head gasket between the block deck and head to keep from scratching the two surfaces. If you don't have the entire gasket—it usually tears during disassembly—use the largest pieces that are left.

Sparkplugs—Start by removing the sparkplug leads. Disconnect them and lay them aside. Use a 13/16-in. socket

Rocking aluminum-engine bearing cage at each end will also loosen it. Five main-bearing caps and oil-seal-housing halves lift off as a unit.

Prying front and rear oil seals from block can make crank easier to lift, but is optional step.

Grasp crank by *snout* in front and place a finger in rear pilot hole. Lift straight up. Don't tilt crank or counterweights will hang up in block.

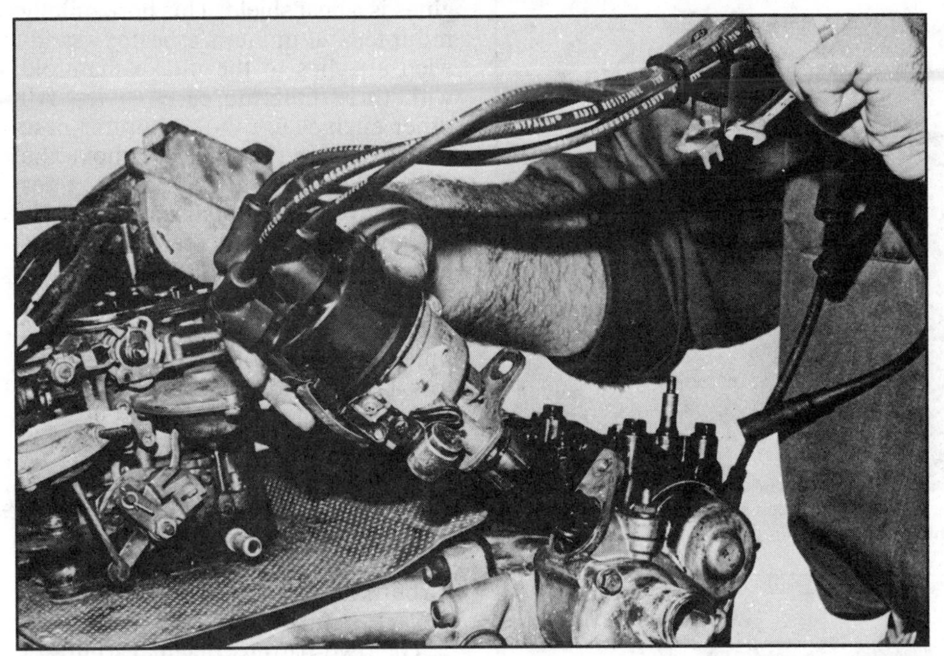

Cylinder-head disassembly. Pull leads off sparkplugs and remove distributor hold-down bolt. Pull distributor from housing on CVCC engine, or head on 1200. Also, remove fuel pump on 1200 head.

to remove the sparkplugs. Unless they are practically new, you won't be using old plugs in the rebuilt engine, so discard them.

Distributor—With a 10mm wrench, remove one distributor mounting bolt, then pull the distributor completely out of its housing.

Thermostat—On CVCC engines, the thermostat-housing bolts clamp the distributor housing to the head. Simply unscrew the two long thermostat-housing bolts and pull them out. The thermostat might fall out. If not, a light pry will dislodge it. The only thing holding on the distributor housing now is the gasket and O-ring sealing it to the head. A light pull will remove the housing.

On 1170 and 1237cc engines, the thermostat housing bolts to the intake manifold. If you are doing a "quick-and-dirty" job, leave the thermostat alone. Otherwise, remove it for manifold cleaning.

Carburetor & CVCC Emission-Control Box—When it comes to the unique CVCC Keihin three-barrel carburetor, every mechanic wants to meet the engineer "who designed this stupid thing!" But when conversation turns to the control box(es), consensis is that it's a clever piece of work. Think of all the connections you'd have to break to remove the carburetor if the control box(es) wasn't there. With it, all you need do is unbolt the four carburetor hold-down nuts, control-box attaching screw and lift off the carb, vacuum hoses and box(es) as a unit. Fantastic!

Clearance around the four *base nuts* is fairly good, but it takes a lot of short swings to remove the nuts. There are *two* washers under each nut, so don't overlook the second one.

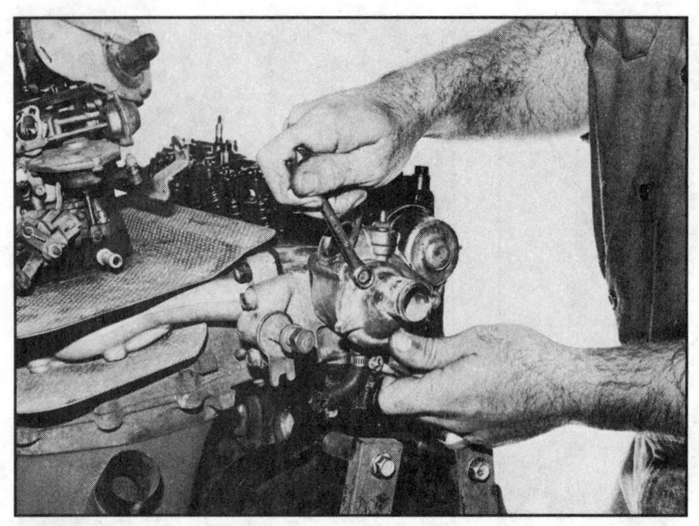

Three bolts retain distributor housing to CVCC head. Two pass through thermostat cover and third is to right of thermostat. Remove these three bolts and pull distributor housing from head. O-ring around distributor drive gear causes enough resistance to require a slight pull.

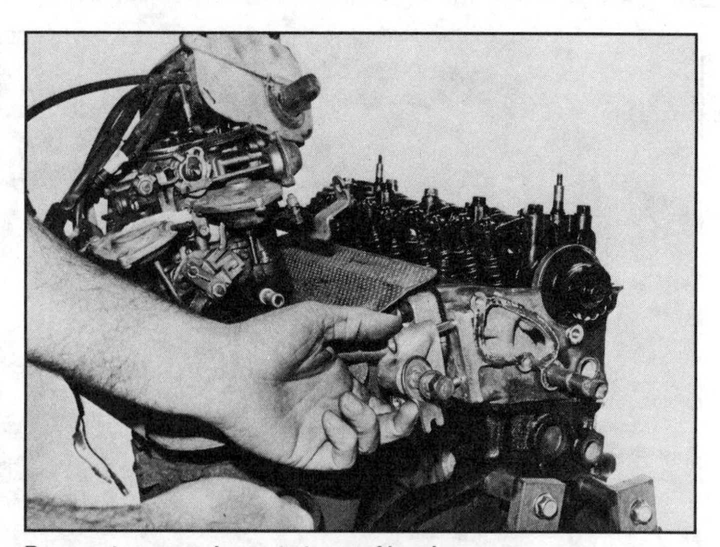

Remove torque-rod mount at rear of head.

After unbolting carburetor base nuts, lift off carburetor and black box as a unit. Separating them would mean reconnecting all those vacuum hoses.

Lift off carb gasket and CVCC-engine heat shield.

Under the carburetor on CVCC engines is a heat shield. This horizontally mounted aluminum/asbestos sandwich attaches to the intake manifold with three bolts on early Civics. All other engines use the carburetor-base nuts to secure the shield. Remove the shield. The 1170 and 1237cc engines don't have a horizontal heat shield.

You may wish to rebuild the carburetor while the engine is apart. This is a good idea. A rebuilt carb will help give new-car performance. But, unless you have rebuilt many carburetors before and don't mind working without a lot of special tools, I don't recommend rebuilding the CVCC carburetor yourself. The Keihin three-barrel carburetor used with CVCC engines is complicated—leave it to a mechanic who specializes in these carburetors. Even with all the correct tools, it's still not easy.

The two-barrel Hitachi carburetor used with 1170 and 1237cc engines is another story. Just buy a rebuild kit at your local import-parts outlet and follow the kit instructions.

Manifolds—As discussed earlier, all CVCC engines have non-crossflow heads. This simply means the exhaust and intake manifolds are on the same side of the head. The 1170 and 1237 heads are a crossflow design—intake and exhaust manifolds are on opposite sides of the head. I'll cover the non-crossflow type first.

CVCC Engines—With non-crossflow, CVCC engines, the exhaust manifold is under the intake manifold. Honda

After removing all exposed CVCC-engine manifold nuts, except center intake-manifold nut, sneak in from side and remove two lower manifold nuts. Use box-end wrench.

Four vertical bolts can now be removed. Be ready for heavy CVCC exhaust manifold as last bolt is unthreaded.

engineers use this arrangement to heat the underside of the intake manifold, resulting in faster warmups for better fuel economy and driveability in cold weather. The idea is to heat the intake manifold immediately under the carburetor, but to insulate the intake-manifold *runners* from exhaust heat. This heats the air/fuel mixture where it's needed for improved atomization.

So much for theory. Without Honda's special S-shaped box-end wrench, it's impossible to reach the bottom-center manifold-to-head bolts from above. So, if you don't have this tool, remove the manifolds separately. First, remove the two or four bolts that pass vertically through the intake manifold into the exhaust manifold. You exposed these bolts by removing the carburetor. Remove the bolts completely, then slide the *riser gasket* from between the intake and exhaust manifolds. The riser gasket looks like a dull aluminum tray with one edge bent slightly. This gasket also shields the intake-manifold runners from exhaust-manifold heat.

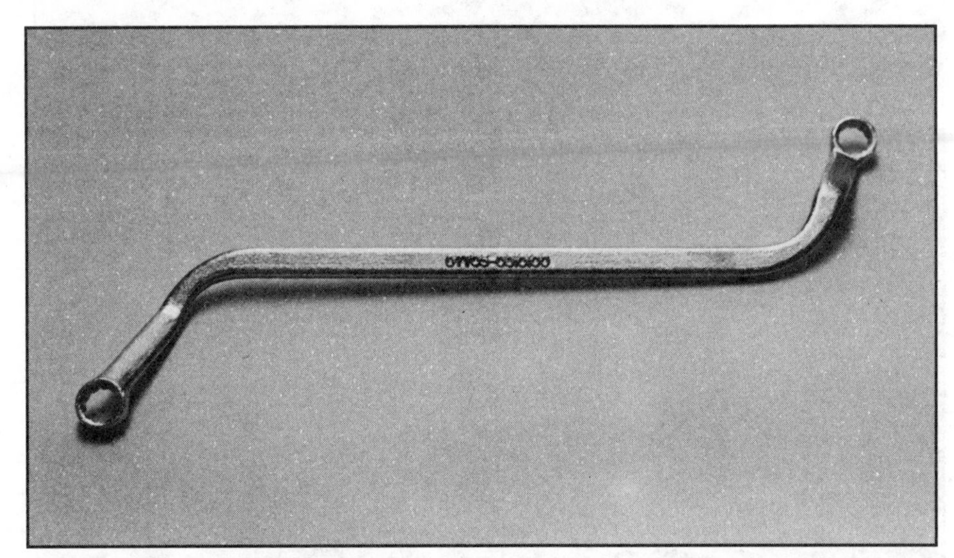
With special S-shaped 12mm box-end wrench from Honda, you can remove CVCC intake and exhaust manifolds from head as a unit. Without it, you must remove manifolds separately. Photo Courtesy of Jackson Racing.

Now, remove the nine or 11 nuts that hold the manifolds to the head. There are two rows of nuts and washers; one at the top and the other at the bottom. The top row is accessible, so you should be able to get them off quickly with a 14mm wrench. The bottom row has four or five nuts. Two at each end are in plain view, but the center nuts are difficult to reach.

Because the intake-manifold flange encircles the mounting studs, you can't remove the manifold to get at the nuts once they're loose. And, you can't easily get at the nuts with the manifold in place. You'll have to snake your hand in from the bottom of the manifolds to reach these two nuts. Guide a 14mm box-end wrench between the exhaust manifold and block until its end reaches the first nut. You can see the wrench and nut by looking down through the intake manifold. Unscrew the nut a little at a time. Let the nut and washer fall on the floor. If your engine has an 11-nut manifold, remove the center nut next. Otherwise, go to the other end of the engine and work your other hand in from that end. Remove the second nut, letting it and the washer drop.

With exhaust manifold off, intake manifold can be removed.

When unbolting cam sprocket, keep sprocket from turning by wedging flat-blade screwdriver between sprocket hole and top of head. Keep broad side of screwdriver tip against head to prevent gouging.

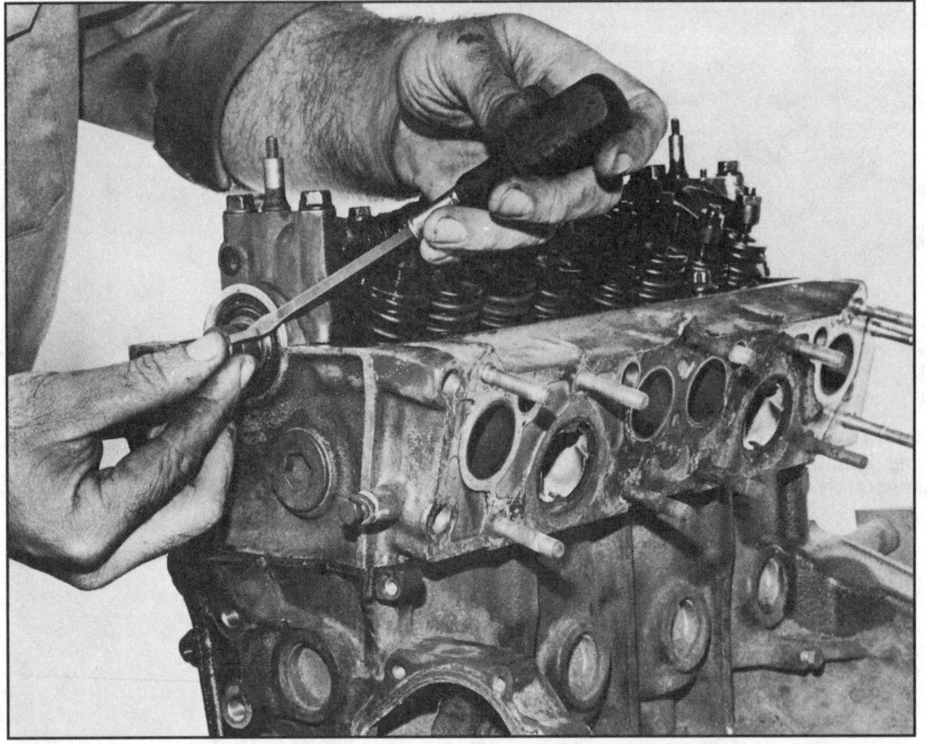

After sliding sprocket off camshaft, remove woodruff key. It fits loosely in keyway and can get lost easily if not removed and stored right away.

Both manifolds should slide off the mounting studs. Set the manifolds and hardware aside, and discard the manifold gasket. Don't forget to pick up the nuts and washers before you accidently kick them into the "fourth dimension." Unbolt the head from the block and set it aside also. Head disassembly is covered in Chapter 6.

1200 Engines—Manifolds are easier to remove from crossflow heads.

Start with the intake manifold. It attaches to the head with six studs, nuts and washers. Four are on top and two are below the runners. Remove all nuts and washers, then slip off the manifold.

On the other side of the head, remove the exhaust manifold in the same way. First, remove the heat stove and its two bolts. Then, remove all eight manifold nuts and slide the manifold off the studs. Also remove the air-injection-pump upper bracket, if so equipped.

Unbolt the head from the block and set it aside. Head teardown and reconditioning is covered in Chapter 6.

Camshaft Sprocket—Remove the camshaft-sprocket bolt and pull the sprocket off the cam. As with the crank sprocket, the cam sprocket uses a straight key. Take care not to lose it.

To keep the cam from turning while breaking loose the sprocket bolt, pass a shaft through one of the sprocket's holes and rest it against the head. It's best to use a round drift so you don't gouge the head. The flat section of a large screwdriver against the head will also work.

Short-Block Reconditioning

5

Let's start the decision-making stage of the rebuild by inspecting and reconditioning parts that make up the short block. The reconditioned short block will be the foundation of your rebuilt engine. Patience and attention to detail will pay off handsomely. Rushing the work and skipping steps will cause problems. You don't want problems to crop up once the engine is back in service. Then, they will be difficult and expensive to fix.

Although you may not have the knowledge and tools to perform all procedures discussed in this chapter, read each description carefully. Knowing what needs to be done will help you select a good machinist. As you can see from the depth of this and the next chapter, machining an engine is not a simple or easy task. Keep that in mind when shopping for a machine shop. The lowest price is often no bargain.

Choosing a machine shop is an important rebuilding decision. A good shop or machinist will save you time and money; even if the bill is higher than the place across town.

CLEAN & INSPECT CYLINDER BLOCK

Head Gasket—What can ultimately be determined from inspecting the head gasket and mating surfaces is the flatness of the head and block. The 1976—'77 Honda CVCC engines are infamous for *blowing* head gaskets. A well-publicized factory recall campaign has repaired hundreds of thousands of these engines. The factory fix involved installing new spiral-shank head bolts, slippery moly-coated head gasket, and head replacement, in severe cases. Later Honda CVCC engines and updated 1976—'77 models don't experience head-gasket problems nearly as often as the old power plants.

Study the head, gasket and block. Look for black streaks caused by combustion leaks. They could lead to another cylinder, water passage, or block or cylinder-head edge. If the streaks are thick and well-defined, the combustion leak was major. You can expect head warpage—possibly block warpage, too. Most minor leaks show as discoloration of the head gasket.

The gasket has a stainless-steel bead that fully encircles each bore/combustion chamber. This bead is shiny where it was sealing. Dark areas, however, indicate combustion leakage. The area between cylinders 3 and 4 is invariably discolored from seeping combustion. Don't worry about discoloration that looks like coffee stains. Cleaning the head and block, plus a new head gasket, will remedy it. However, if you find signs of significant leakage, inspect the head and block deck as described on pages 66 and 91.

Gasket Scraping—Remove all old gaskets with a gasket scraper, putty knife or single-edge razor blade. While you do this, continue to look

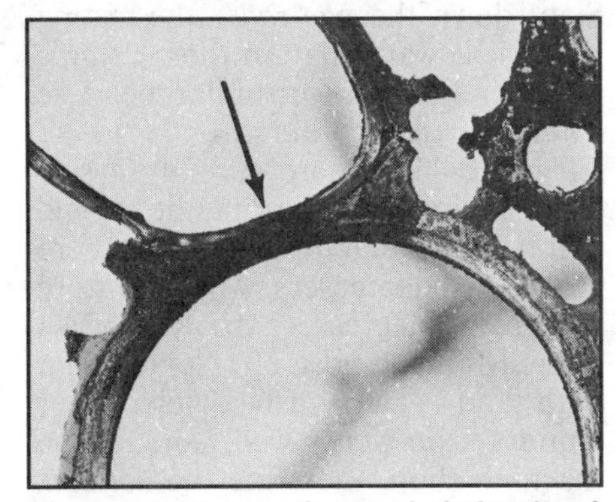

Head gasket was damaged during head removal, but still gives good clues. Dark area (arrow) between cylinder openings indicates combustion leak. New gasket should cure this leak, but more extensive discoloration may require milling head. Check head to make sure.

Core-plug removal is two-step process. First, knock one edge into water jacket . . .

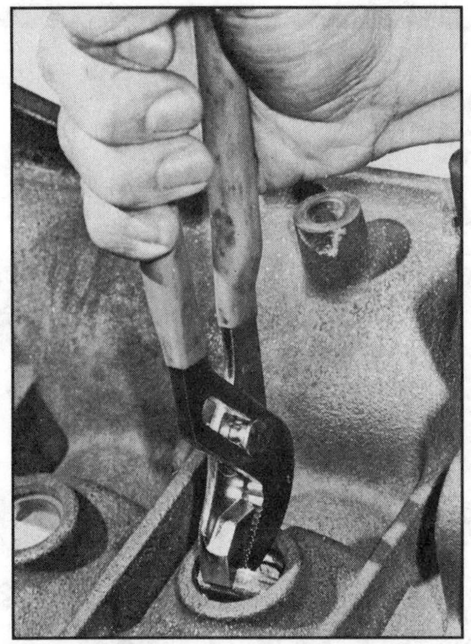

. . . then lever out plug with pliers.

Access to cast-iron-block oil gallery is gained by removing this plug. Aluminum-block gallery and plug is in bearing cage.

Trip through jet tank removes paint, oil crud, metal chips and other grit. Parts rotate on turntable while jets spray hot, caustic solution.

for signs of leakage.

Some Honda gaskets, such as the CVCC manifold gasket, should peel right off. But the typical composition head gasket does not. If you have a choice, use a gasket scraper instead of a putty knife or razor blade. A heavy, wide-blade scraper will make fast work of stubborn gaskets. It's also much safer to use than a razor blade.

Scrape off all gasket material, including material on the pan and sheet-metal covers. Peel off all neoprene gaskets. If you find gasket adhesive around the neoprene seals, remove it. Check sheet-metal parts for flatness when you scrape them. Most Honda engine sheet-metal parts are heavy-gage and not prone to deforming. But, overtightening can cause dimpling around each bolt hole. Sight down the *pan rails*—the mating flanges. If wavy, flatten them using a hammer and dolly or similar tool. Use a punch in tight corners.

Cleaning—There are several ways to clean engine parts. During the rebuild you'll use several of them. First, there are canned degreasers. Any auto store has engine degreasers that you brush or spray on and hose off. Next is plain soap and water—dish detergent or laundry soap works well. And, at the machine shop, there are the *hot-tank* and *non-caustic* cleaners.

Canned degreasers are good for cleaning extremely dirty engines prior to measuring and machining. Do this before delivering the parts to the machine shop.

Ferrous parts, such as iron blocks, oil pans, connecting rods, crankshafts and sheet-metal covers, can be hot-tanked. Hot-tanking is harsh. The heated, caustic solution dissolves dirt, dried grease, and soaked-in oil, along with aluminum, brass, cadmium plate, most core plugs and other soft metal. Hot-tanking is only for iron and steel—*ferrous* parts. Its big advantage is the ability to swirl inside oil passages and clean out hidden goo. If you have a filthy iron-block engine, full of rust, burnt oil deposits, grit, sludge and the like, hot-tanking is for you. However, if your engine is merely worn out, tanking is probably not necessary. You'll save money by not tanking.

Many shops have a *jet tank* or *jet-spray* cleaner that uses hot-tank chemicals specially made for aluminum. Instead of submerging the parts in a tank, the jet tank sprays cleaner on the parts while they rotate on a turntable. It's quick and easy for the shop, and the price of jet tanking is often included with boring and other machine-shop work.

Non-ferrous parts, such as aluminum, magnesium, die-cast metal, plastic and the like, are cleaned with a non-caustic solution. Sometimes non-caustic cleaning is called *cold-tanking,* because the solution is not heated. Carburetor cleaner is a good example. If a jet-spray machine is unavailable, aluminum heads, manifolds and blocks should always be cleaned this way. Cadmium-plated parts—those covers with the yellowish cast—are better off being cold-tanked. Hot-tanking removes cad plate.

The final cleaning is done by you with soap and water. After machining, all engine parts are covered with metal chips and abrasive grit. If the engine were bolted together with this debris inside, rapid engine wear would occur inside of 100 miles. The machine shop *should* clean the parts after machining. If your block was jet tanked after boring and honing, you don't have to scrub it—just rinse it off. If it wasn't, then you *must* soap-and-water it. Dry the block immediately after spraying it with clean water. If you don't have compressed air, use paper towels. Cloth towels leave lint that can clog oil passages. And be

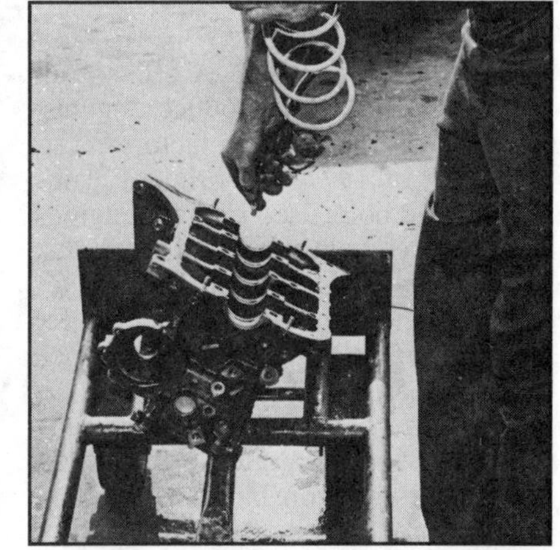
Immediately after rinsing, block is blow-dried with compressed air. Then, machined surfaces are oiled to prevent rusting.

...team.

ly cleaned, machined surfaces unprotected. Just make sure the surfaces are totally dry before you oil them. The major areas to protect on the cylinder block are the cylinder and crankshaft-bearing bores.

Threads—Another cleaning job involves *chasing* the threads. This is done by running a matching tap into the bolt holes. Most important are the head and main-cap bolt threads, but all need a cleanup, too.

You need *bottoming taps,* as opposed to *taper taps* or *plug taps,* to clean threaded holes in the block or head. A bottoming tap is threaded all the way to its flat end. Unlike taper or plug taps, which are tapered at their ends, a bottoming tap cleans the very last thread in a blind hole.

It's suprising how much dirt and grime a tap will drag out of a bolt hole. It's important to remove all of this junk, or the bolts cannot be torqued correctly. Dirty threads cause binding, resulting in less tension being applied to a bolt for a given amount of torque at the wrench.

Correct torque is very important

...d bolts. A head gasket will ...ate inconsistent torque. Not ... there be insufficient bolt ...but tension will vary from ...lt. This increases the chance ...gasket.

...bottom end, proper torque is ...important. An incorrectly ...main-bearing bolt may lead ... *bearing.* Read about *bearing* ...age 67 for detail on checking main-bearing bores.

For all head and main-cap bolt holes, use a 10 X 1.25 tap—10mm diameter, 1.25mm between threads. Also clean the head-bolt and stud threads. Use matching dies from the tap-and-die set, or a wire brush. A wire-wheel brush on a bench grinder does an excellent job of cleaning bolt threads. Remember to wear eye protection. If you use a bench grinder, don't push the bolt too hard against the wheel or you'll damage the threads.

Smart mechanics always buy the best tools. Smart, but poor mechanics always borrow the best tools. This is especially true with tap-and-die sets. A tap or die is designed to cut threads; cleaning threads is secondary. A poor-quality tap or die might mismatch slightly with the threads being chased. For instance, if a tap is too large, it cuts the existing threads, thereby weakening them. Stay with big-name tool manufacturers and beware of "bargain" tap-and-die sets from across the Pacific.

If you are unsure about your taps, inspect the threads cleaned by them. They should be clean, but not bright. Install a cleaned bolt in a chased hole and wiggle it side to side. Then, compare by checking the bolt in a clean, but unchased hole. If it wiggles more in the unchased hole, beware. Inspect the crud removed by the tap—there shouldn't be any metal cuttings mixed in. Those cuttings are from the threads! Finally, beware of any tap or die that's hard to turn. Cleaning threads requires a wrench on the tap, but no more force than running in a bolt.

Head Bolts—Over the years, some Honda head bolts have been improved. First to change were the earliest 1170cc-engine bolts and studs. If your engine is number EB1-1019949 or earlier, check the head-bolt heads for a **9**. Bolts so coded are the *updated* versions, and take 37—42 ft-lb of torque. If your engine's head bolts have no **9**, buy the updated bolts and studs from a Honda dealer. These will help prevent head-gasket problems. From engine number EB1-1019950 on, Honda used the updated hardware as original equipment.

Unlike the bolts, early studs have no identifying mark. The updated studs, however, have a light yellowish tint when new. In normal use, both early and updated studs have the same oil-stained appearance. The bottom line is, if the bolts have no **9**, the studs are probably early versions. When in doubt, change to the late-type bolts and studs.

Another bolt change affects all 1976–'77 1488 and 1600cc engines. As part of the Honda factory-recall campaign of 1977, the original plain-shank head bolts used in these engines were replaced with spiral-shank bolts. The newer bolts are actually weaker and designed to allow the cylinder head to move relative to the block without tearing the head gasket. This movement is caused by the radically different expansion rates of the aluminum head and iron block as they are heated and cooled. If your 1976 or 1977 iron-block engine was not updated—it has straight-shank head bolts—replace them with the spiral variety.

Note that the '75 1488cc engine uses straight-shank bolts and does *not* require updated bolts. This is because it uses the cooler-running eight-port head—and tends *not* to blow head gaskets.

Compressed Air—A real help during the cleaning process is a source of compressed air and a blowgun. You'll then be able to blow dirt and water out of hard-to-reach areas and air-dry washed parts. Be careful that dirt or water doesn't ricochet back into your face or onto bystanders. Always wear goggles or a face shield when using a blowgun.

BLOCK INSPECTION & MACHINING

If your block is crusty dirty, clean it with a degreaser. If not, you are ready to make some measurements.

To measure the pistons and cylinders, you should use 2–3-in. inside and outside micrometers. A telescoping gage can be used in place of the inside mike. Second choice, because it's less accurate, is a *vernier caliper.* If you have neither, try to rent these tools, or let the machine shop do the measuring for you. Also, a *precision straightedge* is needed to check head flatness. Don't forget a set of feeler gages.

Don't be discouraged if you lack many of these tools. In many cases, precision measuring can be dispensed with because the parts are obviously bad. Read through this chapter before going to the machine shop.

Checking Bore Wear—How much cylinder bores are worn *always* determines whether a block should be bored or merely honed. Boring automatically means new pistons. Otherwise, the bores can be honed, or their *glaze broken,* page 66. With honing, you'll be able to reuse the old pistons, *provided that they meet specifications.* Be sure to read the piston section before making a decision, page 76.

There are three ways to measure bore wear. The most accurate and convenient measuring device is a *dial bore gage.* But, this tool can only be found in an engine machine shop. Next best is an inside micrometer, or a telescoping gage and an outside micrometer. The last method is to use an old piston ring and a set of feeler gages. Using the piston-ring method for checking bore, you compare ring end-gap change at different points in the bore. Ring *end-gap* change from the top to the bottom of a bore is directly related to bore wear, or *taper.*

Taper—Although Honda engines are less prone to bore wear than most others, high-mileage engines will develop a slight ridge at the top of the bores. It's usually not enough to require a ridge reamer when removing the pistons, but is noticeable upon close inspection.

Notice how the bores are worn at the top of the piston stroke, but not at the bottom. There is no bottom ridge. The reason is simple; bores don't wear evenly from top to bottom. There is a bell-mouth, or taper, wear pattern. This is because piston rings are forced against a cylinder bore at the top of the piston stroke by compression and combustion pressures. Also, a bore is better lubricated at the bottom. In fact, a bore wears very little midway down—so little that the original *cross-hatch* hone pattern is probably visible at its lower end.

Measuring bore taper involves finding the difference in diameter between the unworn portion of a cylinder and that of greatest wear. Note these measurements for each bore; not all cylinders wear the same amount or in the same places.

Maximum bore wear is *close* to what it will take to *clean up* a cylinder—how much has to be removed from that cylinder so new metal is exposed from top to bottom. Measure all cylinders. The one with the greatest wear governs how much will be removed from the remaining three. On iron-block engines, if you find one cylinder badly damaged to the point that it can't be corrected by boring, consider *sleeving* it before replacing the block. However, aluminum-block engines—1170, 1237 and 1335—*cannot be resleeved* because you cannot remove the factory sleeve without damaging the block. Read about this on page 39.

On 1200s, steel sleeves are cast into cylinders. Barbs (arrow) anchor sleeves in block and prevent resleeving if bore wear limit is exceeded. However, sleeves can be bored 0.04-in. (1mm) oversize on 1170s and 0.02 in. (0.5mm) on 1237s. Photo courtesy of A/T Engineering.

You may notice that the closer a cylinder is to the water pump, the more wear it has. The reason is interesting; the warmer a cylinder gets, the less it wears. The water coming directly from the water pump cools the front cylinders better than those farther downstream. Therefore, the front cylinder wears more than the others because it gets the coolest water. This pattern causes the cylinder to wear into an egg shape. You can sometimes feel it by running your fingernail around under the ridge.

Honda rings have relatively low tension. This keeps bore wear to a minimum. However, bore wear occurs at the top of a cylinder due to the rings reversing direction. Different machinists describe it differently, but the compression rings tend to *buzz* at the top of the stroke. This wears a slight bulge near the top of the bore. Usually, a 0.010 in. (0.25mm) overbore restores the bore.

The final factor in measuring bore wear is the uneven-taper wear pattern we've already seen. Because wear is *not concentric*—uneven around the circumference of the bore—maximum taper measurement will not give you the final overbore figure. That's because the non-concentric bore wear moves, or *shifts,* the bore center. Therefore, you must always bore a little more than wear measurements indicate.

As for eyeball measurements, an experienced machinist will glance at the cylinders and make a snap decision. Perhaps he sees or feels a slight ridge. "No worse than six-thousandths," he says to himself. "I'll bore it ten-over." The machinist knows that if he finds a ridge that indicates the need to bore, the remaining question is how much. Just enough to expose fresh metal the length of the bore is the answer. From experience, the machinist knows it might take 0.008 in. to get fresh *stock* after running his finger through the cylinders. Because the next available piston size is 0.010-in. (0.25mm) over, that's what he'll bore to.

Bore Gage or Micrometer—When checking bore wear directly, with either a dial bore gage or telescoping gage or micrometer, first measure immediately below the ridge. Measure at different points around the bore to find maximum wear. Now, measure near the bottom of the bore. Find the greatest bottom dimension. Subtract this figure from the maximum figure you got at the top. The difference is bore taper. Write down each measurement as you make it.

Ring & Feeler Gage—Unlike reading taper directly with a dial indicator, using a piston ring and feeler gages is an *indirect method* of measuring bore wear. You are comparing the *circumference* of a cylinder at two points, not its *diameter,* so this method doesn't register minor irregularities in bore wear. Although this makes it less accurate than direct measuring methods, it will tell you if you'll have to bore or not.

To measure taper by the ring-and-feeler-gage method, place a compression ring in the bore. Square it up in the bore with a piston turned upside down and without its rings. You can't use a domed piston, such as that used in the EB3 engine, because it isn't flat. Use a flat-top piston, a can

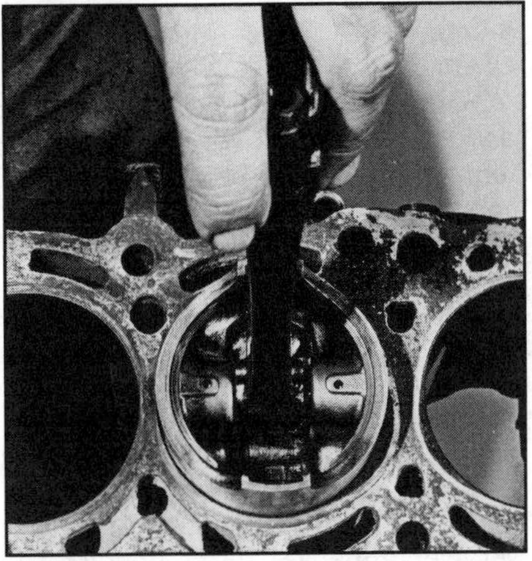

Piston works great for squaring ring in cylinder.

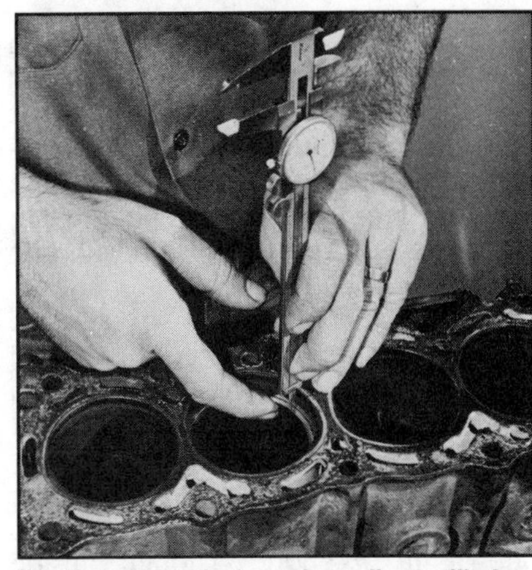

Depth-gage end of vernier caliper will also square ring. Ring can be set to exact depth.

that fits the bore reasonably well, or a 6-in. scale. Square the ring by measuring from the block deck to the ring at several points.

Using feeler gages, measure ring *end gap*—the space between the ring ends. Take two measurements in each cylinder—one immediately below the ridge, another near the bottom of the bore. Do all four cylinders, using the same ring. Record your readings as you go, then use the following formula to determine taper. Taper is approximately the difference between top and bottom ring end-gap measurements multiplied by 0.30.

Allowable Taper—What you decide as *maximum allowable taper* depends on the service you expect from your engine. Honda says no more than 0.002-in. (0.05mm) taper. Any more requires a rebore. That isn't much taper, so little that accurate measurements with the ring-and-feeler-gage method are almost impossible. So, in most cases, if you find *any* taper, you should bore. However, if you hone the bores, don't expect the job to last long. If you want your engine as good as factory new, bore. You'll have fresh bores *and* pistons. You just can't get like-new performance with worn bores and old pistons, so my advice is to bore, unless you can't measure *any* taper.

If a patch 'em-up job is your aim, new rings alone in a 0.002-in. tapered bore will be OK for awhile—maybe 15,000 miles. Then, excessive oil consumption and low compression may become a problem again. It all depends on how badly tapered the bores are.

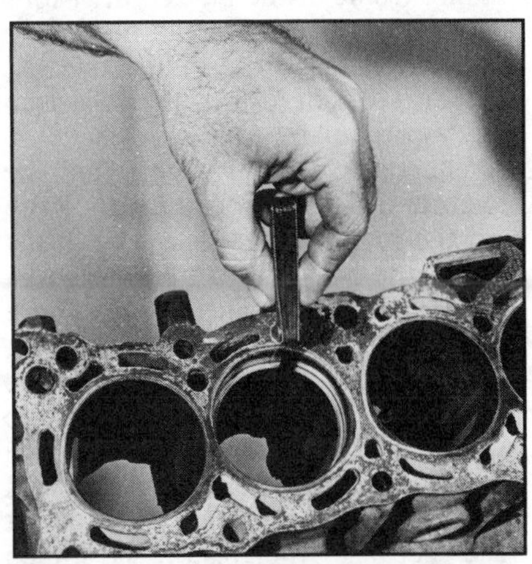

Measure end gap by selecting feeler gages until you find one that just fits between ring ends.

If you decide not to bore, remember that taper will only get worse. Ring durability definitely suffers with increased bore taper. Rings expand and contract on each trip up and down a tapered bore. Therefore, as taper increases, so does ring flexing; the rings fatigue sooner. Piston-ring lands also take a beating. As the rings flex, they rub against the ring lands, accelerating wear and reducing sealing. Therefore, if 0.002-in. taper is exceeded, it's best to bore the block and install new pistons.

Piston-to-Bore Clearance—Before you make the final decision regarding taper and the need to bore, there's another determining factor: *piston-to-bore clearance.* If the pistons are worn out and must be replaced, bore the block. You'll get fresh, straight bores to match the new pistons.

A rare exception might be when

one piston needs to be replaced and there is little bore wear or taper. You might be better off replacing that piston, even though it should be balanced to match the others. Look for a balance number adjacent to the pin bore, and replace that piston with one matching the others.

There are two ways to measure piston-to-bore clearance: direct and indirect. For once, the direct method is easier and cheaper than the indirect method.

To measure directly, select a piston without rings and insert it upside down in *its* bore. Though upside down, the piston must otherwise be in the correct relationship with its bore. The circle should face the intake-manifold side of the block. On pistons with a triangle and a circle, make sure the triangle is on the intake side. As for being upside down, that's just an easy way to hold the piston.

When using feeler gages, find the maximum distance between the bore and the piston *thrust face.* The thrust face is the side 90° to the pin. To be more specific, in an engine whose crankshaft rotates counterclockwise, the thrust face of the pistons is on the right side, when viewed from the timing-belt end of the engine. On Hondas, this is the intake-manifold side. Wear is greater on the thrust face because the piston is forced against that side of the cylinder on the power stroke—as the crank throw swings in the opposite direction. Take several measurements with the piston at the top and bottom of its travel—between TDC and BDC. Record your findings.

The correct reading is given by the feeler gage that slips between piston and bore with slight resistance. Don't try to force the gage. That's too tight.

Indirect measurement requires a 2—3-in. outside micrometer to measure the pistons. The cylinders must be measured with an inside mike or telescoping gage and vernier caliper or outside mike. The difference between piston and bore diameters is piston-to-bore clearance.

Using outside mikes, measure piston diameter across its skirts. Write down this measurement. Next, take an inside mike or a telescoping gage and measure the cylinder bore 90° to the piston-pin center line about 1-1/2-in. down the bore. Record this figure. Subtract the two measurements to find piston-to-bore clearance.

With either method, make sure *both* piston and block are the same temperature. If the block is in the sun and the pistons in a cold garage, your measurements could be off as much as 0.001 in.

How much piston-to-bore clearance is acceptable? The service limit on 1170, 1237, 1335, 1488 and 1600cc engines is 0.004 in. The 1751cc engine can go to 0.006 in. Because original factory piston-to-bore clearance is tight, typically 0.001 in., these service limits are fairly liberal. You shouldn't have any problem with bore clearance unless you have a high-mileage engine. Of course, "loose" pistons promote rapid bore wear and *piston slap,* or *rattle,* page 11. If all other factors are right, a 0.003-in. piston-to-bore clearance is OK with used pistons.

Piston Knurling—Don't be talked into *knurling* your pistons to take up piston-to-bore clearance. Knurling involves *embossing,* or rolling a pattern into the piston skirts. This creates indentations and displaces material to form tiny ridges on the skirt. Excessive piston-to-bore clearance is reduced *at the top of the ridges only.* Now, rather than the full face of the skirts taking the thrust load, thrust is taken by the considerably smaller area formed by the ridges. Rapid wear of the knurled ridges occurs, and you are soon back to square one.

Knurling is backyard engineering. Use it only if you are patching an engine to last no more than 5000—10,000 miles.

Glaze-Breaking—Using a hone to restore cylinder-wall cross-hatch pattern is called glaze-breaking. If you plan to bore the block, there's no need to do any glaze-breaking. This will be done during boring and honing. But, if you are reusing the original bores and pistons, the glaze must be broken. Cross-hatching provides good cylinder-wall oil retention, which is necessary for piston-ring sealing. It also promotes quick ring break-in.

Glaze-breaking is done with a hone, but not just any hone. There are three types of hones: *precision, spring-type,* and *bead.* A precision hone, sometimes called a *rigid hone,* is set to a specific diameter. Because it holds its shape, the precision hone removes a lot of metal. Incorrectly used, such a hone can shift the bore center. Spring and bead hones follow the bore's existing contour, taper and all, but they remove little material.

For glaze-breaking, Honda recommends *only* a 400-grit spring-type hone. Don't use the precision hone—it removes excess material and piston-to-bore clearance could end up way too large.

Block Decking—Another trouble spot is the *deck*—the block surface that mates to the head. If the engine has been overheated, the deck could be warped. Warpage is usually confined to the head. But Honda engines—especially the aluminum blocks—are notorious "warpers," so check the block.

To check deck flatness, use a precision straightedge and a set of feeler gages. The proper straightedge is a large steel bar that has been precision ground—*don't use a yardstick!* If you don't have such a straightedge, ask an engine machine shop to check the deck for you. They won't charge much, if any.

The actual measuring job is simple. Lay the straightedge over the length of the deck, then try to pass a 0.004-in. (0.10mm) feeler gage under it. If the gage fits under the straightedge, there is a depression in the deck. Slip increasingly larger feeler gages under the straightedge until you find one that fits snugly. The snug gage is the amount the deck may be warped. If a deck is warped 0.004 in. or more, the deck requires milling—*decking.*

Double-check that the deck is free of gasket material or dirt before you make a final decision. Take measurements at several different locations. Place the bar both lengthwise and diagonally on the deck at different locations.

A flat deck is doubly important if your machine shop is using a Van Norman-type boring bar; it locates off the deck. If a deck is not flat, the cylinder bores will be machined off-center—they won't be perpendicular to the crank. Also, the deck should be free of any irregularities that could affect the boring bar. Remove any bumps or nicks with a large flat file. Be careful not to nick or gouge the bores.

Decking a block raises compression slightly. To significantly raise compression ratio, you'd have to cut the

Quick method of checking main-bearing-bore alignment is with precision straightedge and feeler gages. You should not be able to pass feeler gage between straightedge and bore. On 1200s, make sure lip at timing-belt end does not hang up straightedge.

Frequent checks ensure machinist doesn't overcut cylinders. Leave 0.002 in. extra material for honing.

By locating off main-bearing bores, boring bar machines bores 90° to crankshaft.

deck so far the pistons would protrude above the deck. The point is, don't deck the block unless it's warped. And, you shouldn't have to remove much material to cure the warpage problem.

Main-Bearing Bores—On rare occasions, the roundness and alignment of the main-bearing bores will change. Honda main bearings have proven quite strong, so you should have no problems with them. But, just to be sure, check main-bearing alignment and *out-of-round.*

Check alignment first. Make sure you do this if the crank didn't spin freely when you checked it during teardown. One good way to check alignment without a lot of special equipment is to bolt in a *known* good crankshaft. First, inspect the crankshaft for *runout,* page 73. Then read how to install the crank, page 114.

If the crank has no runout, and the main-bearing bores are straight, the crank will spin freely. If the crank sticks, the bearing bores may need *align boring* or *align honing.* All nominal main-bearing-bore diameters are 2.13 in. (54mm). Similar to rod reconditioning, align honing or boring the main-bearing bores is a machining process that ensures that the bearing bores are in perfect alignment. This is an expensive process—expect to pay at least $100. Usually, align honing or boring is required only when the original main caps are not used, or when a bearing bore is out of spec or damaged. Align honing is preferable because align boring removes too much material.

A precision straightedge and feeler gages can also find an out-of-line main-bearing bore. Lay the straightedge in the main-bearing bores and try to pass thin feeler gages under the bar. You should not be able to get any feeler gage under the straightedge.

A more accurate checking method is with a special dial indicator that accompanies an align-boring or align-honing machine. The problem is few machine shops have these, so they won't have the special dial indicator either.

Actually, there's no need to worry about the main-bearing bores unless the crank didn't spin freely, or the bores appear to have been pounded. It's a rare Honda that needs align boring or honing. If your engine does, you might be best off with a new block because of the cost and difficulty in finding a shop to do the job.

Just in case you want to check main-bearing bores for out-of-round, you'll need a 2—3-in. mike and telescoping gage, or a dial bore gage. Install the main caps, less bearing inserts, in order and in the right direction. Lightly oil the main-cap bolt threads, install and torque them. Tighten 1170, 1237 and 1335 main-cap bolts to 29 ft-lb, 1488s and 1600s to 32—35 ft-lb, and 1751s to 44—51 ft-lb.

All main-bearing bores should measure 2.128—2.129 in. all the way around—they should be perfectly round. If a bore is not to spec, the bearing insert will be crushed too much or not enough. Either way, the bores should be align bored or honed.

Main-Bearing Caps—Make sure your block has its main-bearing caps installed before sending it to the machine shop. They're needed to complete honing the bores. But don't torque the main caps in place just yet because the machinist will probably remove them.

The caps are installed so the cylinders will be round once the engine is assembled. This is because the tightened main-bearing caps distort the cylinders. It isn't much, especially with cast-iron blocks and Honda's low torque specs, but the distortion is enough to allow for.

Don't be alarmed if you see your block being bored on a machine that locates off the main-bearing bores. The one pictured is that type. You don't need to bore with the caps torqued in place, just during the final hone.

Crack Check—*Magnafluxing* and *Spotchecking* are two machine-shop odds and ends that need mentioning. Both are crack-checking processes, useful for finding damaged parts that may be unusable *before* you spend money reconditioning them.

Magnafluxing works only on ferrous parts, because it relies on magnetism. Iron dust is sprinkled on the part, then a powerful magnet is attached. The dust will congregate around any cracks. It's a quick-and-easy test.

Spotchecking can be done on ferrous and non-ferrous parts. A spotcheck kit has a cleaner, dye and developer. The part is cleaned, then sprayed with the dye. The dye is cleaned off, then the developer is applied. If a crack exists, penetrant

Precision-hone remaining 0.002 in. of material. Honing with 220-grit stone leaves correct oil-carrying finish for Honda OEM rings without *fracturing* metal—as does boring.

Because cylinder was lit from underside, only half of honing marks stand out. Look harder to see other half of *crosshatch*. Also visible is chamfer at top of bore.

that has *penetrated* the crack will bleed through the white developer and show up as a bright red line.

Cracks usually mean the part must be replaced. However, cracks in aluminum parts can sometimes be Heliarc-welded. Talk to your machinist about any cracked part. Impurities in Honda's aluminum makes it hard to weld. Make sure an experienced welder performs any repairs, and uses high-quality rod.

Bore Chamfer—When you get your block back from the machine shop, inspect the tops of the bores. Each should be *chamfered*—angled by machining, filing or grinding—so there's no sharp corner at the top of the bore where it meets the deck surface. Some machinists cut the chamfer with the boring bar after machining the cylinder. If your block came back without being chamfered, do it yourself. Removing the 90° corner will provide a lead-in for the piston rings during installation. It also eliminates a sharp corner that may overheat and cause preignition.

Chamfer the bores with a *rat-tail* or *half-round* file. Don't use a flat file or you'll be making a lot of sharp edges. Use a light stroke to arrive at a 45° angle cut about 1/32-in. wide. Try to keep the file at an even 45° around the entire edge. And above all, be careful. Keep the file end from gouging the cylinder wall! You sure don't want to gouge a fresh cylinder bore.

Suds Time—The block might have to be cleaned *again* when it comes back from the machine shop. Most machinists won't let a block out of the shop without jet-tanking and spraying it with oil. Make sure yours has been cleaned and oiled. Run a finger down the bores. You shouldn't feel any grit. Inspect the crankcase area the same way. If the block is bare-metal clean and the only thing inside is fresh oil, bag it and set it aside.

If not, you *must* clean the block. Otherwise, grit will embed in the bearings, get between cylinder walls and pistons, valves and guides and in other critical, close-tolerance areas of your engine.

The final cleaning is a little different from previous solvent sessions. This time, soap and water is the cleaning agent. You'll need a bucket with dishwashing detergent and a stiff-bristle scrub brush. If your engine was dirty to begin with, a rifle bore brush—.22 cal.—and rod are needed for the oil passages. After you brush out the oil passages, follow up with a shot of high-pressure water from the hose.

With the block in the driveway, turn on a garden hose to a fast trickle. Hose down the block, then vigorously scrub every inch with the scrub brush and soapy water. A wipe through each bore with a white paper towel should leave the towel perfectly clean. If dirt shows on the towel, keep scrubbing until the bore passes the towel test.

Drying with compressed air is best, but toweling works almost as well on most surfaces. The oil passages, if opened, really need to be blown out. Remember, don't use cotton shop rags or other cloth for drying. Cloth threads or lint could clog lubrication passages. Paper towels are better; paper fragments dissolve in engine oil with no ill effects.

As soon as you finish cleaning each bore, rinse it again, dry it quickly and give it a shot of water-dispersant oil. This will keep that bore rust-free while you're scrubbing the others. Do the same to all machined surfaces. After it is completely scrubbed, dry and oil the block once more.

Whatever you do, don't walk away and let the block air-dry; the bores will rust quickly—and rust is abrasive. An oil-free, freshly cut-and-honed bore can rust in less than five minutes if humidity is high.

When the block is cleaned and rust-protected, that oily surface becomes a dust magnet. Set the block on its rear face and slip a plastic trash bag over it. Tie off the end to seal the bag. Store the block out of your way until assembly time.

CRANKSHAFT

Along with the auxiliary valve and matched bearings, Honda crankshafts

Laying CVCC-engine main-bearing caps on pan rail is best way to arrange bearings for reading.

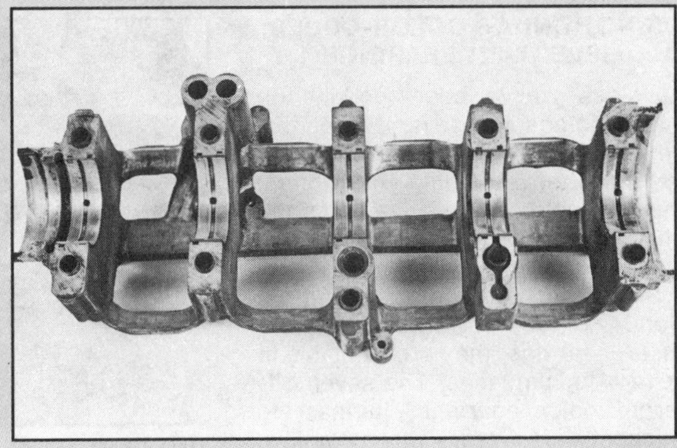

Aluminum-block main-bearing-cap *cage* keeps bearings in order and makes reading them easy.

READING BEARINGS

As with other worn engine parts, the main and rod bearings can provide good clues as to engine condition and past use.

First, a lesson in bearing construction: Like most modern engines, Hondas use what is called a *precision-insert* bearing. A steel shell gives the bearing its strength and shape. Over the shell is the *babbitt.* Babbitt is a soft material that allows the bearing to "absorb" small, hard dirt particles to prevent bearing-journal damage. This feature is called *imbedability.* A thin tin-plate *overlay* is spread over the babbitt.

Place the bearings in order for inspection. With rod bearings, this means having the rod and caps *loosely* bolted together in their correct relationship. With main bearings, place each cap next to its mate in the block.

All bearings will show some *scoring*—scratching. The dirtier the oil and air filters were, the worse the scoring. Be alarmed if you see large, individual grooves. These mark the passage of big chunks of foreign material. Such large pieces could have come from outside, if the air filter was defective or not in place. More likely, they came from inside. A broken ring or piston is usually the culprit. If you find a badly scored bearing, check its crankshaft journal for damage.

Worse than dirty lubrication is no lubrication. If a bearing was oil-starved, it will be heavily scored and probably *blued* or blackened from heat. To check for a blocked oil passage, use a length of coat hanger or welding rod to *rod out* that journal's passage. If more than one bearing has been oil-starved, check

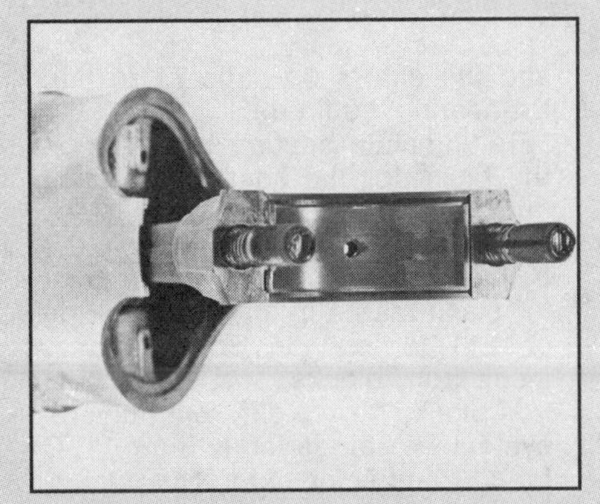

Don't forget to look at rod bearings. Dark area in bearing center is babbitt. Wear pattern is typical of high-mileage Hondas.

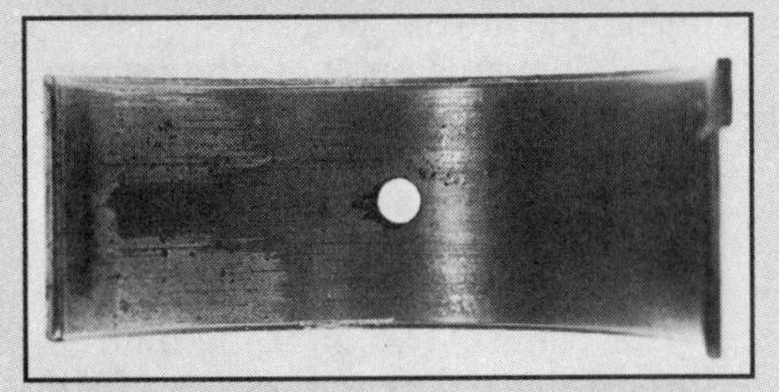

A closer look at same rod bearing shows hydraulic erosion around oil hole. Dark spots are embedded dirt and pitting from missing overlay.

the oil pump. See page 82 for oil-pump overhaul. If the pump is OK, the engine may have been low on oil.

You may notice that rod bearings seem more worn than the mains. This is typical of high-rpm engines such as the Honda—1200s in particular. Two factors cause the rods to wear quickly. First, they get their oil after the mains, so if there isn't enough oil, or oil pressure, the rod bearings are the first to suffer. Second, normal crankshaft movement swings the rods, resulting in more oil *sling-off.* When an engine has large oil clearances, the rods wear very quickly because centrifugal force slings off a lot of oil.

A worn-out bearing shows its babbitt through the tin-plate overlay. The babbitt on stock Honda bearings is copper-colored, which contrasts sharply against the gray overlay. If you see copper, you are seeing the babbitt of a very tired bearing. Aftermarket bearings may have a dark-gray babbitt—look closely to see the light-gray overlay contrasted against it.

Don't be surprised if the rod bearings show babbitt and the mains don't. Rod bearings usually show babbitt at their top and bottom—where reciprocating motion and compression and combustion pressures impart the greatest loads.

Other than the uneven wear from top to bottom, uneven wear from front to back around the full circumference of a bearing is abnormal. This means the bearing journal is *tapered*—larger in diameter at one end than the other. Although taper should show up when you mike the bearing journal, this bearing wear pattern is a warning sign.

Also, a badly worn thrust bearing allows the crank to *walk*—move forward and backward. With this excessive *end play,* the main-bearing journal *fillet*—the radius at the ends of each journal—contacts the main bearings. This will show up as a thin line of exposed babbitt around the bearing edge.

Compare the bearings to each other. Honda's five-main-bearing bottom end is tough, but problems can occur. If a main bearing is worn in mirror image to its neighbor bearing, the crank could be bent. There are a lot of possible combinations, including more than two involved bearings. To check for a bent crank, see page 73.

USING HONDA'S COLOR-CODED, SELECT-FIT BEARINGS

Unless you've been working on auto engines for several decades, *select-fit* bearings are probably new to you. After rebuilding your Honda, they'll be old hat because that's the only type of engine bearing Honda uses.

Select-fit bearings, as used by Honda, differ from other *precision-insert* bearings; they are available in several thicknesses. The seven different color bearings—thinnest to thickest—are red, pink, yellow, green, brown, black and blue. Thicknesses are in 0.0008-in. (0.02mm) steps. The bearing thickness range extends below and above the crankshaft's supposed standard, or *nominal,* diameter, from -0.0003 in. (-0.008mm) to +0.0005 in. (+0.013mm). A negative number means that the bearing is smaller than nominal; positive means it's larger.

These minute differences in bearing thickness compensate for equally minute variations in *both* crankshaft-journal diameter and housing-bore diameter. If both the housing bore—block or rod big end—and crankshaft journals are at their nominal dimensions, there is no variation to compensate for. In that case, the bearing that covers the nominal range is used—yellow color code. However, if the housing bore is 0.0003-in. larger while the journal diameter remains nominal, a brown bearing is used because the *combined* housing bore and journal deviation is 0.0003-in. larger than nominal. If the housing bore is 0.0003-in. larger than nominal and the journal is 0.0002-in. smaller than nominal, then the combined deviation is 0.0005 in. from nominal; a blue bearing is used.

Now, don't let all the figures scare you. All you need to do is use the bearing-selection numbers or letters and the charts on page 71 to find the bearing required.

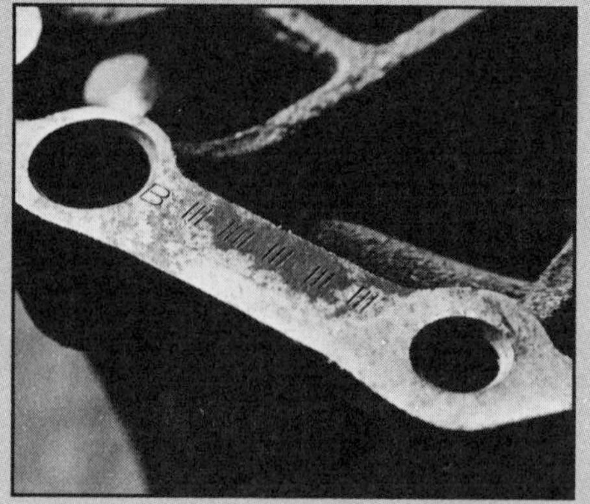

Main-bearing-bore diameters are stamped in block. On iron blocks, look on bellhousing mating face as shown. On aluminum blocks, check at bottom of block near left rear pan rail.

To find main-bearing size, look on the block for the bearing-selection numerals. On most iron-block engines, the numerals are on the bellhousing mating surface. Look for slash marks in any combination of 1, 11, 111 or 1111. Some replacement iron blocks and late-model iron blocks have a different numeral system—A stands for 1, B for 11, C for 111 and D for 1111. The numerals are arranged with the first, or pulley-end, bearing near the dowel and the fifth, or flywheel-end, bearing next to the bolt hole.

On aluminum engines, the numerals are along the left pan rail, near the flywheel end. As with late-model iron blocks, all aluminum-block engines have the numeral system using A for 1, B for 11, C for 111 and D for 1111. The first bearing is toward the pulley end of the block, the fifth, toward the flywheel.

Find the numeral for the bearing you are checking and write it down.

Now, go to the crankshaft. The bearing-selection codes are on the counterweights. See the drawing for the exact location. Find the code—numeral 1 through 4—for the journal you are checking. Write down the numeral next to the numeral from the block. Now, go to the chart and find that combination. Where the two columns intersect, you'll find the color bearing necessary for that combination of bearing-bore size and journal diameter.

Although the explanation is lengthy, finding the bearing color takes only seconds.

Rod bearings are done the same way. Match the connecting-rod numerals with the crankshaft letters. Then, use the chart to find the correct color bearing.

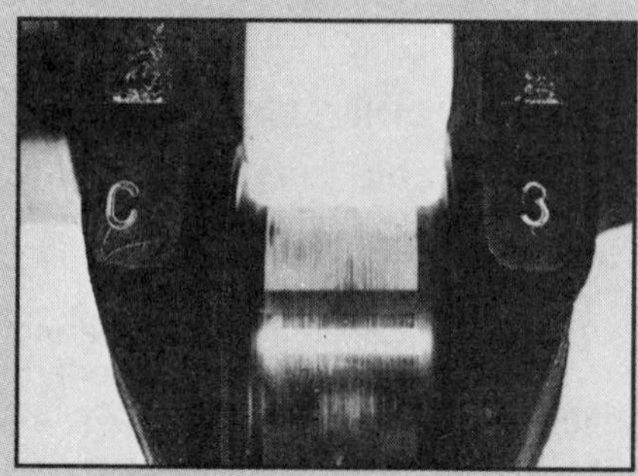

C is rod-journal-diameter code. The 3 is main-journal identification number for main journal to right. Rod-journal fillets are evident.

With new bearings, the color is a faint paint swatch on the edge of the bearing. Unfortunately, you can't use the color swatch as a guide on old bearings because engine oil washes it off.

Never, never mix different-color bearing halves on the same journal. For example, if the second main bearing needs a green insert, be sure to use two green main-bearing halves—one upper and one lower. Honda sells bearings by the half shell—one half of what you need to service one journal. Keep that in mind when pricing their bearings. Make sure the price quoted is for a *complete set*—otherwise, it can be twice that amount.

scare off many would-be rebuilders. They've heard many doom-and-gloom stories about destroyed cranks and engines. It isn't that bad.

Construction—Honda *forges* steel into the crank's shape, *machines* the critical surfaces, then *surface hardens* the finished product.

This process is not unusual. What is different is the *softness* of the original metal—before it is surface hardened. It doesn't seem so soft if you drop a crank on your toes, but as metal goes, Honda cranks are *soft.* So soft, that without the hardening, the journals can be damaged by normal engine operation.

Crankshaft hardness is a consideration during rebuilding. Normally, cranks with little wear are *polished* and returned to service. If crank wear is measurable, it is *ground* to an *undersize.* Thicker bearings are then used to retain the proper *oil clearance*—distance between the bearing and crank journal. With Honda cranks, polishing is OK—it doesn't remove an appreciable amount of metal. Grinding, however, removes the *surface hardness,* exposing soft, unhardened material. The crankshaft then needs to be rehardened, even though some rebuilders seem to get

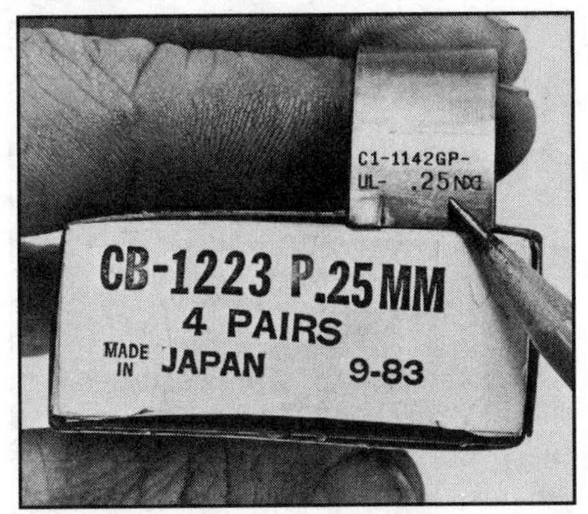

Aftermarket bearings will be marked for size, such as this TRW rod bearing. Always check bearings against box before leaving store.

by without doing it.

Several processes can be used to reharden a crank. The original method used by Honda is *nitriding,* commonly called *Tuftriding.* Another method is *cold quenching.* Hardening is not part of a normal rebuild, so it's not available from machine shops—not even very large ones. Only the largest crankshaft specialists or metal shops can reharden a crank. In most parts of the U.S., you'll have to ship the crank for the work. Then you are paying for the crating, shipping (two ways) and hardening.

Another consideration when grinding a Honda crank is bearing availability. Honda does not supply undersize bearings. Aftermarket sources, such as TRW and Federal-Mogul do. The Federal-Mogul catalog shows Honda bearings in 0.25, 0.50, 0.75 and 1.0mm undersizes. In English units, those are 0.010, 0.020, 0.030 and 0.040-in. undersizes. So, if you grind a Honda crank, you *must* use aftermarket bearings. Also, this means all main or rod journals will be the same size. The crank grinder will start with the journals' *nominal* or stated diameter, not their actual diameter.

So when you get your crank back from the grinder, the journals will all be the same diameter. There will be no need for Honda's blue, pink, brown or whatever color-coded bearings. An aftermarket undersize bearing set will be needed. The crank must also be rehardened.

Of course, grinding the crank journals does nothing for the bearing bores. Aftermarket manufacturers ignore any difference in bearing-bore

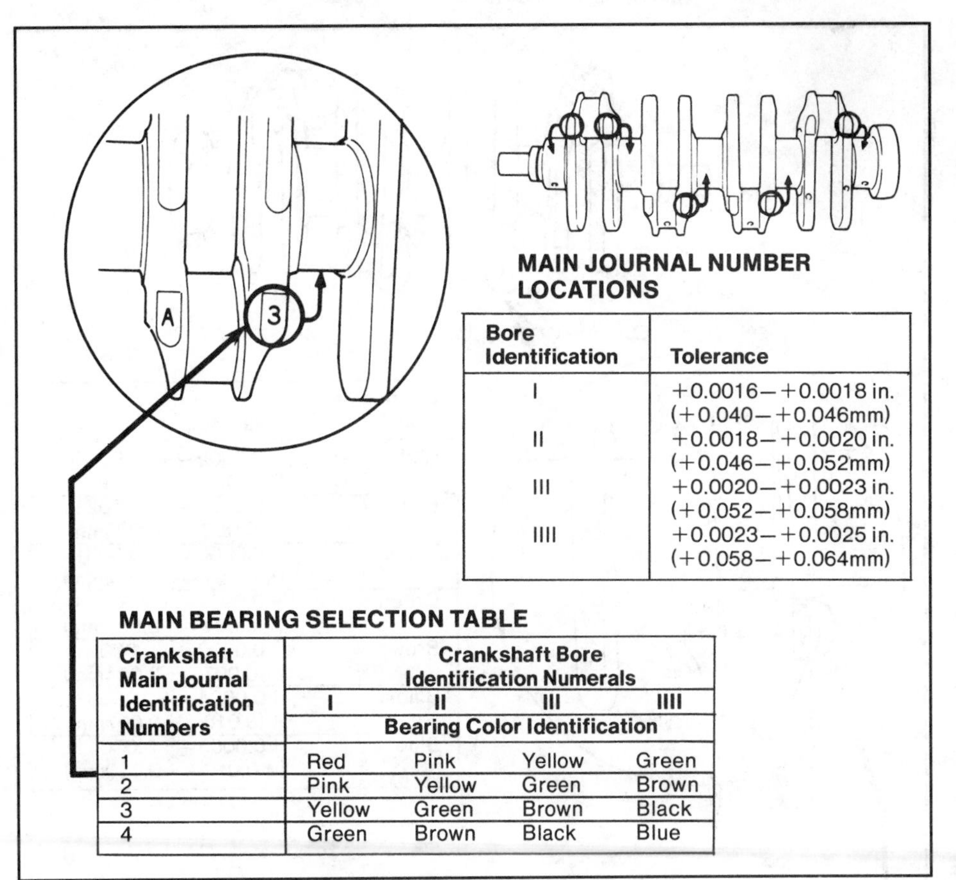

Bore Identification	Tolerance
I	+0.0016–+0.0018 in. (+0.040–+0.046mm)
II	+0.0018–+0.0020 in. (+0.046–+0.052mm)
III	+0.0020–+0.0023 in. (+0.052–+0.058mm)
IIII	+0.0023–+0.0025 in. (+0.058–+0.064mm)

MAIN BEARING SELECTION TABLE

Crankshaft Main Journal Identification Numbers	Crankshaft Bore Identification Numerals			
	I	II	III	IIII
	Bearing Color Identification			
1	Red	Pink	Yellow	Green
2	Pink	Yellow	Green	Brown
3	Yellow	Green	Brown	Black
4	Green	Brown	Black	Blue

Main-bearing selection; all engines except 1335. Courtesy of American Honda Motor Co., Inc.

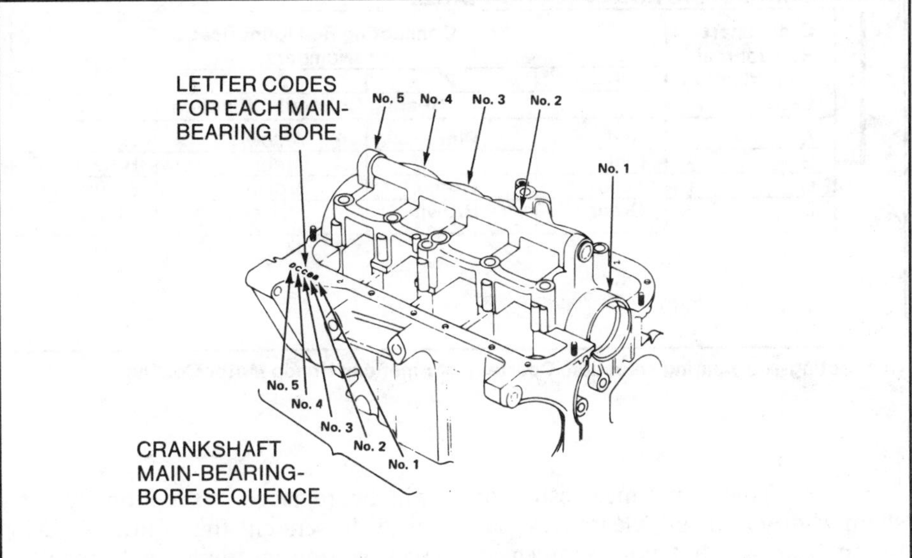

For No. 1, No. 2, No. 4 and No. 5:

Bore Letter	Tolerance
A	0 to +0.006 mm (0 to +0.0002 in.)
B	+0.006 to 0.012 mm (+0.0002 to +0.0005 in.)
C	+0.012 to +0.018 mm (+0.0005 to +0.0007 in.)
D	+0.018 to +0.024 mm (+0.0007 to +0.0009 in.)

For No. 3:

Bore Letter	Tolerance
A	–0.010 to –0.004 mm (–0.0004 to –0.0002 in.)
B	–0.004 to +0.002 mm (–0.0002 to +0.0001 in.)
C	+0.002 to +0.008 mm (+0.0001 to +0.0003 in.)
D	+0.008 to +0.014 mm (+0.0003 to +0.0006 in.)

Main-bearing selection; 1335 engine. Courtesy of American Honda Motor Co., Inc.

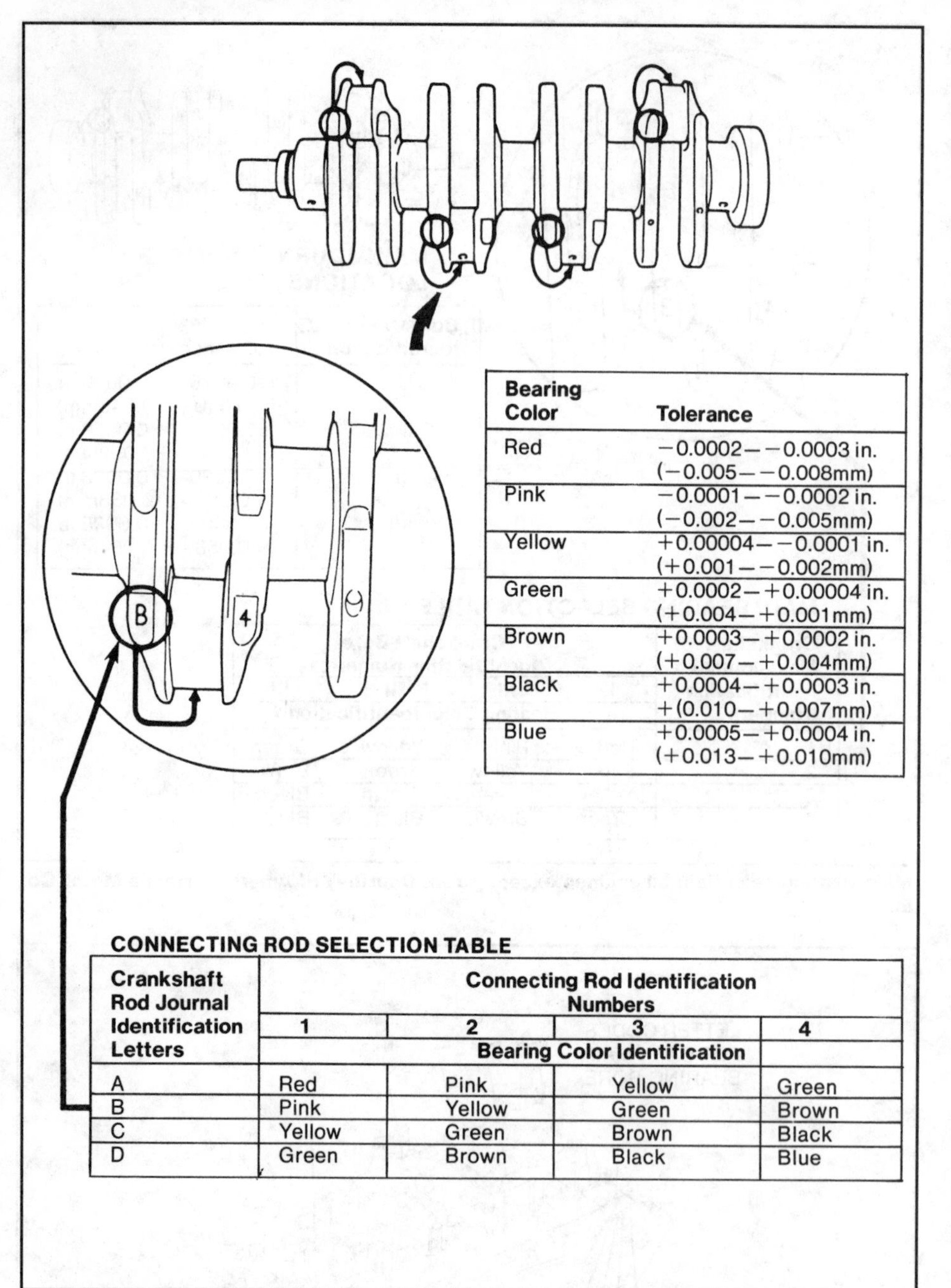

Bearing Color	Tolerance
Red	−0.0002−−0.0003 in. (−0.005−−0.008mm)
Pink	−0.0001−−0.0002 in. (−0.002−−0.005mm)
Yellow	+0.00004−−0.0001 in. (+0.001−−0.002mm)
Green	+0.0002−+0.00004 in. (+0.004−+0.001mm)
Brown	+0.0003−+0.0002 in. (+0.007−+0.004mm)
Black	+0.0004−+0.0003 in. +(0.010−+0.007mm)
Blue	+0.0005−+0.0004 in. (+0.013−+0.010mm)

CONNECTING ROD SELECTION TABLE

Crankshaft Rod Journal Identification Letters	Connecting Rod Identification Numbers 1	2	3	4
	Bearing Color Identification			
A	Red	Pink	Yellow	Green
B	Pink	Yellow	Green	Brown
C	Yellow	Green	Brown	Black
D	Green	Brown	Black	Blue

Connecting-rod-bearing selection. Courtesy of American Honda Motor Co., Inc.

diameters. This sometimes results in slightly different oil clearances between bearings, but not as much as you would think. Honda allows 0.0008-in. (0.020mm) variation in bearing-bore diameter. Usually, any variation will be clustered at the middle or one end of the scale—not all over the scale.

Crankshaft Problems—Most crank problems are caused by dirty oil. Although dirty oil is harder on the bearings than the crank, there is a limit to what a crank journal can take. This is as it should be because the bearings will be replaced. Exceptionally dirty oil will scratch the journals. Deep scoring and grooving will result if large fragments are circulated through the oiling system.

Besides being scored, scratched or grooved from dirty oil, crank journals could be worn *undersize, out-of-round* or *tapered.* Additionally, the thrust faces could be worn, especially if crankshaft end play was excessive. Or, there might be cracks hiding anywhere on the crank.

Start crankshaft inspection by checking the texture of the journals.

When a fingernail catches on scoring, Honda says crank should be replaced. Most independent garages will have crank ground and rehardened.

Let the bearing inserts be your guide. Match each insert with its journal. If an insert shows wear, then its journal may also be worn. Drag a fingernail the length of each journal. If there is any roughness, you'll feel your nail catch on it.

Don't think you are doing something wrong if you can't feel anything. Usually, there's no roughness to feel. This will be true if the engine was well cared for with regular and frequent oil changes. Assuming journal sizes are OK and the crank is straight, all you need to do is polish the journals to recondition the crankshaft.

If your fingernail catches on deep scratches or scoring, Honda says to *junk* the crank and buy a new one! However, if you prefer to grind the crank, this is your signal to do so. How much to grind off depends on how deep the scratches or scores are. Also, the taper and out-of-round checks discussed later will show how much it will take to clean up the journals.

Even if only one journal is bad, *all* the main or rod journals must be ground. Example: The crank is OK except for number-2 rod journal; it's deeply gouged. You should grind all rod journals to the size it takes to clean up number 2. Bearing inserts are sold in sets; all inserts in that set are the same size. Therefore, it doesn't make sense to grind only one journal. True, precision-fit Honda bearings are sold individually, but if the crank is ground, you *cannot* use them.

Machining just one journal doesn't save money, either. The cost of setting up the crank in the grinder represents most of the cost. And, it's the same for grinding one or four or five journals. You'd have trouble getting the machinist to do only one anyway. However, you can have the mains ground and not the rod journals, or vice-versa. This is standard practice. Or, you might grind the rods 0.020 in. and the mains 0.010 in., and so on.

Out-of-Round—Bearing journals do not necessarily wear evenly, especially rod journals. They tend to wear into an elliptical, or oval shape. It's easy to see why when you consider the load applied to a rod-bearing journal. Load changes occur as the piston goes through its four strokes, or cycles.

The rod and piston push against the journal during the compression and exhaust strokes, and particularly the power stroke when near TDC. And, the rod and piston pull against the journal during the intake stroke. Inertia loads change with the angle of the crankpin, or rod journal, and the relative movement of the piston in its bore. In short, varying loads cause uneven journal wear.

To detect journal wear—even or uneven—you need a 1–2-in. outside micrometer. Measure the journal at several positions around its circumference, but on the same plane. If your measurements vary, the journal is *out-of-round.* Ideally, you should be able to make a measurement, then move the mike around the journal without changing its setting.

To determine how much a journal is out-of-round, subtract the *minor dimension* (smallest measurement) from the *major dimension* (largest measurement). The resulting figure must not exceed 0.0004 in. (0.010mm). If it does, Honda recommends that you replace the crankshaft. Or, if you are so inclined, have it ground undersize, then rehardened.

It's all well and good to talk about 0.0004 in., but in the real world, it's difficult to measure. To have a good chance at it, you need a micrometer that reads to 0.0001 in. (0.0025mm), and some experience using such a precision tool. Go ahead and use a 0.001-in. (0.025mm) mike on the crank, but if you think a journal has an out-of-round problem, have a machinist double-check your measurements.

Don't forget to measure the main

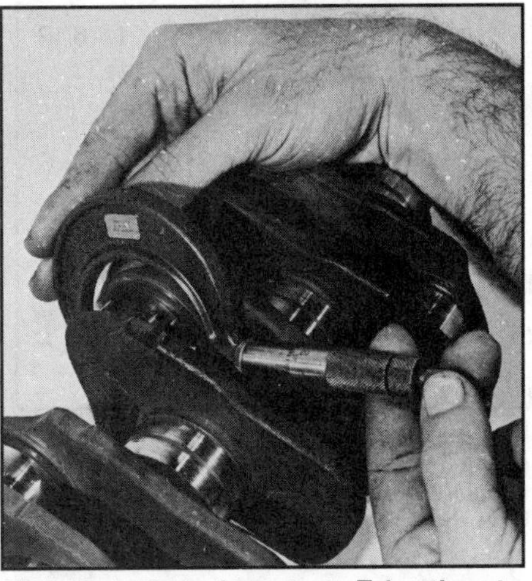

Miking crank is important. Take time to record the figures. You'll never remember them if you don't.

journals. Although they are not as prone to out-of-round wear, check them anyway.

Taper—When a bearing journal wears front-to-rear, or along its length, it is said to be *tapered.* Measure the journal along its lengthwise axis. If these measurements vary, the journal is tapered.

Excess taper must be corrected. If not, the bearing insert will be loaded more at the big end of the journal than at its smaller, worn end. In effect, the load that should be distributed evenly over the length of the bearing will now be concentrated in a smaller area. Bearing design load will be exceeded, resulting in rapid wear.

Taper is expressed in *thousandths-of-an-inch per inch.* So, make your first measurements at one end of the journal, then move to the other end and take another. Subtract the smaller from the larger to find taper. Maximum-allowable taper is again a very small 0.0004-in. (0.010mm) for both main and rod journals. The taper limit on 1200s is only 0.0002-in. (0.005mm).

Again, Honda recommends correcting excessive taper by replacing the crank. Grinding and rehardening will also do the job.

Journal Diameter—Now that you've checked taper and out-of-round, check journal diameters. Find the journal's letter or numeral, then check the charts for that journal's diameter. I think you'll be surprised how close the journals measure. The 0.0008-in. spread is not much. Nominal diameter for all mains is 1.969 in. (50mm). The 1170, 1237 and 1335 rod journals should be 1.575 in. (40mm); the rest should be 1.654 in. (42mm).

Be aware that the crank could be undersize if the engine has been rebuilt. This is probably the case if the main or rod journals are grossly undersize by the same amount and in increments of 0.010 in. (0.25mm).

You are looking for consistent measurements. Your findings should be close to what Honda says they should be, unless the crank has already been ground. If the journals are slightly different, try the appropriate bearing, no matter what the letter/numeral code says.

Also, if you polish the crank, remeasure the journal diameters. Depending on the duration and intensity of the polishing, a slight amount of material may be removed. Even slight metal removal may mean a different-color bearing. I went to all-blue main bearings for this reason. Remember, we are really "splitting hairs" when it comes to the difference between a pink or yellow bearing, for example. What you ultimately want to end up with is a specific *oil clearance*—the distance between journal and bearing.

Typically, bearings of the size used in Honda-car engines can run an oil clearance of 0.0007–0.0035 in. (0.018–0.090mm). Granted, when a 0.003-in.-or-larger oil clearance is used, oil pressure and bearing life drop. Don't expect more than 50,000 miles from such clearances. Optimum clearances would be 0.0015–0.002 in. (0.038–0.050mm). So, don't get too excited about 0.0002-in. differences as long as the oil clearance is between 0.0015 and 0.002 in. on a non-tapered, and non-out-of-round journal.

If your journal measurements show the bearings may end up with excessive oil clearance, either replace the crank or grind and reharden it.

Runout—Before declaring your crank OK or sending it off to have it ground, check its *runout.* Runout is simply how much a crankshaft is bent. You'll need a dial indicator and supporting base to measure runout. Place the block upside down. Install the front and rear main bearings, coating each with oil. Lay the crank in the block and install the two end main caps. Tighten the cap bolts to 25 ft-lb. Set up the dial indicator next to and in

Five main bearings do a great job of keeping crank in line, so runout problems are rare.

contact with one of the center three main-bearing journals. The indicator plunger must be 90° to the journal and positioned so the journal oiling hole will not go under the plunger tip as the crank is rotated.

Rotate the crank slowly by hand and watch the dial indicator for movement. When the indicator is at its *lowest* reading, zero the dial face. Turn the crank again until you find the *highest* reading. Read runout directly. Check runout at the other two center journals. Pay special attention to the center main—if there's any runout, it should have the most.

Honda allows new cranks as much as 0.002-in. (0.05mm) runout. Maximum permissible runout is 0.004 in. (0.10mm). If your crank yields a runout greater than this, don't worry. A crank may bow slightly when unbolted from the block. You may be able to correct this by bolting it back in. Rotate the crank so its point of maximum runout is away from the block. With only the cap half of the center-bearing insert installed, torque the cap bolts. Leave the assembly undisturbed for a couple of days, then remove it and remeasure.

If runout is now within limits, great. If not, replace the crank. Or if you prefer to keep the original crank for some reason, search out a shop that uses a hammer technique to straighten cranks. This job is for experts only, not the local blacksmith. If you don't live in a large city, you'll probably have to ship your crank away

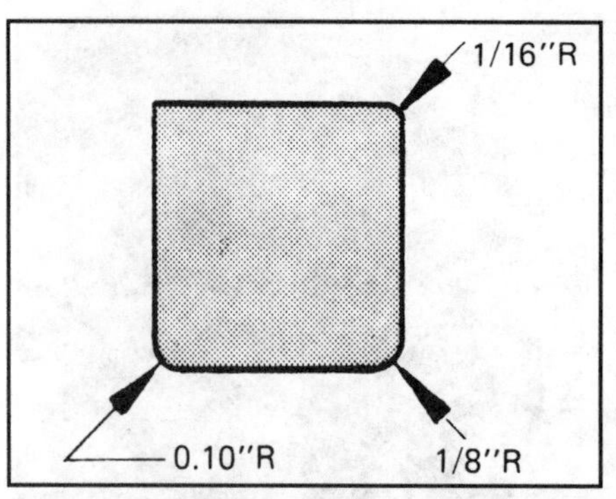

Use pattern for making a template to check bearing-journal radii. Journal radius should not exceed 0.1 in.; otherwise, it will edge-ride bearing.

for this service. Above all, don't use a hydraulic press to do the straightening. Although this will straighten the crank, using a press may also crack it.

Speaking of cracks, Spotcheck or Magnaflux the crank. This is a must if you straightened it. If a crack is found, replace the crank.

Purchase Bearings—After a crank has been ground, undersize bearings must be used. Ask your parts supplier for undersize-bearing inserts to compensate for the metal removed from the crank. For Honda engines, the books usually show undersizes in metric units. These would be 0.25, 0.50mm, etc. The crank will be marked in English units—0.010, 0.020 in., etc. The replacement bearings could be marked with either system. It doesn't matter. If the crank was ground 0.010 in. and the bearings are 0.25mm, you have the right combination.

When you buy bearings, open the box right there at the counter. Inspect the bearing back for a size marking. All but standard-size bearings will be marked. Make sure the box marking is correct. More than once, the right bearings have been put in the wrong box or vice versa.

Honda bearings come in a plastic bag. Don't forget to check their color code before walking out the door.

Post-Regrind Checks—After you have the crank back from the machine shop, make some more checks. Begin with the oil holes. The edge of each oil hole must be chamfered so it blends smoothly into the journal surface. A sharp edge will cause trouble. It could cut a groove in the bearing. If the oil holes are not chamfered, return the crank and voice your complaint.

Another area where improper regrinding can cause trouble is at the journal fillets—where the journal blends into the counterweight or throw at each end. These fillets serve a purpose—to reduce the possibility of cracks developing at the ends of the journals. If a fillet radius is too small, a crack could start in the sharp corner. The crack eventually causes a break. In this respect, the bigger the fillet radius, the better. However, a fillet radius that's too big will *edge-ride* the bearing. The flat surface of the journal would then be too narrow to accommodate the width of the bearing—the edge of the bearing would ride up on the fillet.

You may have noticed old bearing inserts that showed a line of babbitt around their edges. This was caused by edge-riding, but not necessarily due to an overly large fillet. On high-mileage engines, it is usually caused by excessive crankshaft *end-play*—front-to-rear crank movement. Although it could be caused by crank thrust-face wear at the number-4 journal, it's usually caused by mating thrust-bearing wear.

High-mileage edge-riding is usually found on engines mated to manual transmissions. This is a result of the force of the release bearing against the clutch and, consequently, the thrust journal and bearings.

The industry standard for fillet radii is 3/32 in. (0.0938 in. or 2.38mm). If your crank was ground, check its radii with a template made from the pattern on this page. An alternate, but cruder method is to fit a half of a new bearing insert onto the appropriate journal and check for interference between the journal and fillet. If there is interference, the fillet is too large. Return the crank to the machine shop for corrective action.

Final Polishing—If your crank doesn't need to be ground, reconditioning it is simple. Polish the journals and smooth any small scores or dings. There shouldn't be any dings unless the crank was abused while the engine was apart. However, a *needle-file* works well to remove any projections caused by nicks or dings. While you have the file in hand, check the oil holes. Smooth any sharp edges.

Even if the crank was ground, polish its journals. Use narrow strips of 400-grit emery cloth as shown.

Polish journals with even pressure. Tug only hard enough to cinch strip around journal. Reposition crank often to avoid working one spot too much.

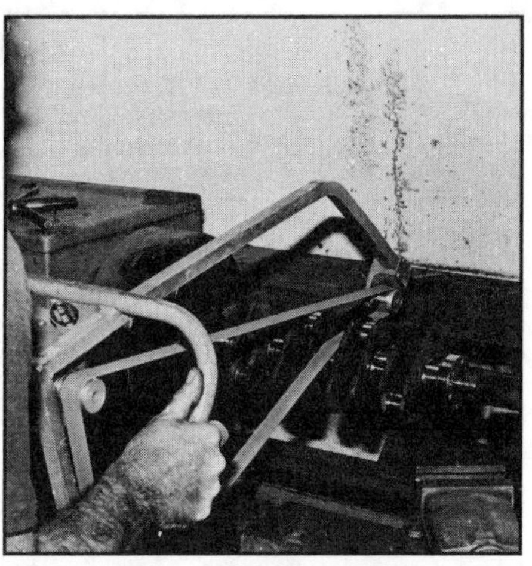

Machine shops use motor-driven polisher and lathe to get even polish in minutes.

Clean oil passages with .22-caliber bore brush. Push brush completely through passage before reversing direction. Otherwise, bristles will fold over, ruining brush. Finish with solvent flush and blow drying.

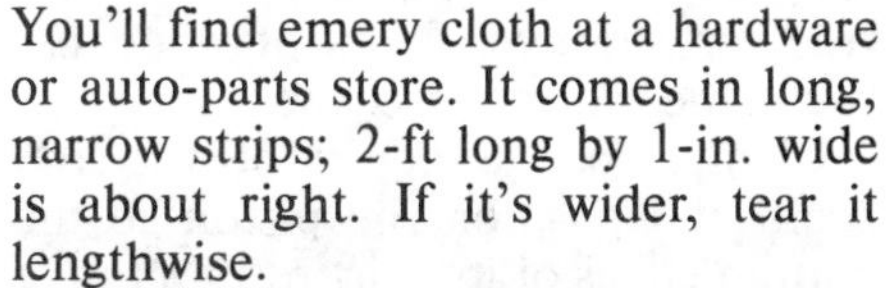

You'll find emery cloth at a hardware or auto-parts store. It comes in long, narrow strips; 2-ft long by 1-in. wide is about right. If it's wider, tear it lengthwise.

Position your crank on the bench so it won't roll around, then loop the strip over a journal. Start at one point on a journal and move around its circumference, ending up where you started. Use a light, even pressure and take care not to polish one spot too long. You don't want to flat-spot the journals. The idea is not to remove metal, but to clean and resurface the journals. Do the oil-seal area as well, but don't bring it up to a bright polish. Like the cylinder walls, the seal area should have some *tooth* to carry oil and lubricate the seal. This increases seal life and improves sealing.

Final Cleaning—Now that crankshaft reconditioning is complete, make sure it's clean. Even if you had the crank hot-tanked, brush out the all-important oil passages. A .22-cal. gun-bore brush works perfectly. Dip the brush in solvent, then push the brush completely through the oil passage before reversing direction. Otherwise, you'll ruin the brush. Continue this process until all crankshaft oil passages are clean.

Use a stiff nylon or brass brush on the counterweights if they're still dirty. After scrubbing, rinse the crank in clean solvent. Pay special attention to flooding the oil passages. Air dry, if possible, or use paper towels. Avoid linty rags.

When the crank is clean, a paper towel wiped over it will come off clean. Then you are ready to spray the crank with water-dispersant oil.

Once crankshaft is completely prepped, protect with spray-on oil.

CRANKSHAFT SPECIFICATIONS

in. (mm)

Engine	Journal Diameter Main	Journal Diameter Connecting Rod	Stroke	Taper & Out of Round	Runout	End play
1170	1.9676-1.9685 (50.006-50.030)	1.5739-1.5748 (39.976-40.000)	2.99 (76)	New: 0.0002 (0.005) SL: 0.004 (0.010)	New: 0.0013 (0.03) SL: 0.002 (0.05)	New: 0.004-0.014 (0.010-0.35) SL: 0.002 (0.05)
1237	1.9676-1.9685 (50.006-50.030)	1.5739-1.5748 (39.976-40.000)	2.99 (76)	New: 0.0002 (0.005) SL: 0.002 (0.05)	New: 0.0013 (0.03) SL: 0.002 (0.05)	New: 0.004-0.014 (0.010-0.35) SL: 0.002 (0.05)
1335	1.9676-1.9685 (50.006-50.030)	1.5739-1.5748 (39.976-40.000)	3.23 (82)	New: 0.0002 (0.005) SL: 0.002 (0.05)	New: 0.0013 (0.03) SL: 0.0024 (0.06)	New: 0.004-0.014 (0.010-0.35) SL: 0.002 (0.05)
1488	1.9676-1.9685 (50.006-50.030)	1.6526-1.6535 (41.976-42.00)	3.41 (86.5)	New: 0.0002 (0.005) SL: 0.002 (0.05)	New: 0.002 (0.06) SL: 0.004 (0.10)	New: 0.004-0.014 (0.010-0.35) SL: 0.002 (0.05)
1600	1.9676-1.9685 (50.006-50.030)	1.6526-1.6535 (41.976-42.00)	3.66 (93)	New: 0.0002 (0.005) SL: 0.002 (0.05)	New: 0.002 (0.06) SL: 0.004 (0.10)	New: 0.004-0.014 (0.010-0.35) SL: 0.002 (0.05)
1751	1.9676-1.9685 (50.006-50.030)	1.6526-1.6535 (41.976-42.00)	3.701 (94)	New: 0.0002 (0.005) SL: 0.002 (0.05)	New: 0.0012 (0.03) SL: 0.0024 (0.06)	New: 0.004-0.014 (0.010-0.35) SL: 0.002 (0.05)

SL—Service Limit

Numeral 10 near piston-pin bore is weight code. If you replace just one piston, use one with *same* weight code. Otherwise, engine will be unbalanced.

Pushing lightly on oil ring yielded no movement. Ring was stuck, resulting in high oil consumption, increased combustion-chamber deposits and contaminated oil. Frequent oil and filter changes, and longer trips—resulting in higher oil temperature—would have helped.

Finish by slipping the crank into a plastic trash bag and storing it out of harm's way.

PISTONS & CONNECTING RODS

If you are boring the block, you'll need new pistons. Throw away the old ones without a second look. However, if you can get by with honing the bores, make sure the pistons are reusable.

Don't make a snap decision about boring solely on cylinder-wall condition. Before you can reuse old pistons, they have to be in good condition, *along with the cylinders.* If the pistons are poor or even borderline, go ahead and bore and get new pistons. The major cost of "new cylinders" is the cost of the pistons. Then you'll have the best possible combination—new pistons in fresh bores.

Dark patch centered below oil-ring groove is scuff mark. General vertical scratches are scoring. Shiny area around scoring shows piston-skirt wear pattern—scores are too small to show in photo. Dark areas near pin bore are oil residue and illustrate how only piston thrust faces contact cylinder wall. This amount of wear isn't excessive, but piston should be replaced to gain maximum life from rebuilt engine.

Boring not only increases engine displacement, it raises compression slightly. For example, if you bore an 1170cc engine 0.030-in. (0.76mm) oversize, you increase its displacement 1.36 cu in. (22.28cc). Compression ratio (CR) will increase slightly because the volume displaced by the larger piston—*swept volume* (SV)—increases, but the area the displaced air is squeezed into—the combustion chamber or clearance volume (CV), does not. Expressed in mathematical terms, $CR = SV/CV + 1$. You might gain one tenth of one point of compression by boring 0.030 in.

The point to remember is that used bores and pistons are just that—*used.* Some of their working life has already been spent. You'll get the longest possible use from your rebuild with fresh bores and pistons.

INSPECT PISTONS

If you are not going to rebore, inspect the pistons before removing them from the rods. If the pistons are OK, you can rebuild the engine without disassembling and assembling the pistons and rods. It is one less job and will speed the rebuild. It may also mean one less trip to the machine shop to have the piston pins pressed out. One less bill too, and that's what this book is all about—reducing those bills!

Remove the rings. There are two ways to do this: by hand or with a *ring expander.* The ring expander is a plier-like tool that grips the rings by their ends and spreads them to allow you to lift them off the piston. The advantage of using a ring expander is it lessens the chance of scratching the pistons. However, Honda rings are small and supple, so spreading them with your thumbs works fine.

Ring removal is easier when the rod/piston assembly is held steady by something other than your own hands. A vise is ideal, but be careful not to mark or bend the rod. Either use soft lead, brass or aluminum vise jaws, or slip two pieces of wood between the rod and vise jaws.

Clamp on the edges of the rod beam and position the piston so the bottom edges of its skirt rest on top of the vise jaws or wood blocks. This keeps the piston from flopping back and forth.

If you don't have a vise, clamp the rod and piston to a bench with a C-clamp. Use a wood block between the clamp and rod. If the bench is metal, put a piece of wood between it and the rod, too. Clamp the assembly so the piston rests against the table edge, thereby steadying it.

Piston Damage—After you have the rings off, check for the following: general damage to the dome, skirt or ring lands, or excessive ring-groove, skirt or pin-bore wear. If any of these exist, the piston must be replaced.

General Damage—Obvious damage is done to a piston when a valve drops into the cylinder, or when a broken ring destroys its groove. You'll have to look harder for other types of damage. Scuffing, scoring, a collapsed skirt, ring-land damage and dome burning can be found through a visual inspection.

Scuffing refers to abrasion damage to the piston skirt. Scoring is like scuffing, but is limited to deep grooves. Both types of damage are caused by insufficient lubrication, a bent rod, or very high operating temperatures—causing skirt-to-bore contact.

The common denominator is excessive pressure and temperature at the piston and cylinder wall. The most common cause of scuffed pistons is overheating. A stuck thermostat, blocked water passage, inoperative fan or low coolant level can all cause severe overheating. Honda cars, like many high-revving, small-engine cars, can't tolerate the severe service American owners are accustomed to.

Honda engines don't have the reserve power and cooling capacity of a big V8. If you are going to "bomb" your Civic down the interstate from Omaha to San Antonio in the summer, its cooling system better be 100% efficient.

In addition to visible scuff damage to a skirt, the shape and resulting piston-to-bore clearance may have been adversely affected. The loads that caused the scuffing may also have bent or broken—*collapsed*—the skirt. Discard a piston with a collapsed skirt and start over. Additionally, a high-mileage piston will have lost some of its controlled-expansion qualities, even if it doesn't show bad scuffing. It's simply a matter of heating and cooling—*heat cycling*—thousands of times.

Examine the bent and twisted connecting rods in the drawing, page 80. If a rod is bent or twisted, it will cause the piston skirt to wear or scuff unevenly from side to side. A little variation is normal, but when a piston skirt is scuffed left of center on one side and right of center on the other, you can be sure the rod is bent or twisted. Checking for bent rods is part of normal machine-shop procedure, but it never hurts to advise them of your findings. You can't check the rods accurately for straightness because you don't have the equipment—but you may have evidence that indicates they need checking.

Piston-Skirt Diameter—Measuring piston-skirt diameter is not so much to determine wear as to check for collapsed skirts. But, before measuring and judging the results, you should fully understand what piston-skirt collapse means.

Practically all modern pistons, Honda's included, are widest at the bottom of their skirts. This wide-skirt design allows the piston to fit tightly in its bore for better piston stability. Stability makes for quieter running and less hammering of the cylinders.

In normal service, piston and cylinder walls warm to their operating temperatures and the piston skirts expand to a given *size* and *shape*. But, if the engine overheats, the piston skirts continue to expand until piston-to-bore clearance is gone; they are forced against the cylinder wall and begin scuffing. This scuffing action increases piston heat even more, until the cylinder wall doesn't permit further expansion. The skirts are then overstressed, as they try to expand further, causing the skirts to break or collapse.

After the engine cools, the piston skirts contract smaller than before because they have been permanently deformed—they have collapsed. The result is greater-than-acceptable bore clearance, both during warmup and at normal operating temperature. The piston will slap during warmup and, possibly, after the engine is at operating temperature. The skirt may even break off.

If you were to reuse pistons with collapsed skirts, they would be noisy. The oil-ring grooves may have been deformed as well, providing poor oil control.

Heavy scuffing is a common sign of collapsed piston skirts, but to be sure, measure the piston in two places and compare the readings. First, measure across the skirt immediately below the oil-ring groove and 90° to the piston pin. Then, move to the bottom of the skirt and measure again. The skirts should be about 0.0005-in. (0.013mm) wider at the bottom. If the measurements are the same, or less at the bottom, one or both skirts has collapsed. The piston is junk.

Old compression rings make great scrapers. Guard piston edges from gouging when cleaning with a ring.

Inspect Domes—Because the melting point of aluminum is lower than cast iron, the piston dome is first to suffer from heat caused by preignition or a lean fuel mixture. Such damage removes aluminum from the dome and deposits it on the combustion chamber, valves and sparkplug, or blows it out the exhaust. This leaves the dome with a spongy-looking or porous surface that carbon adheres to.

Clean the piston domes so you can inspect them. If carbon buildup is heavy, remove it with a dull screwdriver or old compression ring. Make sure the screwdriver is old and worn. One with a sharp blade will scratch the piston easily. Don't use a sharp-edged chisel or gasket scraper on the pistons.

Follow screwdriver cleaning with a wire brush, or wheel, if you have one. Be extra careful to keep wire bristles off the piston sides, particularly the ring grooves. Light scratches on top of a piston won't hurt, but they will cause trouble in the grooves and on the skirts.

Piston-Pin-Bore Wear—All Honda car engines use pressed-in pins—the pin has an *interference fit* in the rod. The pin is slightly larger than the rod's pin bore, so removing the pin from the rod requires a press.

Pressed-in pins make checking pin-bore wear difficult for two reasons. First, the pin is usually galled during pressing, and second, it costs several dollars to press out a pin. Fortunately, there's a reasonably accurate and easy way to check for excessive pin and

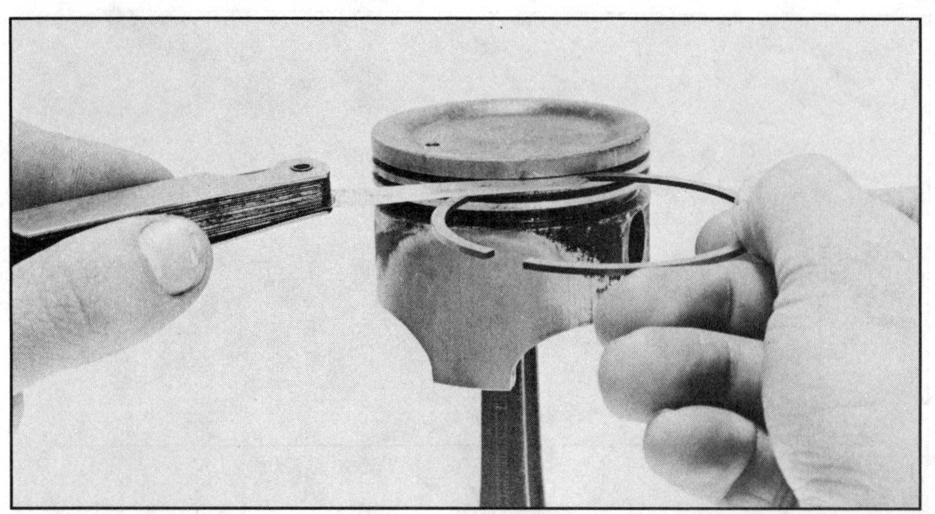

Checking ring-groove wear with ring and feeler gages. Ring-to-land clearance can be as much as 0.005 in., but 0.0008–0.0018 in. is best. Check for consistent ring-groove width by sliding ring and feeler gage around entire ring groove. If they bind or loosen, ring land is bent or groove is worn unevenly. Replace piston.

To resize rod big end, some of cap's mating surface is first ground off. Disc grinder is one way of doing job. Most shops use a special fixture and grinding wheel to remove metal.

pin-bore wear without removing the pins.

To check pin-to-piston clearance first, soak each piston in solvent to remove all oil. Work the rod back and forth while the pin is submerged in solvent. This removes the oil cushion from between the pin and bore. The piston-and-rod assembly should be at room temperature to get the most accurate check.

Hold the assembly upside down on a bench, grasping the piston with one hand and the big end of the rod with the other. Hold the piston with the end of your thumb on one end of the pin and finger on the other. Now try to rock and twist the rod without moving the piston. Rock the rod in the direction of the pin center line. Next, twist it in a circle concentric with the piston circumference. If wear is excessive, you'll be able to feel pin movement at your thumb and finger tips.

Again, if you feel *any* movement, the piston-pin bore is worn. Chances are the remaining piston-pin bores are worn if the first one checked is. Check them anyway. The cure for worn pins and piston-pin bores is piston replacement.

Ring-Groove Cleaning—There's no accurate way to measure ring-land wear if the grooves are dirty. You have a choice of two ring-groove cleaning methods. One is free, the other costs a little more. The free method uses the time-honored broken-ring method. The other uses a *ring-groove cleaner.*

Unless you rebuild an engine every now and then, use the broken ring. Break a compression ring in half and grind or file one end. Round off all ragged edges except the one cutting edge. This will keep you from removing metal from the ring lands. These surfaces must remain as tight and damage-free as possible for proper ring sealing.

When using a broken ring, be very careful not to gouge or scratch the ring grooves. Use steady, relatively slow strokes. Keep the ring centered as much as possible at all times. This is important because piston rings rely greatly on their grooves for sealing. Once a ring groove is damaged, the ring cannot seal well against the piston.

If you can afford some special tools, a ring-groove cleaner is nice to have. This will allow you to clean the pistons much faster with less chance of damage.

Measuring Ring Grooves—The easy way of measuring ring-groove width or *ring side clearance* is with feeler gages and a new ring. However, don't lay out the money for new rings just to discover that your old pistons are bad. Assuming you'd do the logical thing and rebore and install new, oversize pistons and rings, you'd be stuck with a set of rings that you had no use for. Therefore, use an old ring and compensate for its wear.

To measure ring-groove width, mike the thickness of an old compression ring. Subtract this from the *original* thickness measurement of that ring. See page 43 for original ring thickness. Add the difference to the specified ring side clearance. The result is the ring-groove *checking clearance.*

Now, insert the edge of the ring in the groove and gage the distance between it and the groove with feeler gages. You have the checking clearance right away, but you must also check around the groove.

Slide the ring and feeler gage around the piston and you can easily tell whether the ring lands are bent. The feeler gage will tighten up in any constricted area and loosen where the ring lands are worn or bent. You've just killed two birds with one stone, ring side clearance and ring-land condition.

When you find the gage that fits snugly—not tight or loose—subtract this measurement from the checking clearance. The difference is *piston-ring side clearance*—the dimension used to determine if the piston-ring grooves are serviceable.

The side clearance for new Honda rings is 0.0008–0.0018-in. (0.020–0.046mm). The service limit is 0.005 in. (0.13mm). So, if you have more than 0.005-in. ring side clearance, get new pistons.

Oil-Ring Grooves—Because oil rings and their grooves are so heavily oiled—compared to compression rings—they wear little, if any. Therefore, if the compression-ring grooves are OK, visually inspect the oil-ring grooves. If they look OK, they should be OK.

Special dial indicator quickly reveals out-of-round big end.

Rod is made round again with special precision hone. Dial indicator is used to check progress. Several rods can be honed at one time.

INSPECT CONNECTING RODS

Checking the Big End—Think of the rod's big end like the main-bearing bores in the block. Both can change shape. The conventional rebuilding fix for these problems is called *reconditioning.* It is similar to align boring or honing the main-bearing bores. The rod-cap mating surface is ground, removing several thousandths of material. The cap is then installed on the rod and the nuts torqued to specification: 20 ft-lb for all engines except the 1751; 22–25 ft-lb for the 1751. The bearing bore is now egg-shaped. Next, the big end is honed round on a special precision hone. The machinist checks his progress with a dial-indicating fixture.

Reconditioning, or *sizing,* the rods ensures that the bearing bore is truly round, not tapered, and the *correct diameter.* Therein lies a problem with Honda engines. When the machinist is done reconditioning the rods, the big ends will all be one size. It's like grinding cranks; the machinist uses the nominal diameter. This is fine, but the big ends are not going to be within 0.0002 in. (0.005mm) of their original diameter, unless they were nominal to start with.

This means you will either have to accept a slight difference in oil clearance using the same-color bearing, experiment until you find the desired Honda bearing, or use an aftermarket bearing. All methods are acceptable, but understand that the original-size bearings might not work.

Honda doesn't acknowledge rod reconditioning. They'd rather have you replace the rod if it's not to specs. Actually, *most* Honda rods don't need reconditioning. But, early 1170 and 1237cc rods are another story. With four-speed manual transaxles and numerically high gearing, the small aluminum engines must rev high to keep up with highway traffic. The rods suffer accordingly. All later engines use five-speed manual or automatic transaxles, which reduce engine rpm, even at highway speeds. On the other hand, many Honda mechanics complain of over-revved CVCC engines with deformed rods. The rods are a good example of a part that works fine if properly operated and serviced, but is unable to tolerate any abuse. If you took advantage of the high-rpm capabilities of your Honda engine, rod reconditioning or replacement might be necessary.

The best way to accurately check the rod big end is with a dial indicator. Any shop with a rod-reconditioning machine can tell you quickly if the rods are out-of-round. The cap is torqued in place without bearings. Then, the big end is checked on the machine's dial-indicating fixture. Any out-of-round shows up on the dial indicator.

Checking the Small End—It is a rare rod that needs replacement because of small-end problems. Unless you detected movement during the piston-pin check, the pin bore should be just fine. Actually, this is something the machinist will bring to your attention because he's the one with the press. If, after the machinist presses out the pin, a bad pin bore is found, replace the rod.

Visually inspect all rods for cracks.

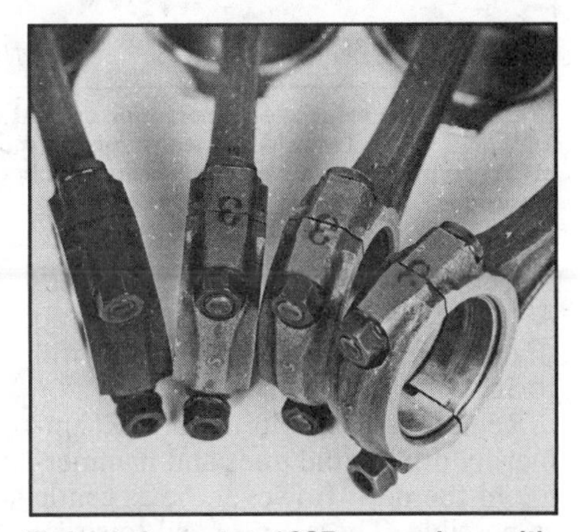

Rods belong to 1237cc engine with clogged oil screen, page 53. Rod at left is burned black from spun bearing; resizing it would not repair extensive damage. Numeral 3 on other rods is for bearing selection. Don't use carb cleaner—it will wash off numeral.

Magnafluxing is good, but generally not necessary. Rods can also be overheated—usually the result of poor oiling. If a rod has been overheated, it will be darker in the affected area. Severe examples will be black or dark blue. Although overheating is not a cause for replacement, overheated rods usually have other problems that require it.

Rod Bolts—Along with beating the rods out-of-round, another effect of high-rpm operation is *stretched* rod bolts. Again, the principal offenders are the 1200s. Don't bother miking these bolts, replace them. Constant high-rpm use requires strong rod bolts.

Replace rod bolts before a rod is reconditioned. Otherwise, there might be a slight mismatch between the rod and its cap. This is because the

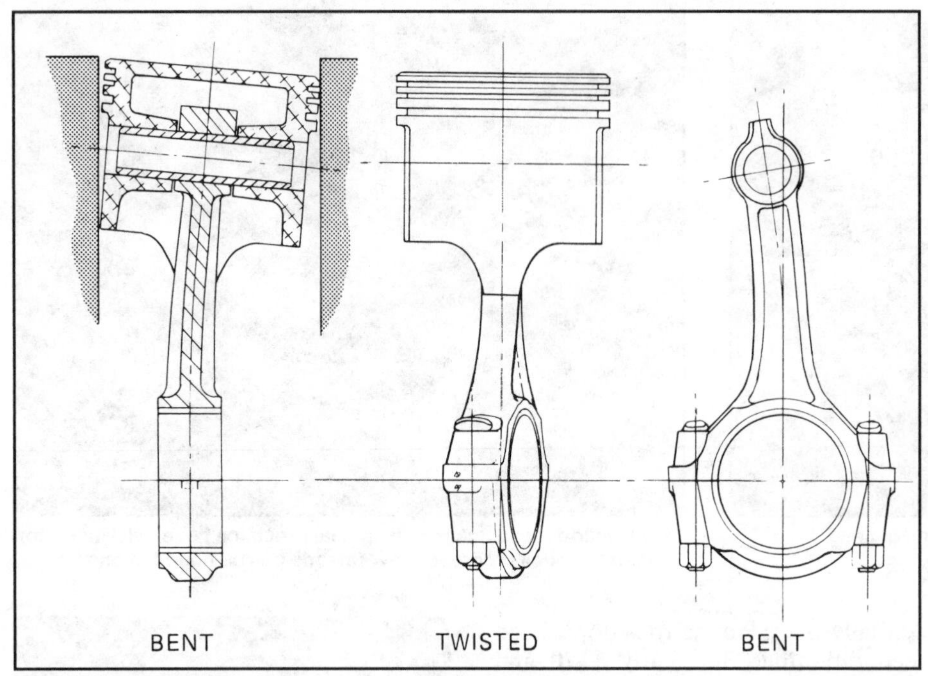

Uneven piston wear is one indication of bent or twisted rod. First rod wears piston most at TDC and BDC. Twisted rod wears piston most at mid-stroke. Third rod, however, has no effect on piston wear and can be reused if bend is not visible to naked eye. Drawing by Tom Monroe.

Gage shows 900 psi used to push out piston pin—a graphic illustration of why you should *never* try to hammer out a piston pin. You'd destroy the pin, piston and probably the rod. It would take all day, too.

When pressing out a pin, give it somewhere to go. Piston is on blocks centered over gap between steel plates.

rod bolt positions the cap on the rod. A different bolt can shift the cap slightly; reconditioning the rod will correct the mismatch.

Replacing rod bolts involves hammering out the old ones and hammering in the new. However, be as gentle as possible. There's no need to stress the rod. Hold the rod upside down, with the small end against the bench or table. Or, use a block of wood on the floor, if it is the only solid surface you have. With firm backing, the rod bolt should budge with a couple of moderate blows.

To seat the new bolts, turn a short piece of 2x4 upright in a vice and straddle the rod big end over the wood. Set in the new bolt and tap it "home." Watch the offset bolthead; the flat side should line up with the flat shoulder of the rod.

If you used an impact wrench on the nuts and chewed up the hex, replace the nut. Otherwise, the original nuts are OK. If you are going to change rod nuts, buy them ahead of time so you don't have to stop in the middle of engine assembly.

PISTON/ROD DISASSEMBLY & REASSEMBLY

Align Rods—In the machine shop, rod *alignment* is checked *before* the piston is removed from its rod. You are actually checking the rods for bending. Two fixtures are commonly used for this test. One uses a vernier-type scale to detect a bent rod; the other, light, passed by a flat plate. This job is best done at the machine shop because you can use the checking fixtures.

Or you can check rods visually by stacking them one on top of another. By switching the order of the stack, a bent rod can often be detected.

After removing its piston, a bent rod can be straightened with a large pry bar. The rod is not heated, and bending the rod while cold does not harm it. It sure gives me the willies to see someone clamp my rods in a vise and yank on them with a yard-long bar, though!

Disassembly Tools—A press with the appropriate mandrels is required to remove pressed-in pins. If you have a press and the proper tooling, by all means do the job and save yourself some money. Otherwise, leave the removal-and-replacement work to your machinist.

Above all, *do not hammer on a piston pin!* Even though you can get them apart, you'll probably ruin the pin, piston and rod in the process. The same advice goes for power tools. Trying to blast a press-fit pin out with an impact, or *zip,* gun is criminal laziness. *Always* use a press.

Remove Pin—To remove a piston pin correctly, rest the piston on a hollow mandrel, squarely on its pin boss. The hole must be large enough to allow the pin to pass through. There must also be a hole in the press bed for the same reason. Always use a punch between the press ram and pin. Best is a *stepped* punch—one that butts against the pin end and has a reduced diameter to center the punch in the pin. Center the pin, mandrel and punch under the press ram. Before starting to press out the pin, make doubly sure the pin boss rests squarely on the mandrel. Now, push out the pin.

Install Pin—Assembling the pin and piston can be done two ways: with a press or with heat. Pressing a pin is easy enough to understand. The pin is forced into a smaller bore, creating au-

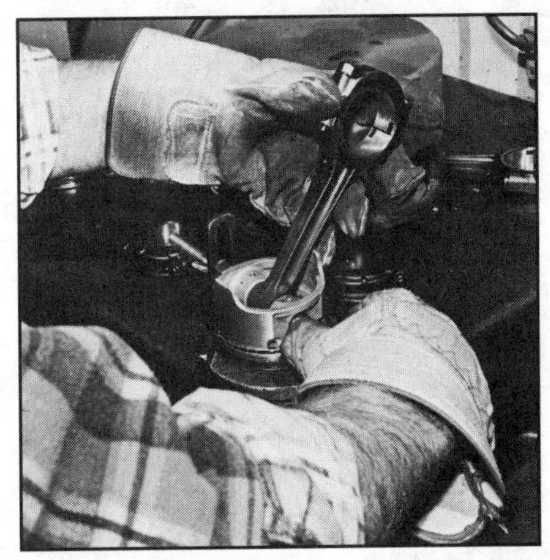

I prefer heating rods for piston-pin installation. You need a fixture to limit pin travel so it centers in rod. You also must be very quick because pin and rod temperatures quickly equalize, cinching the pin. After installation, pin and rod are oil-quenched.

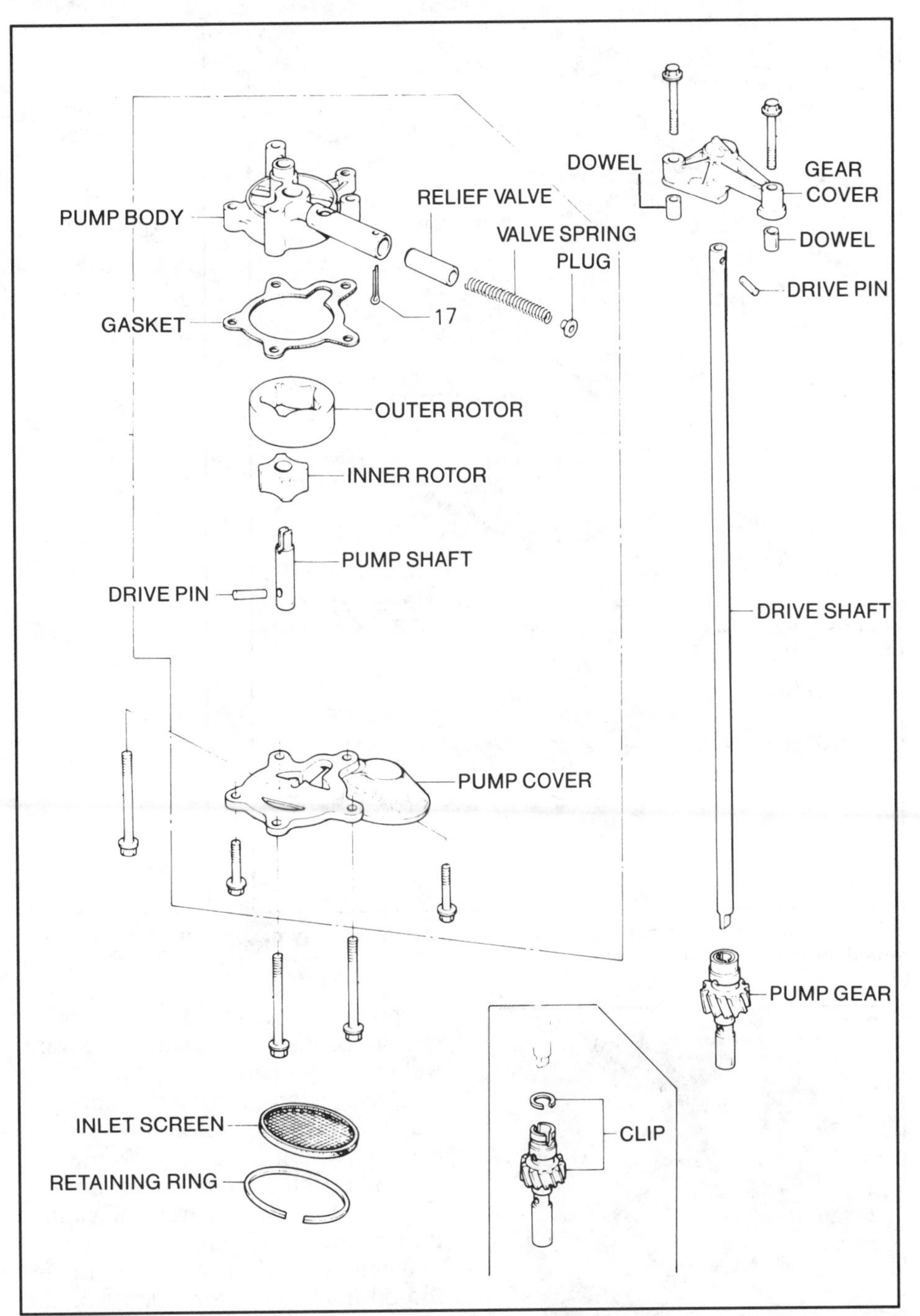

Exploded view of single-stage oil pump; 1335 shown, 1200 similar. Courtesy of American Honda Motor Co., Inc.

tomatic retention. However, when the small end of the rod is heated, the bore expands, allowing the pin to slide in by hand. When the rod cools, it clamps tightly on the pin.

I prefer heating the rod. When done by a machine shop, you are practically guaranteed a perfect, non-galled pin installation. When heating is not possible, the pin should be pressed in.

Pressing in a pin is basically the reverse of pressing it out, but you must be extra careful to line up pin, piston and rod. First, inspect the ends of each pin. If they are not chamfered, grind one on them. Otherwise, the sharp 90° edge will gall the rod.

Assemble the rod and piston in their correct relationship. Remember, a rod can be easily ruined if the pin is not aligned with the rod pin bore while pressing it in. It won't do the piston any good either. A little lubricant on the pin—moly or anti-seize compound—reduces the chance of galling. It also reduces the force needed to press in the pin. Finally, the pin must be installed so it projects equally from both sides of the rod.

Using heat to install piston pins helps prevent galling—if you successfully complete each installation. To use the heat method, a heat source such as an electric rod heater, acetylene or propane torch is needed. It is very difficult to monitor the amount of heat applied to the rod, or how far in to install the pin—you have to be *very quick*. If you take too much time, the rod cools and clamps onto the pin. A press is required to move it any farther. Then you'll probably gall the rod and pin.

If this is your first time installing rods and pistons, count on finishing about half of them with a press. Also, Honda rods have been known to fail due to overheating during piston-pin installation. Leave this job to the experienced hand.

Some machine shops have an electric rod-heating device that does a beautiful job of heating connecting rods. If you want the best possible piston/rod assembly, look for a machine shop with a rod heater—or a machinist who is proficient with a torch.

With either heat or press, make doubly sure the pistons are positioned correctly on their rods. Honda pistons are easy: The circle or triangle on the top of the piston goes on the same side as the rod oil hole. If the piston has *both* a circle and triangle, put the triangle on the oil-hole side.

Aftermarket pistons have their own mark. Instructions with the pistons should show which side the mark

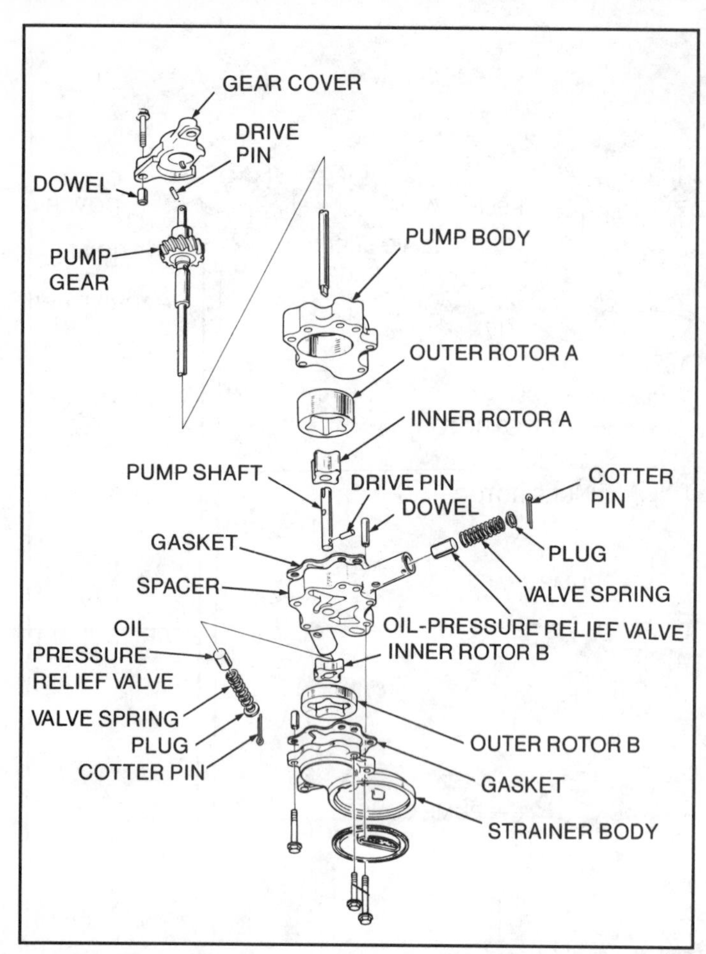

Exploded view of two-stage oil pump; 1751 engine. Courtesy of American Honda Motor Co., Inc.

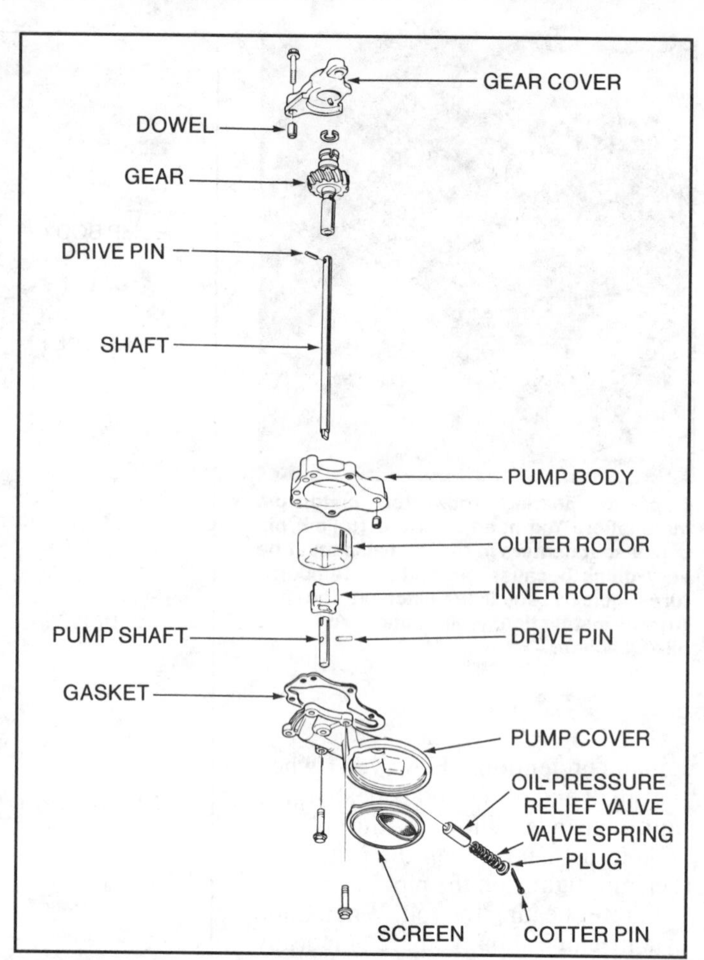

Exploded view of single-stage oil pump; 1751 shown, 1488 and 1600 similar. Courtesy of American Honda Motor Co., Inc.

Inner and outer rotors have dimples on one side only. Note their positions on disassembly and reassemble rotors on same side as originally installed. If dimples can't be seen, mark rotors with felt-tip pen.

should face. Pistons are directional because the pin bore is offset. Therefore, the installed piston is not perfectly centered on the rod. This offset geometry loads the piston against the thrust- or sparkplug-side cylinder wall. Pushing the piston against the wall stops the piston from slapping, providing quieter engine operation. Pistons installed backward will knock in operation, and may have incorrect piston-to-valve clearance.

If you install a piston 180° out of position, you'll have to push the pin out and turn the rod or piston around. This doubles the chances of galling the pin, rod and piston.

When correctly installed, both the rod oil hole and piston mark face the intake manifold. It's a good idea to check your work by laying out the piston/rod assemblies as installed in the engine.

OIL PUMP & DRIVE SHAFT

Most Honda oil-pump components are durable. Unless your engine has ingested large amounts of dirt or you have run the same oil for five years, the oil pump should be OK. Nevertheless, even if the engine has seen light duty, the pump should come apart for inspection.

Start by placing the pump pickup in a pan of solvent. Rotate the shaft counterclockwise until clean solvent gushes from the pump outlet. This is an easy way to flush most of the remaining oil from the pump's innards. Then scrub the exterior of the pump with a stiff-bristle brush and solvent. This keeps grit out of the pump when you open it up.

Disassemble & Inspect Pump—Turn the pump upside down, watching for spilling solvent, and undo the remaining bolts. Remove the cover and completely disassemble the rotors, pin and drive shaft. Don't let the rotors fall out onto the floor. This will damage them. Note that the rotors are marked with dots. These are manufacturing marks, not alignment dots! They must be reinstalled in the same relationship, so note the dot's relationship at disassembly.

By same relationship, I mean only that if both dots are on the same side, they should be reassembled that way. Sometimes, the outer-rotor dot will be on one side and the inner-rotor dot on the other. If you find them like that, reassemble them like that. There is no need to align the dots next to each other. All that matters is that the gears are not flipped end-to-end.

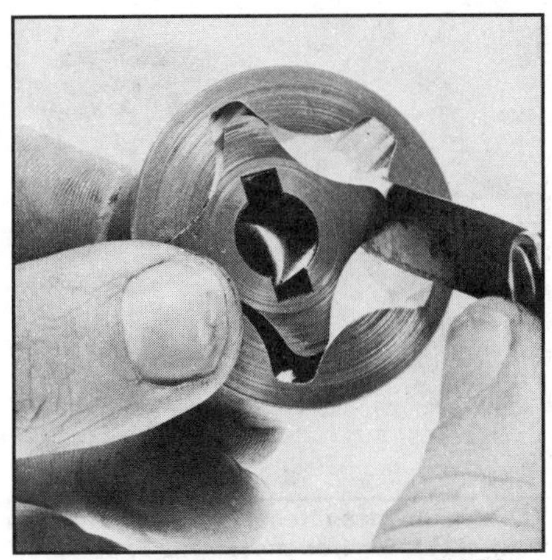

Inner-to-outer rotor clearance is 0.006 in. on new pumps. Service limit is 0.008 in. Radial scoring is from dirt or metal particles in oil—pump gets oil *before* filter. It's best to use a wire-type feeler gage, not the flat type shown.

Measuring end clearance: Reduce clearance by lapping pump body—light-colored piece on bottom. Note splayed cotter pin at top of photo. Remove pin to reach pressure-relief-valve components.

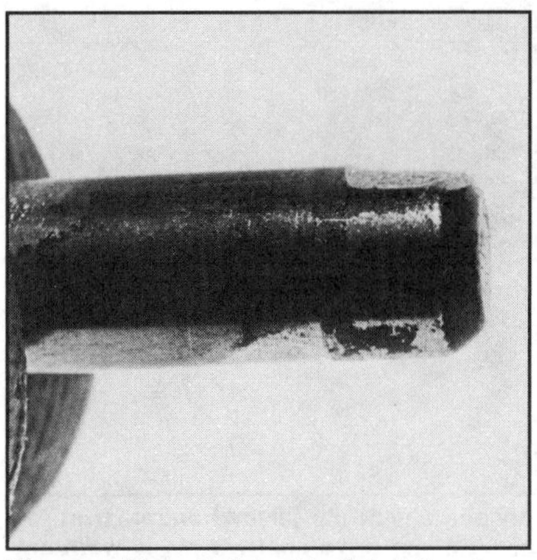

Wear on drive pin is not bad, but it got replaced anyway. Pin doesn't rotate, so surging rotational forces hammer the same spot constantly. Replace pin.

Visually inspect the pump body and rotors. Deep scoring of the pump body and chipped rotor lobes indicate that large dirt or metal particles passed through the pump—get a new pump. Minor scratching and scoring from dirty oil is normal—reuse the pump.

Visually check the oil-pump drive shaft. It should not be grooved where it contacts the oil-pump body.

Feeler-Gage Checks—Set the inner and outer rotors back inside the pump body. To accurately measure the outer-rotor-to-inner-rotor clearance as well as the outer-rotor-to- pump-body clearance, you need a *wire-type* feeler gage. A *flat* feeler gage takes up too much room, so the clearance is actually greater than the thickness of the gage. Experiment with wire-type feeler gages until you find one that just fits between the installed inner and outer rotors. Measure between one of the inner rotor's points and the outer rotor's softly rounded points. Then measure between the outer rotor and pump body. The service limit for both is 0.008 in. (0.20mm).

Typically, these clearances measure a couple-thousandths less than the service limit. If the pump is badly worn, an 0.008-in. gage will probably be lost in the excessive clearances. If the rotors are worn, replace the pump.

Next check *end clearance*—distance between the pump cover and rotors. You can use a *flat* feeler gage for this. After removing any remaining gasket material, lay the pump cover partially over the pump body. This way, you can slip a flat feeler gage between the cover and rotors. Find the gage that fits; it should be 0.006 in. (0.15mm) or less.

Two-stage pumps used in 1751cc engines have two sets of rotors. However, both the thick "A" rotor and thin "B" rotor use the same wear limits as single-stage pumps. When measuring end clearance on two-stage pumps, do so with the gasket in place. Otherwise, measurements will be inaccurate.

You can also check end clearance with a depth gage or Plastigage. The depth gage will give the most accurate readings, but a depth gage isn't found in every toolbox. Plastigage is used in a manner similar to checking bearing clearance, page 114.

Chances are end clearance is close to normal if you kept the oil clean. If end clearance is too much, but the rest of the pump is OK, you can reduce it. In fact, new pumps measure between 0.001 and 0.004 in. (0.025—0.100mm). So, if end clearance is approaching 0.006 in., it's a good idea to reduce it to 0.003 in. (0.075mm) or so. The pump will prime quicker and operate more efficiently.

End clearance is reduced by *lapping* the pump body on abrasive paper backed by a *flat,* rigid surface—a thick piece of glass works well. To lap the body, buy one sheet of 220 Wet-or-Dry emery paper. Find a hefty piece of glass; a glass shop might have a broken piece you could use. A marble slab works great, too. The surface supporting the paper *must* be flat and thick enough not to break or bend when you press on it with the pump body. Lay the paper down on the glass, grit-side up. Lightly flush the paper with water, cutting oil or automatic transmission fluid (ATF). Place the pump body on the paper, cover-side down and without rotors.

Move the pump back and forth with light pressure. Don't bear down or you'll remove too much material from the cast-aluminum body. After several passes, rotate the pump 90° and continue. Periodically rotate the pump so the lapping marks go in different directions. Don't lap too long; it doesn't take much to remove a thousandth or two.

Make frequent feeler-gage checks. You should be able to get end clearance right where you want it. It's better to have slightly more clearance than not enough. If you overdo it, lap the rotors the same as you lapped the body until end clearance is the desired 0.001—0.004 in.

When you reinstall the rotors to check end clearance, be sure to remove grit from the pump-body mating surfaces. If you don't, the tiny granules will likely affect your measurements. You'll be tricked into removing too much material.

After lapping, clean the pump in solvent. Remove every last particle of lapping grit. Otherwise, it will be pumped through your newly rebuilt engine.

Drive Pin—Earlier I said *most* Honda oil-pump parts were quite durable. The exception is the drive pin that transfers drive-shaft rotation to the inner rotor. This pin wears during normal operation, and has been known to shear. When this occurs, the drive shaft turns, but not the

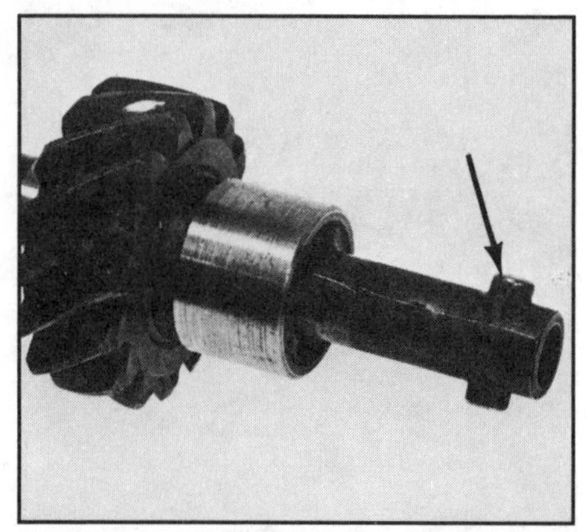

Always check pin (arrow) and gear at top of oil-pump drive shaft. If pin is worn any more than this, replace it. If you replace the camshaft, also replace this gear. If cam is OK, make sure gear-tooth wear pattern on gear and is *centered*—midway between ends of gear teeth. Wear pattern on gear looks fine.

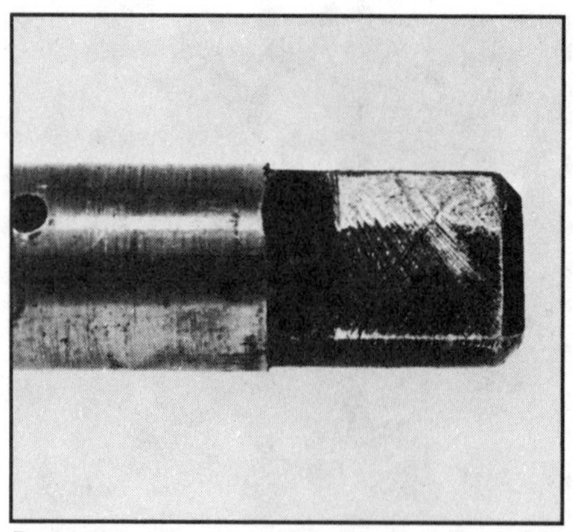

Problems are rare at bottom end of oil-pump drive shaft, but check it anyway. This one shows some wear, but is in good shape.

Thoroughly clean relief valve and bore, then check for free movement of valve in bore. If valve sticks, pump might develop only partial pressure—say 20 psi. That would be enough to keep oil-pressure warning light off, but insufficient for high-rpm, high-load operation. Use your little finger to bottom valve in bore.

Shiny area on relief valve is where valve wears in bore. Sealing end of valve is angled section at top. Make sure angled section is not worn concave. If it is, double-check all pump parts. Pump replacement is recommended.

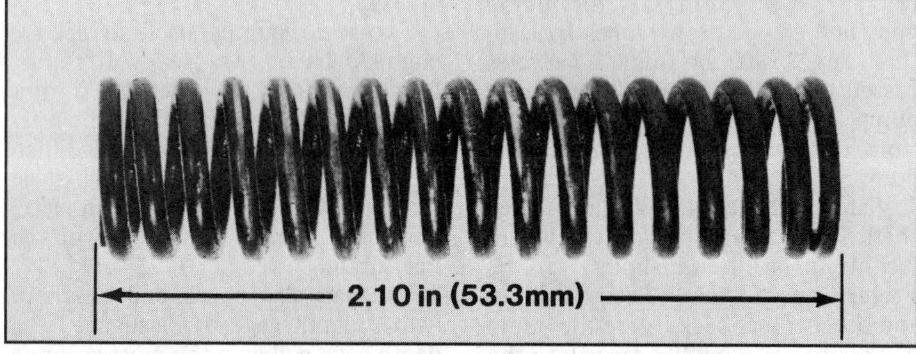

Use vernier caliper or ruler to check relief-valve-spring free length. Keep dual springs of two-stage pumps separated so they can be returned to their original bores.

rotors—presto, no oil pressure. Severe bottom-end damage results shortly thereafter. Considering this pin is inexpensive, replace it. You don't even need tools to do it because it slips right out of the drive shaft.

Pressure-Relief Valve—Here's a part that rarely needs much attention. Disassemble the valve by removing the cotter pin from the pressure-relief-valve housing. This is the boss that juts from the oil-pump body. Two-stage oil pumps have *two* pressure-relief-valve assemblies. Keep the valves and springs from each bore *separated*—they have different calibration.

With the pin removed, the retainer, spring and valve slide out of the housing. Slip the valve back and forth several times in its bore. It shouldn't hang up, but slide smoothly from one end to the other.

If there are burrs in the pump housing, hone them out. A brake-spring-type hone is the tool you'll need. Use fine-grit stones and hone the relief bore for no more than 30 seconds. Use light load on the stones and lubricate with cutting oil.

Inspect the sealing end of the valve. You'll see the mating surface, which shines a little differently than the rest of the valve. The mating, or sealing surface should not be grooved from long-term wear. If it is, replace the valve.

On single-stage pumps, measure the spring *free length*—length of spring when unloaded. It should be 2.10 in. (53.3mm). If shorter, replace it. On two-stage pumps, merely inspect each spring for damage.

After completing these checks, reassemble the relief valve. Use a new cotter pin. Make sure there is no grit in the bore or on the removed parts. This warning also applies to rag lint. A small piece of debris can hold the valve open, routing oil back into the pan instead of the oil galleries. That's one problem you don't need.

Normally, the oil pump needs no work other than checking. If you find considerable wear on the gears, and the relief valve is scored, buy a new pump. Don't replace a lot of little parts and put them in a worn-out body.

Final Assembly—Pack the pump rotors with light grease or petroleum jelly during final assembly. This ensures that the pump will develop sufficient suction, or prime, to draw oil from the sump during initial engine start-up.

Discard the old oil-pump pickup screen. You'll be installing a new screen after installing the pump in the engine.

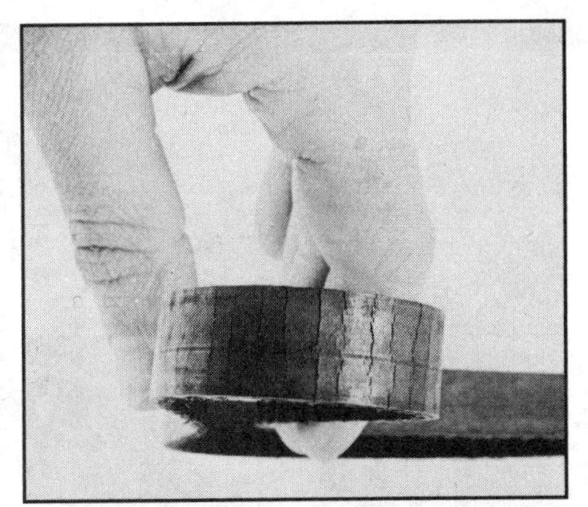

I'm bending timing belt to make a point. Cracks mean belt must be replaced. Rubber parts don't last long in hot, dry climates. Each side should be parallel. Bending like this harms belt backing.

CAUTION: ALUMINUM THREADS

Because all Honda engines covered in this book have aluminum cylinder heads, and some have aluminum blocks, some words about aluminum-thread care are in order.

When a steel bolt is threaded into aluminum, the softer aluminum threads are easily damaged. Common problems are crossthreading, overtightening and galling. Crossthreading is probably the number-one cause of destroyed aluminum threads. However, if the threads are clean, of the right size, and reasonable care is taken starting the fastener in its hole, the chance of crossthreading is small.

When a fastener is overtightened, something has to give. In the case of a steel bolt in an aluminum head, the softer aluminum threads usually give.

Sometimes, aluminum threads strip even without crossthreading or overtightening. This is a result of galling. It's common with sparkplug threads. The problem is heat-expansion rates. When an aluminum head is hot, it grows several thousandths of an inch in all directions. Thus, the pressure between the sparkplug threads and those in the sparkplug hole increases. This forces the aluminum into voids in the sparkplug threads, locking the plug in its hole. When the plug is subsequently removed, its threads tear or gall the aluminum threads, particularly in the case of a hot engine. To protect against this, always use anti-seize compound on sparkplug threads; and don't change plugs on a hot engine.

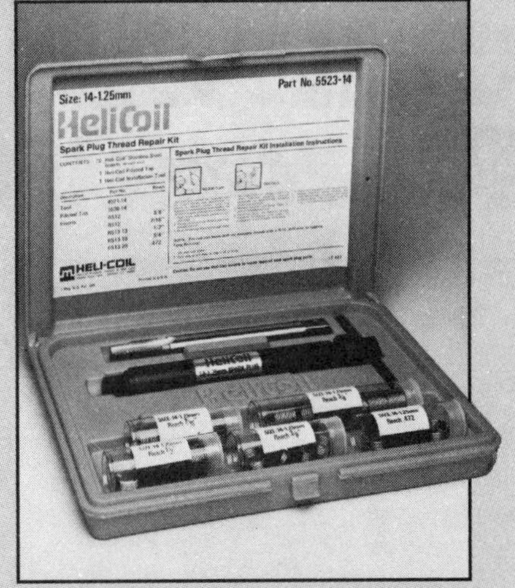

Damage to sparkplug threads in aluminum head can be repaired with thread inserts. Shown is popular Heli-Coil 14-1.25mm thread-repair kit. Photo courtesy of Heli-Coil Products Division, Mite Corp.

How do you restore stripped threads? It can be done with a steel thread insert. Thread inserts fall into three groups: steel-wire coil, threaded sleeve and locking threaded sleeve.

The procedures for installing inserts, sometimes called *thread savers,* differ. Basically, the old threads are drilled out, the hole is tapped oversize and the thread insert is installed. The new steel threads are actually stronger than the aluminum ones they replace. For this reason, race-car builders often install thread inserts in light-alloy components as a matter of course. It's a good preventive measure in any component that is removed and installed frequently.

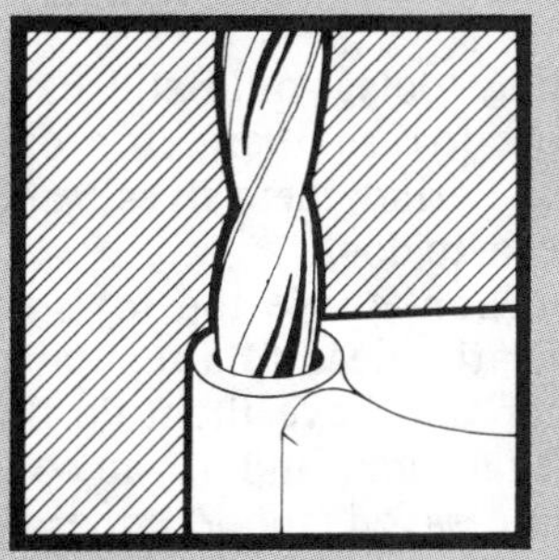

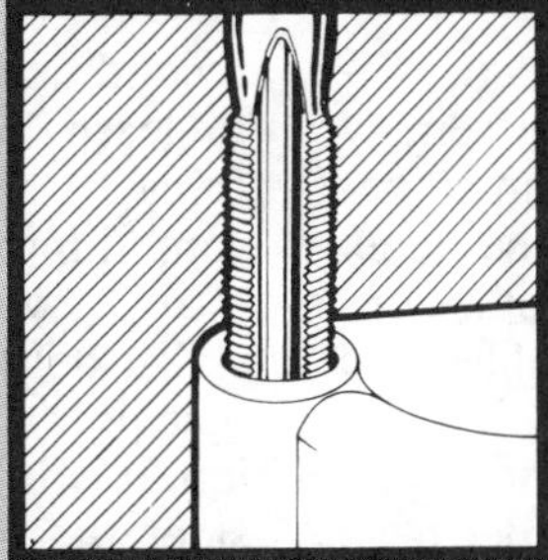

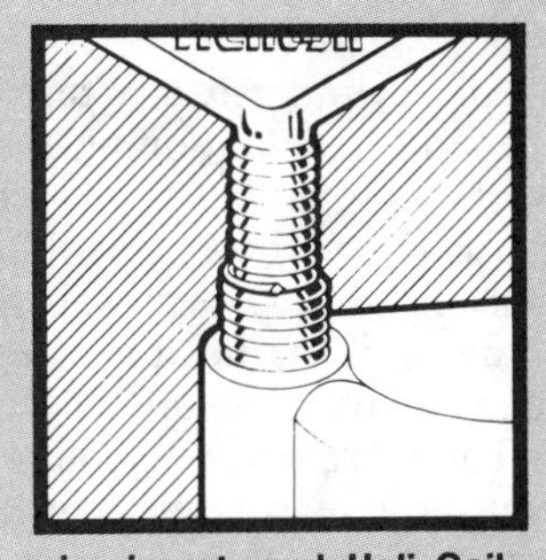

Damaged threads are drilled oversize, hole tapped for wire insert, and Heli-Coil inserted. Inserts are stronger than original aluminum threads. Drawing courtesy of Heli-Coil Products Division, Mite Corp.

TIMING BELT & SPROCKETS

About the worst thing you can say about a camshaft belt-drive system as used in Honda engines is it wears silently. As timing belts go, Honda's are very durable. Performance deteriorates gradually from incorrect valve and ignition timing, but without any telltale noise to get your attention.

Check the belt. Its back should be free of cracks. Oil- or solvent-soaking will quickly destroy a belt, and is cause for replacement. On the cogged side, inspect for wear in the area shown, page 127. A very slight rounding of the corners is normal, but sloping sides mean the belt should be replaced. Regardless, I suggest that you change the belt now just for the peace of mind. Plus, it's a lot easier to do now than with the engine in the car.

Sprockets are obviously more durable than the belt. Inspect both the crank and cam sprockets for wear. Look for rounded corners on the cogs. Usually, wear is easily detectable on this type of sprocket because a small portion will not mate with the belt. This unworn band provides the necessary color contrast to detect wear. Sprockets are not normal replacement items, so don't be surprised if they don't look worn.

Check the fit of the sprocket keys in their respective shafts. If you detect lots of slop and can see shiny wear marks, replace the keys. In rare cases, the sprocket keyway will get pounded by a loose key. Then, you must replace the sprocket.

Servicing the belt tensioner means checking it for noisy operation. Spin the tensioner drum. It will make a little noise, but if it growls and rumbles like a bad alternator bearing, or if it feels loose or *notchy* as you rotate it, get a new one.

When washing the tensioner, remember the sealed bearing that supports the drum. If you submerge the tensioner in solvent, you'll wash away the bearing lubricant. If the bearing wasn't noisy before, it will be shortly.

Cylinder-Head Reconditioning 6

Honda's CVCC cylinder head presents some unique challenges to head reconditioning.

All cylinder-head teardown, inspection, reconditioning and assembly work is covered in one chapter because this is how it should be done. Once head work begins, it should continue until the head is assembled and ready to install on the block.

There are several good reasons why you should farm out the cylinder-head work. First, there's the equipment you'll need: valve grinder, valve-seat grinder, valve-spring compressor and spring tester. With the possible exception of the valve-spring compressor, these tools are so specialized and expensive that no one rents them. And you sure can't justify buying them for doing one engine job.

Second, even if you had access to a machine shop full of head-reconditioning tools, you'd still have to learn how to use them. It's one thing to say "grind the valves" and something else to do it.

So, if you don't have the tools or skills, deliver the head to a reputable machine shop and have them do the job. It won't take long and, if you consider the time saved, the job doesn't cost that much.

A word to the wise: Don't shop for a machine shop on price alone. Money should not be the main consideration. It usually pays to go to a somewhat more expensive, but faster, cleaner and more professional shop. The extra care and time spent on *your* head costs money—so don't worry about spending a little extra now. Better a little more now than lot more later.

If you're on a tight budget, save money by disassembling and cleaning the head yourself. If time is more important, drop the head off at the machine shop after removing the manifolds, rockers and camshaft.

To make the first stages of cylinder-head work more pleasant, clean it. Loosen crud with engine degreaser, then rinse it off with a garden hose. Or, if you don't like making a mess at home, use the local car wash. Spray the head with degreaser and blast off the dirt with pressure spray.

DISASSEMBLY

If the oil-pump-drive-shaft cover is in place, remove it. Don't lose the two dowels. Unbolt the ten rocker-

Begin head service by unthreading rocker-arm-assembly bolts.

On 1200s, also remove two Phillips-head screws at rear of rocker-arm assembly.

arm assembly hold-down bolts. You'll need a 10mm socket for one row of bolts and a 12mm socket for the other. Completely remove all hold-down bolts before attempting to remove the rocker arms. Because the rocker-arm assembly is doweled to the head, you'll have to rock it slightly while pulling up. Watch for loose dowels.

Pulling the rocker-arm assembly removes the rocker arms, rocker-arm shafts and cam-bearing caps in one step. It also loosens the oil seal at the front of the camshaft. Pry it from its register in the head. Lift off the camshaft. All that's left are the valves.

You can now deliver the head to the machine shop, or if you have the tools and equipment, continue on.

Valve Removal—The first step in head reconditioning requires a *valve-spring compressor.* The most common type is the *C-type,* so called because it looks like a huge C-clamp. It is designed for use with the head off the block, unlike a *fork-type,* best used with an installed head.

A C-type compressor fits around the head. One end butts against the valve head, the other end straddles the valve-spring retainer. A lever on the compressor moves its two jaws together against the spring with an over-center effect. This compresses the spring, forcing the spring retainer down the valve stem to expose the two *keepers, collets, locks* or *keys*—whatever you prefer calling

Rotate cam one turn to lift rocker-arm assembly off dowels. When removing rocker-arm assembly, don't lose dowels.

them. Their purpose is to lock the spring retainer to the valve stem with a wedging action, so the retainer holds the valve spring in compression.

A common problem when using a spring compressor is breaking the retainer loose from its keepers. If you try to force the compressor, you may bend its frame. To prevent this, lever down with the compressor and tap the retainer/compressor with a soft mallet. This will break the retainer loose, allowing the spring to be compressed.

Be careful when doing this; parts tend to fly out of sight. It's best to point valve springs at the wall behind

Lift camshaft from lower bearing halves. Discard front oil seal.

your workbench. If you have safety glasses, wear them. Keep your head and body out of the way, but watch the keepers' trajectories. Find the keepers before continuing.

There's no need to arrange the keepers and retainers in order. Just dump them into a container so they won't get lost. You must keep the valves, springs, spring seats and any shims in order, however. You could make a special holder such as the one in the photo on page 88, or use a 2x4. With the 2x4, bore eight shallow holes large enough for the eight valve stems. Space the holes at least 1-1/2 in. apart. Arrange the valve springs in

Compress valve spring with compressor and fish out keepers with magnet.

After removing keepers, slide retainer, outer and inner springs . . .

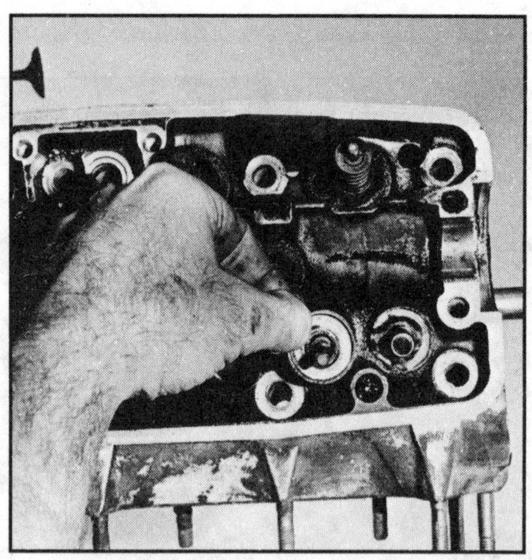
. . . valve-stem seal and spring seat off valve and guide.

Holder helps keep valves and springs in order. One shown is two pieces of wood with holes drilled and dowels set. You can do the same by merely drilling holes in a 2x4.

"quiet" corner of the bench. Or, you could punch the valve stem through cardboard and place the spring over the stem. The trouble with cardboard is it bends with all that weight, allowing the springs to go flying.

If you're working on a CVCC head, also make a board for the auxiliary valves.

A word about shims. Honda uses "shims," or valve rotators, as original equipment on the '75 1488cc engine only. But, if your non-'75 1488 head has been to a machine shop, it may have shims. If you find any, just keep each with its valve for now. Don't confuse the spring seat—that washer-like piece with the upturned edge—with a shim. All valve springs have a seat to keep the springs from damaging the aluminum head. Shims install *under* the spring seat.

Some shops use the '75 1488 "shim" under all exhaust valves, regardless of year. Honda calls this piece a *thrust washer;* machine shops call it a *valve rotator.* Regardless of the term, the washer or rotator turns the exhaust valve a little each time it opens and closes. This helps exhaust-valve cooling and extends durability. This is why machine shops add the washers to each Honda head they recondition.

Auxiliary Valve—On CVCC heads, the auxiliary valves must be dealt with. The valve fits inside the *auxiliary-valve holder*—it contains the valve seat and guide. A *spring washer* fits over the outside of the holder so the valve seal, spring, spring-seat retainer and keepers can work against it. An O-ring seals the bottom of the valve holder to the cylinder head. The *valve-holder nut* threads over the holder and into the head, cinching the complete assembly into place.

Underneath the auxiliary-valve assembly is the auxiliary chamber. It consists of a steel bucket—*chamber collar*—and two sealing rings or gaskets. The chamber collar is held in place by the valve-holder nut, via the valve holder. I know it sounds complicated, but once you've taken one apart, it's nothing more than a miniature valve assembly with some special parts.

No special tools are required for auxiliary-valve disassembly. Unscrew the valve-nut holder with an adjustable wrench or ratchet and deep-wall socket. Once completely unthreaded, the nut holder will pass over the valve spring. The valve holder, spring and all related parts should now come out as a unit. If you are lucky, the assembly will wiggle out between your thumb and index finger. If it doesn't pop out at the first try, it's time to be resourceful.

On auxiliary valves with large, single-hole prechambers, turn the head upside down, insert a punch through the combustion-chamber hole and tap the assembly out. A light tap should get it moving. Set the assembly aside for the moment.

With the multi-hole prechamber design, you must remove the auxiliary valve from the rocker-arm side of the head. Honda has a special tool that resembles a slide hammer. Mechanics report a high incidence of bent auxili-

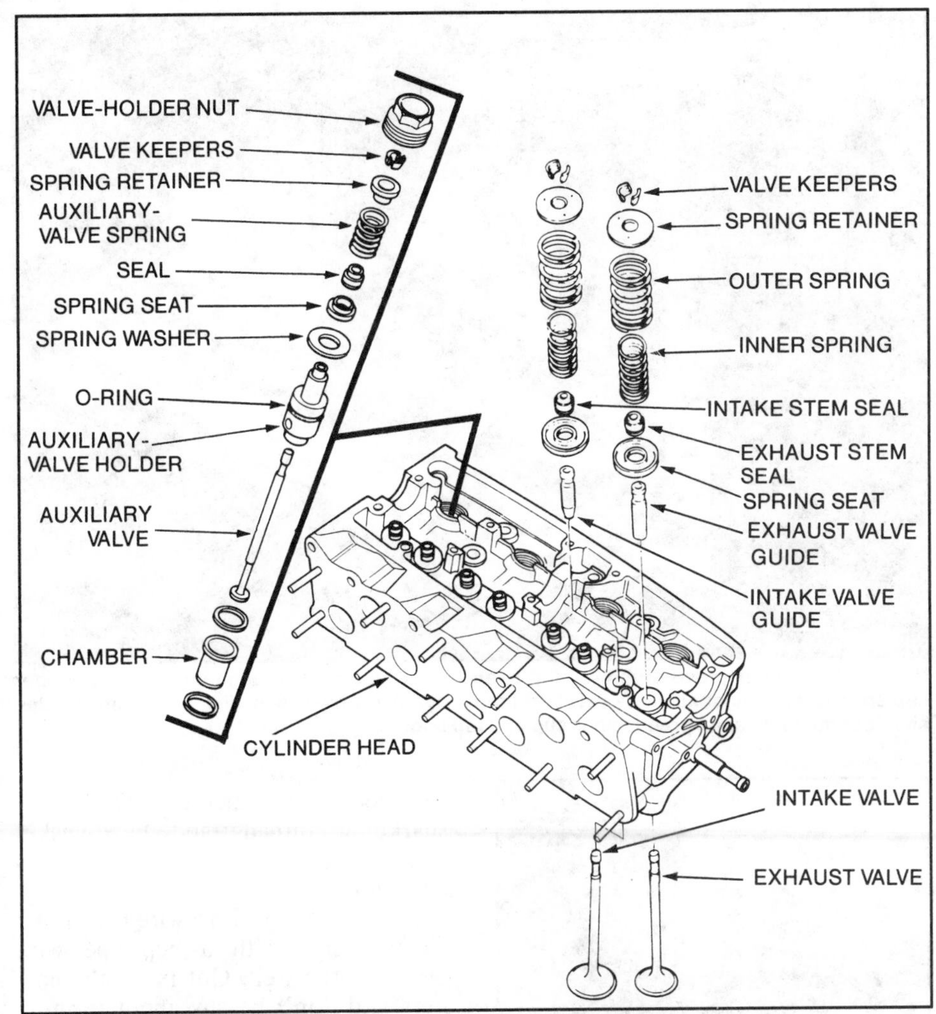

Exploded view of CVCC cylinder head. Courtesy of American Honda Motor Co., Inc.

Cylinder block makes good fixture to hold head during disassembly. After loosening auxiliary-valve holder nut, slide nut over auxiliary-valve spring and off head.

Either wiggle complete auxiliary-valve assembly out of head or disassemble it in place. Low-rate spring can be compressed by hand or with a compressor. It has keepers just like a main valve spring. Valve-stem seals were so brittle, they shattered when I tried levering them off.

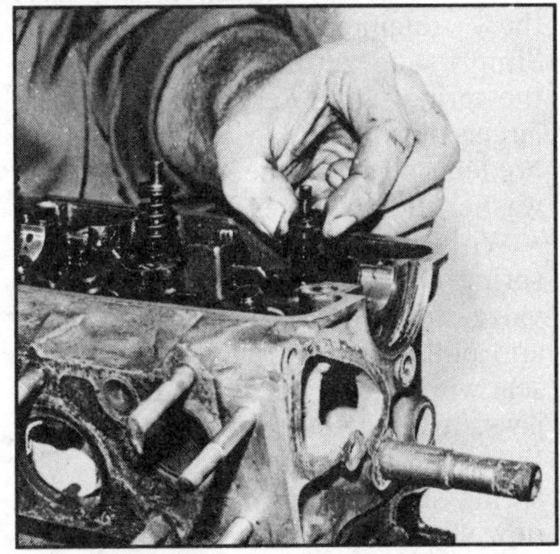
After valve-stem seal, slip off spring seat.

ary valves with this device. If you do try the slide-hammer tool, be careful. All you are working against is the valve-holder O-ring, so force should be minimal and straight up from the head.

Another method is to disassemble the auxiliary-valve keepers, retainer, spring, spring seat and stem seal. Grasp the top of the valve holder with pliers and wiggle it out. The problem here is the pliers will leave marks on the holder. So, you are better off wiggling it out by hand.

You could drive out the prechamber collar and auxiliary-valve assembly by hammering against the collar. But, this method will likely result in a bent and mangled prechamber.

Use a punch resting against the metal piece visible through the combustion-chamber hole. This is the side or bottom of the chamber collar. Lightly tap it out of the head. Heavy pounding will distort the collar. There are two sealing rings, one above the chamber collar, another below. The first pops out with the chamber collar, the second is fished out after chamber-collar removal.

Back to the valve-holder assembly: If you have a helper, compress the valve spring while your helper gets the keepers with a magnet. Use partially opened pliers resting sideways against the retainer to compress the spring. Or, use a valve-spring compressor—it'll need a lot of travel in its adjusting screw to compress this tiny spring. Try your best not to over-compress the auxiliary-valve spring. Little force is required to compress it, and a little more will overdo it.

If you do a lot of Honda CVCC head work, or just enjoy making things, try this: Find an old pair of pliers with enough throat to open the length of the auxiliary-valve spring. Weld a 1/2-in.-ID flat washer to each jaw. One washer should be cut so it

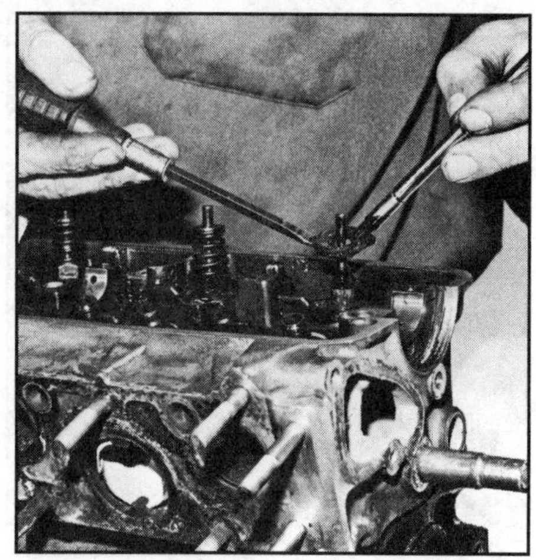

Valve-spring washers are hidden beneath goo. A magnet helps.

It takes a King Kong grip to wiggle auxiliary valve and holder from head.

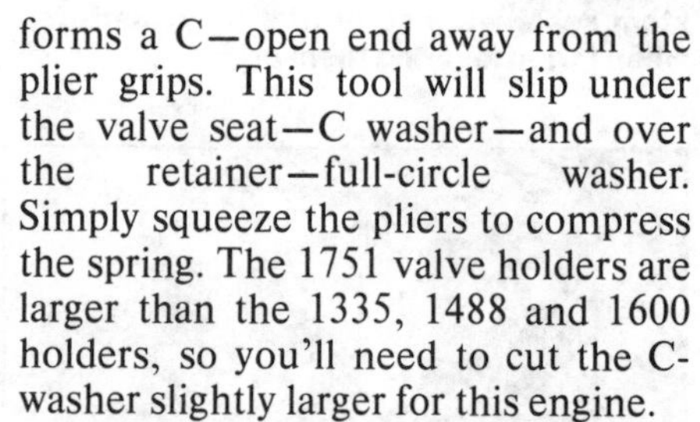

forms a C—open end away from the plier grips. This tool will slip under the valve seat—C washer—and over the retainer—full-circle washer. Simply squeeze the pliers to compress the spring. The 1751 valve holders are larger than the 1335, 1488 and 1600 holders, so you'll need to cut the C-washer slightly larger for this engine.

With the keepers, retainer and spring out of the way, break off the valve-stem seal. Grasp it with pliers and pull it off the valve holder. The seal will probably crumble in the plier jaws, making removal easy. Or, lever the seal off with a screwdriver. The spring seat and spring washer should now slip off the valve holder; and the valve will slide out the bottom of the holder. Finish disassembly by prying the O-ring from the valve-holder exterior.

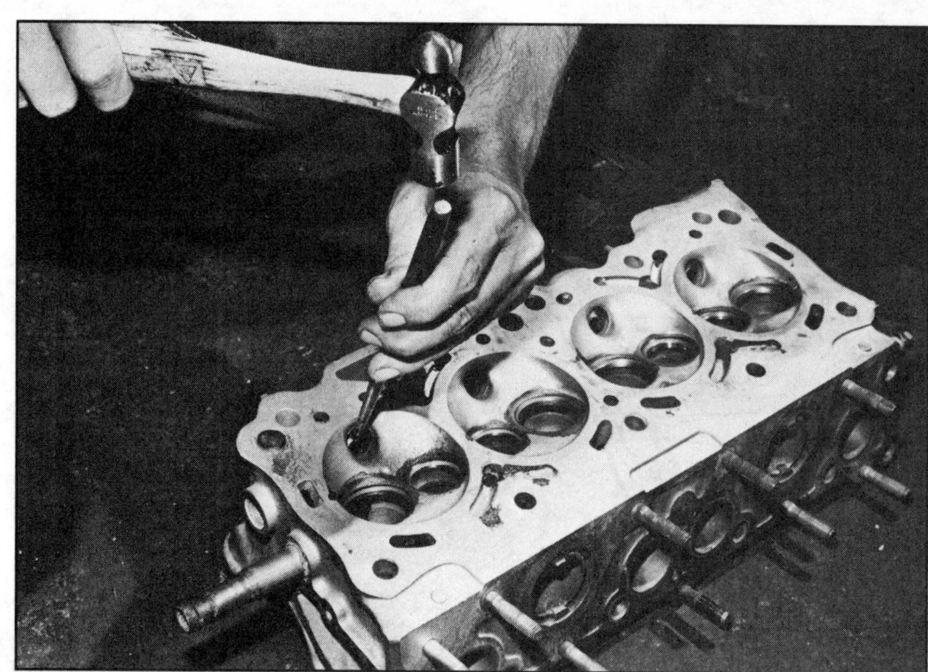

If valve and holder prove stubborn, flip head over. Slip a drift through hole in prechamber cup so it butts against auxiliary valve and tap it out. After freeing auxiliary valve and holder, shoulder punch against prechamber cup and tap it loose.

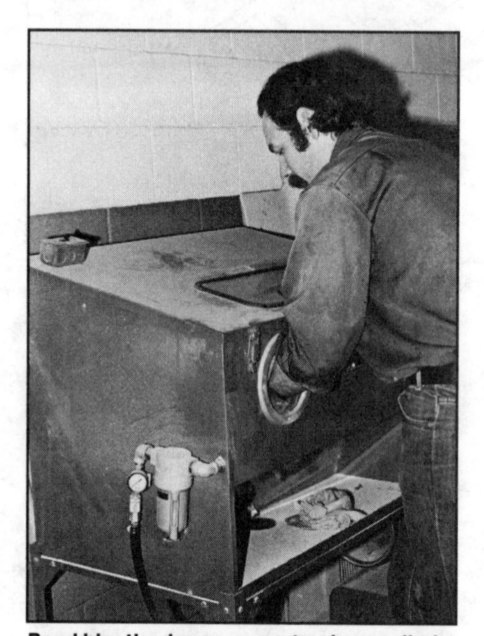

Bead blasting is easy way to clean cylinder heads. Unfortunately, bead residue gets everywhere—oil passages, coolant passages and threaded holes. After bead blasting, head *must* be thoroughly solvent blasted and air blasted to remove all bead residue—especially from rocker-arm bolt holes. If you leave beads in these blind holes, you may strip threads in aluminum head when torquing bolts.

Cleaning—Wash all parts in solvent, but don't worry about carbon in the combustion chambers. It is too baked-on for the solvent to do much good. Knock the largest carbon pieces off with a blunt screwdriver or chisel, but be careful not to nick the valve seats, sparkplug threads and head-gasket surface. Remember, the head is aluminum!

Finish the job with a wire brush. An electric drill with a cup-type wire brush works well. Get in all the corners and don't be shy about running the brush into the valve seats. Use the gasket scraper on the gasket surfaces: head-to-block, and intake- and exhaust-manifold surfaces.

Be extremely careful while scraping the head-gasket surface. Gouge it and you'll provide a path for combustion pressure to blow past the gasket.

The edges of all the water passages also need cleaning. Use a pocketknife or dull screwdriver to remove the large deposits, then finish with a few passes of a rat-tail file. Don't remove any metal—just the scale. To remove the junk that fell into the water passages, hold the head right-side up over a trash can and shake it around until all debris falls out. You could also use a shop vacuum.

In the machine shop, the head can be bead-blasted. This quickly removes carbon deposits without danger of gouging critical surfaces. If the head is bead-blasted, make sure it gets a *thorough* cleaning afterwards. Blasting residue collects in bolt holes and corners, and it takes quite a while to get it all out. Force it out with high-pressure water, steam or solvent.

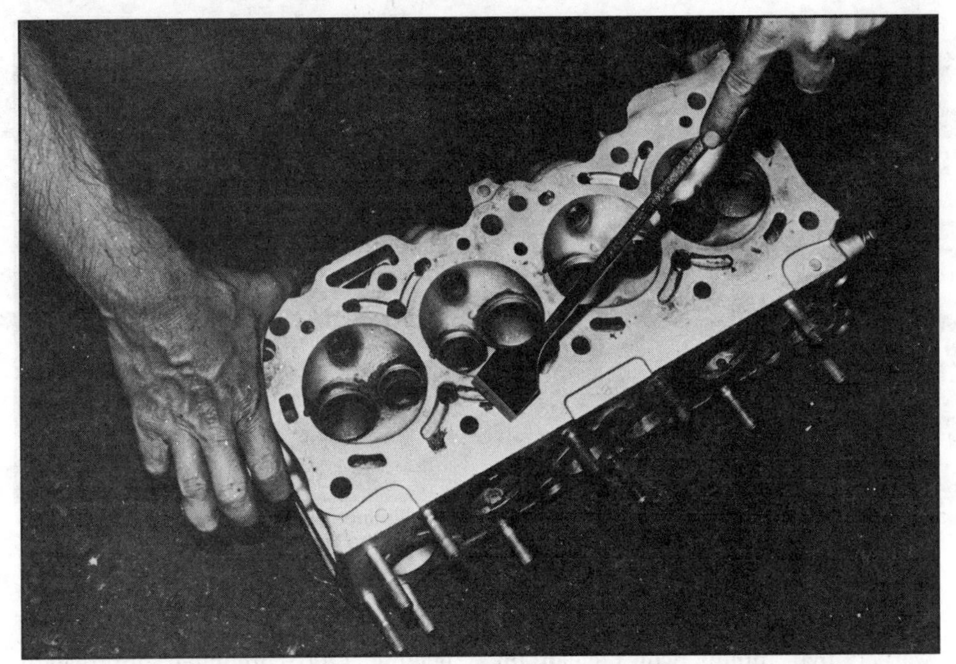
Make sure all gasket material is scraped off, especially on head/block mating surfaces.

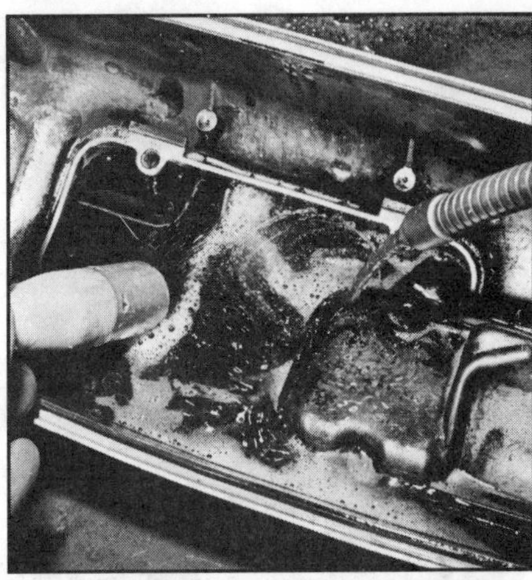
Clean oil separator with solvent. Remove four screws and lift off separator cover. Be prepared for sludge and caked-on deposits—oil trapped here gets baked.

Chase all threads to ensure all grit is removed. If this junk is not washed away, the rocker-arm bolts will bind up during assembly, plus grit will eventually end up in the oiling system.

Because it's difficult and time-consuming to get bead-blasting residue out of a head, I recommend cold-tanking the head. As detailed earlier, cold-tanking is a cold, chemical cleaning process that is not corrosive to aluminum.

Head Warpage—Aluminum cylinder heads warp easily from overheating, or experiencing many thousands of *heat cycles*—warming and cooling. If your previous engine experience has been with cast-iron heads, especially the large American variety, you might be surprised at the narrow warpage service limits for Honda heads. Typically, this is 0.0012 in. (0.03mm) for 1200s and 0.0020 in (0.05mm) for CVCC heads.

This is due to several factors. First, all Honda heads are cast aluminum. This soft material is more prone to warping, so you can't mill it, run it hot again and expect the head to remain flat. Second, Honda heads don't have much reserve material. That would add weight, going against Honda's design philosophy of maximum mileage and good power-to-weight ratios. Third, Honda engines are overhead-cam (OHC) designs. This means that every time the head is cut, you decrease the distance between cam and crankshaft, which upsets cam timing. So, *milling*—machining off head material with a rotary cutter—is kept to a minimum; a mere 0.008 in. (0.20mm) is allowed on CVCC heads, 0.004 in. (0.10mm) on 1200 heads.

Another factor is cam-bearing-bore alignment. When a head warps, it pulls the aluminum out of shape—including cam-bearing-bore alignment. Milling restores flatness to the head-gasket surface, but not bearing-bore alignment. To get them straight again, you'll have to replace the head or have the head heated and straightened in a press. Align boring or honing Honda heads just isn't feasible because of the bore sizes involved and valve-timing considerations.

Measure for warpage first at the gasket surface, then at the cam-bearing bores. To measure the gasket surface, you'll need the same precision straightedge and feeler gages used on the block deck. The procedure is the same. Clean off old gasket material, dirt and carbon so you'll get accurate readings.

Lay the straightedge across the head, then try to pass a 0.002-in. (0.05mm) feeler gage under it. Measure in several different directions, both straight and diagonally across the head. If the 0.002-in. gage slips underneath, try thicker gages until

Check head for flatness. Head must be perfectly clean or dirt will raise straightedge, giving false readings.

you find one that just passes under the straightedge. This is the amount of warpage. If warpage is below the limit, the head can be used as-is. Warpage above the allowable limit, but below the milling limit means the head can be milled. Warpage in excess of the milling limit requires head replacement unless the head can be straightened by bending.

On the bottom of each head is a protruding dot that acts as a head-thickness indicator. It is cast slightly lower (0.008 in. or 0.20mm on CVCC heads and 0.004 in. or 0.10mm on 1200 heads) than the mating surface of the head. When a head is milled more than the minimum allowable thickness, the dot is also milled. This is a quick way to check a used head. If the dot is milled flat, the head is too thin. Don't use it. If the dot is

As with other OHC designs, keep minimum head thickness in mind when milling Honda heads.

Check cam-bearing-bore alignment with straightedge and feeler gages. A 0.002-in. deviation is the limit.

untouched, the head should be usable, provided it is not cracked, badly warped or has other problems.

Cam-Bearing Bore Alignment—You can measure camshaft-bearing bores two ways. The first is with a straightedge and feeler gages. Lay the straightedge in the cam-bearing bores, then try to pass the thinnest feeler gage you have under it. You shouldn't be able to pass a 0.001- or 0.002-in. (0.025 or 0.050mm) gage under it. The second way is to install a known straight camshaft and the rocker-arm assembly; the valves and springs should be removed. This allows you to spin the camshaft without rocker-arm drag.

The cam should turn smoothly, without binding. If the cam-bearing bores pass this test, they're OK. If they don't, check them with the straightedge and feeler gage. Or, you could measure cam-bearing oil clearance with Plastigage. It should be 0.002—0.006 in. (0.05—0.15mm) for all but the two 1200s. The 1200s should measure 0.002—0.0035 in. (0.05—0.09mm). If you find one bearing with very tight oil clearances, it's the troublemaker.

If you still don't know for sure where the bind is, measure camshaft *runout.* See camshaft inspection, page 105. If the cam yields too much runout, replacing it should get you a free-turning camshaft.

Usually, cam-bearing-bore alignment is no problem—especially if the engine was well maintained and not overheated. If the bores are out of alignment, the head is likely so warped that milling won't clean up the head-gasket surface anyway—the head will need replacing. And, in almost all cases, as you'll soon see, the camshaft will also be replaced, so don't concern yourself with camshaft runout.

Water Damage—Another Honda cylinder-head phenomenon is water damage. This is *pockmarking* of the combustion chambers and head-gasket surface due to coolant seepage. Coolant gets into the combustion chambers in one of two ways: through the head gasket or through the aluminum. This last condition is caused by *porosity* in the head casting. If aluminum is cast so it's full of *voids,* or holes, coolant eventually erodes its way through the head.

Typical porosity trouble spots are in so-called *cross-drilled* CVCC heads. These heads date from about 1978 and feature a water passage between the combustion chambers. This small passage may eventually leak into a combustion chamber. Aside from the obvious problems with compression leakage into the cooling system, coolant can erode the side of a valve seat. Because the passage is adjacent to the exhaust-valve seat, no welding can be done. High temperatures from welding distort the valve seat. Also, the passage is curved, so it can't be welded and then ground back into shape. Replacement is the only cure.

Other possible leak areas are the core plugs under the camshaft, especially on 1200s. A leak here can cause camshaft and rocker damage due to coolant-diluted oil. If you suspect a leaking 1200 core plug, unthread it with a 14mm Allen wrench and clean the threads with a wire brush. Reinstall the plug using non-hardening or silicone sealer on the threads.

More common is coolant damage to the head-gasket mating surface resulting from a leaky gasket or eroded coolant passages. The large, V-shaped coolant passage between CVCC combustion chambers can become unrecognizable from coolant erosion. Also, the area between combustion chambers 3 and 4 on CVCC heads is a known coolant-damage spot. Some coolant damage is acceptable, but when the edges of damaged material border the areas sealed by the head gasket, you must cure the problem. Try milling out the damage, without exceeding the milling limits. If that doesn't correct the damage, replace the head.

All this points out the need to use antifreeze compatible with aluminum. Needless to say, it's cheaper to use the correct coolant than replace a cylinder head.

Don't worry about raising the compression ratio when milling the head or block. You'll raise compression slightly, but hardly enough to matter, assuming you follow Honda's milling limits. You'd have to remove two or three times the milling limit to make a difference.

VALVE GUIDES & STEMS

Let's consider valve-guide wear and how to measure it. As a rocker arm pushes its valve open, a side force is created that forces the valve

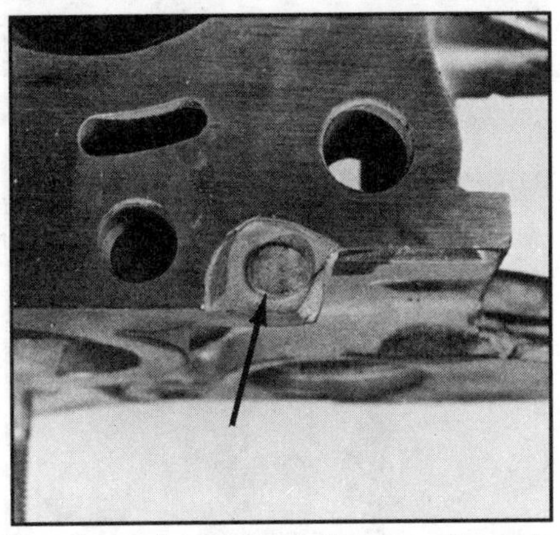

Dot (arrow) indicates minimum allowable head thickness. This head has been milled, but dot is still lower than gasket surface, so it doesn't show shiny metal. If head was cut much more, dot would be milled, also—machinist would know head was milled beyond limit.

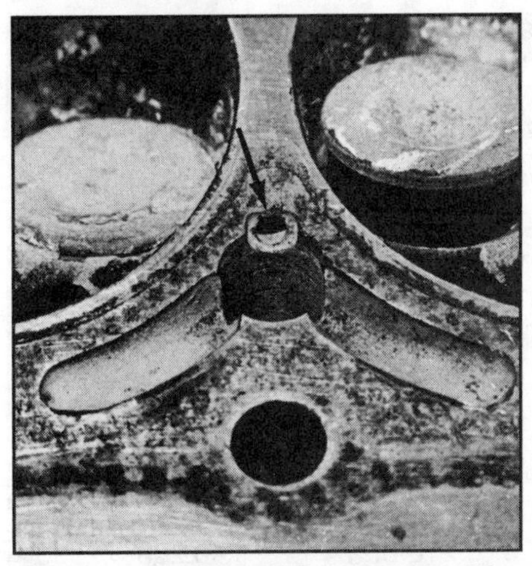

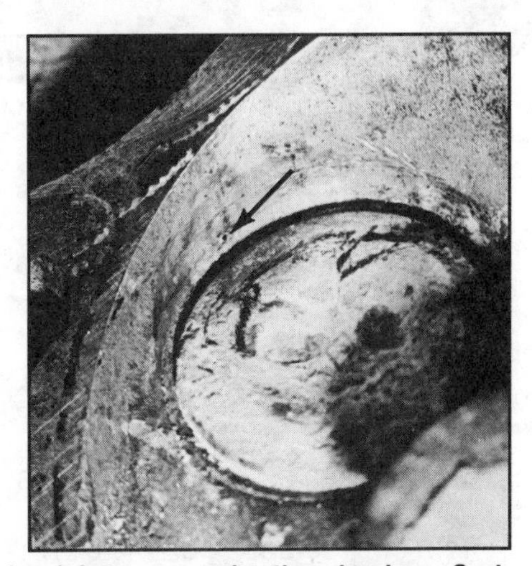

Small water passage leads to other side of head, between combustion chambers. Such *cross-drilled* heads, dated from 1978, were designed to prevent blown head gaskets. Unfortunately, lack of material and porosity between combustion chamber and passage led to many water leaks (arrow). Symptoms include overheating, poor power and mileage, and water in oil. Trouble area is too deep in head to reach from inside passage. Welding from outside may fix leak, but disturbs exhaust-valve seat. Replacement is only sure fix.

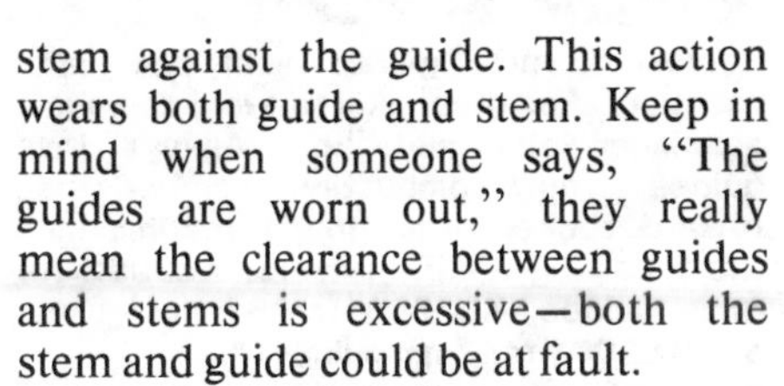

stem against the guide. This action wears both guide and stem. Keep in mind when someone says, "The guides are worn out," they really mean the clearance between guides and stems is excessive—both the stem and guide could be at fault.

A quick and fairly accurate wear check makes use of a valve in the guide. Insert the valve in the guide so the valve is off its seat approximately 3/8 in. While holding the valve head, rock the valve back and forth. Move it in different directions until you find the one that gives the most movement. Measure this movement, or *play* with a 6-in. scale, the depth-gage end of a vernier caliper, or a dial indicator, if you want the most precise reading. Maximum allowable play is 0.006 in. (0.15mm) for intakes and 0.008 in. (0.20mm) for exhausts. Or you can divide the play measurement by 3.5 to get *approximate stem-to-guide clearance.* This method does not tell you if the valve stem or guide is the culprit; it could be either one.

Guide & Stem Measurements—To tell exactly which part is worn, you'll have to measure them. Start with the valve stems—they're the easiest. Use a 0—1-in. outside micrometer or vernier caliper to measure the valve stem. The point of maximum wear, or highest point on the stem wiped by the guide, is immediately below the wear mark at the top of the stem. Measure stem diameter there and compare your findings with the chart, page 94. If wear exceeds the service limit, the valve must be replaced.

Measuring guide ID is more difficult. For this job, you need a *small-hole gage,* sometimes called a *ball gage,* and a 0—1-in. outside micrometer. Insert the gage into the guide to find the point of maximum wear. As you expand the gage, rotate it and run it up and down until you find the widest point in the guide. Maximum wear occurs at one end of the guide. Remove the gage and mike it. Standard guide diameter less this figure is *exact* guide wear.

In the real world, you aren't going to find a bunch of machinists running around with small-hole gages measuring every last valve guide. From experience, they know that the exhaust valves will be replaced unless they have very few miles on them—less than 20,000. Honda CVCC engines—especially 1976—'79 models—run hot, taking a toll on exhaust valves. Replacing them saves inspection time and ensures against valve failures. Wiggling the valve in the guide, then measuring stem OD, by eye or with a mike, is the usual way engine machinists check guide wear. If the valve wiggles a lot in the guide, but has no apparent wear, the guide is bad. If the valve stem *is* worn, then they'll recheck the guide with a new valve to see how much the valve wiggles. If most movement disappears, they know the valve stem is *badly* worn and the valve will have to be replaced.

It's not unusual for a Honda head to get both new exhaust valves and guides.

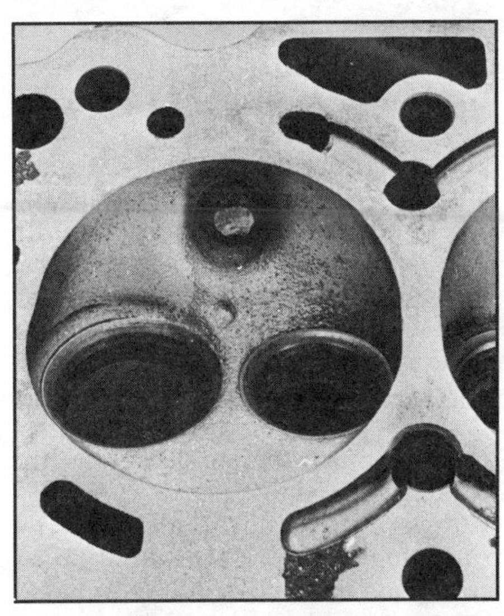

Erosion shown is acceptable if it hasn't exposed side of valve seat. More-extensive erosion may require head replacement.

If you have the tools, measure the valve stems and guides, and subtract diameters to get *exact* stem-to-guide clearance. If you don't have a small-hole gage, measure stem OD and rewiggle the stems in their guides. Knowing how badly worn the valves are will help you *guesstimate* guide condition. Using a new valve will help considerably. Then, any excessive wiggle can be attributed to a worn guide.

VALVE-STEM DIAMETERS

in.(mm)

Engine	Year	Intake	Exhaust	Auxiliary
1170	73	0.2592-0.2596 (6.58-6.59)	0.258-0.259 (6.55-6.56)	N/A
1237	74-79	0.2592-0.2596 (6.58-6.59)	0.258-0.259 (6.55-6.56)	N/A
1488	75-83	0.2592-0.2596 (6.58-6.59)	0.258-0.259 (6.55-6.56)	0.2162-0.2166 (5.48-5.49)
1600	76-78	0.2592-0.2596 (6.58-6.59)	0.258-0.259 (6.55-6.56)	0.2162-0.2166 (5.48-5.49)
1335	80-83	0.2592-0.2596 (6.58-6.59)	0.2574-0.2578 (6.537-6.547)	0.258-0.2593 (6.572-6.587)
1751	79-83	0.2748-0.2751 (6.98-6.99)	0.273-0.274 (6.94-6.95)	0.258-0.2593 (6.572-6.587)

N/A—Not Applicable

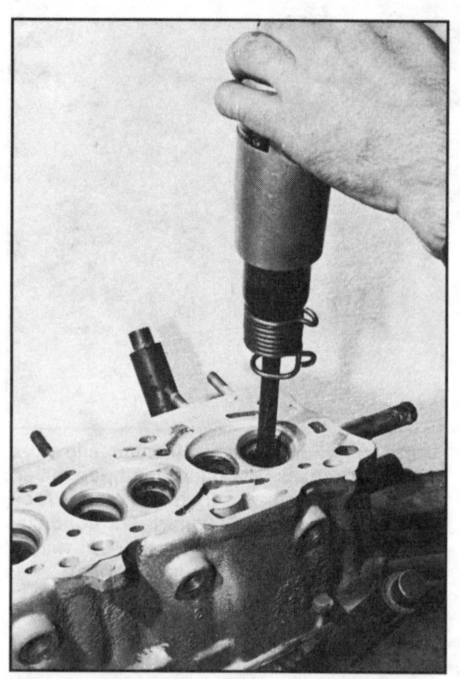

Removing valve guides with pneumatic hammer takes great skill to avoid damaging aluminum valve-guide bore. Always drive guides from combustion-chamber side toward rocker arm. Honda recommends the following K-Line drivers: KL2992 for 6mm guides, KL2993 for 6.5mm guides and KL3001 for 7mm guides.

Guide Replacement—If your stem-to-guide measurements indicate worn guides, they must be replaced. Guides are replaced by driving them out with a hammer and stepped punch or drift. The drift must have a shoulder that butts against the guide end, but small enough to pass through the guide bore in the cylinder head; about 10mm or 3/8-in. diameter should do. Before you drive out the guides, measure how much they project above the rocker side of the head.

Always drive out guides *from the combustion-chamber side.* New guides are driven in *from the rocker-arm side.* Otherwise, you'll be driving against the tapered OD of the guides. Honda guides are difficult enough to install without driving them in the wrong direction.

If the guides won't budge, heat the head to 300F in an oven for about one-half hour and try again. *Don't burn yourself!* Another trick is to use a pneumatic hammer, but only if you are an experienced machinist. This little pneumatic tool takes a deft touch. It's way too easy to "pull metal"—gall the guide bore. If you use one of these, be sure to use a stepped drift and keep your weight against the gun so it won't jump around.

Installing guides is easier if they are chilled and the head heated. Again, if you can put the head in the oven without domestic complaint, bake it for one-half hour at 300F. This will expand the guide bores. Have the guides go in the freezer for the same time. Don't overheat the head! Anything over 300F may warp the head. If the valve seats fall out, you know you overheated it!

A word of caution: Don't bake the head unless it is *perfectly clean* or you'll stink up the house for a week.

Many machinists don't take the time to heat and cool the parts—they just hammer harder. I prefer the heating-and-cooling method. No matter which way you do it, use some sort of lubrication on the guide ID and bore. A thin coating of general-purpose grease or anti-seize compound works well.

Drive in the new guide until it projects from the rocker-arm area the same distance as the one it replaced. You can save time by fitting a section of tubing over the guide during installation. Cut the tube to the guide's proper installed depth, then drive in the guide until it's flush with the tube. Or, use a machinist scale to measure your progress—just don't get carried away and hammer too much without measuring.

Reaming—After the new guides are installed, they are reamed. As sold, their ID is smaller than the valve stems. This allows you to ream the guide for an exact fit with its valve. Reaming is done with a long, drill-motor-driven reamer. Cutting oil makes a cleaner cut and carries away spent material.

After reaming, try to fit the valve in its guide. It probably won't go. Oil, then ream the guide again for several seconds, then try the valve again. Ream as necessary until the valve *barely* wiggles in the guide. About 0.0015-in. (0.04mm) stem-to-guide clearance is best for the intakes and 0.0025 in. (0.06mm) for exhausts. Flush out the guide with oil and a small brush, if possible. Check for burrs on the valve tip or guide opening. If the valve still won't start after all that reaming, look for a burr holding things up.

Obviously, you must have the valves in order from here on so they'll end up in the guides that were reamed for them.

INSPECTING & RECONDITIONING VALVES

In this section, I cover conventional intake and exhaust valves. CVCC auxiliary-intake valves are dealt with later on.

Next on your list after valve guides is valve inspection. You may have already declared some valves unusable because of obvious damage; they were burned, warped or badly worn. However, if you haven't given the valves the once-over, do it now.

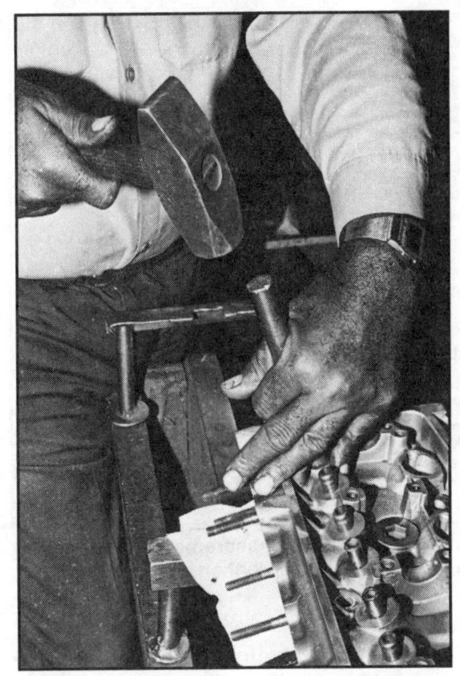

Hammer and punch are usual tools for guide removal and replacement. Guides must go in from rocker-arm side. Heating heads—see text—is highly recommended to avoid *galling* aluminum. This fellow's watch sure took a pounding.

Rushing guide replacement can be fatal to cylinder head. This one cracked when guide was driven in without heating head first. Crack was welded, but guide bore had not been drilled to size before photo was taken. I don't know if fix was successful.

Inspect the valve-stem tips for damage. If a tip was severely pounded by its rocker arm, it may be *mushroomed*—it'll look similar to the hammer-end of an old chisel. Replace the valve. Check the rocker arm if you find such a valve tip.

More common than mushroomed valve tips are pitted tips. Look for pitting at the center of the tip where the hardened material may have broken away. Most of the time, pitting can be corrected by grinding. But, if it won't clean up after removing 0.010 in. (0.25mm), replace the valve.

Valve-Stem Wear—You'll need a 0—1-in. micrometer to compare worn and unworn portions of the valve stem. Mike the unworn portion of the stem between the deeper groove and the *wear-line*—where the stem stops at the top of the guide when the valve is fully open. You may be able to feel a small step here with your fingernail. Record your findings. Then measure the valve stem immediately below the wear-line. Subtract this last measurement from the first to get *valve-stem wear.*

Because the greatest valve-stem wear occurs farthest from the valve head, you should always take the second micrometer reading right below the wear-line in the swept area. The farther away you get from the wear-line, the less accurate your reading will be.

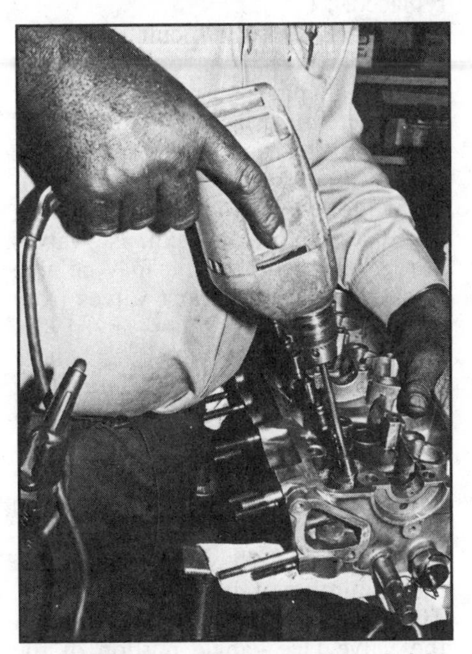

Once guides are driven in to proper height, ream them to size. Check fit with valve for that guide. It takes experience to feel when valve-to-stem clearance is just right.

Pitted and galled valve-stem tips are common. If pits don't disappear after removing 0.010 in. from stem, replace valve.

How Much Wear is OK?—If you realize that you have only about 0.002-in. (0.05mm) *additional* stem-to-guide clearance to play with, you can see there isn't much left for valve-stem wear. For instance, if you replace the valve guides, but run valves with stems worn 0.001 in. (0.025mm), you've already used up about half of the allowable wear. So even though you could use a stem worn as much as 0.002 in. and be within limits, I don't advise it. It wouldn't be long before your engine would start burning oil. Use 0.001 in. as your stem-wear limit—less for optimum durability.

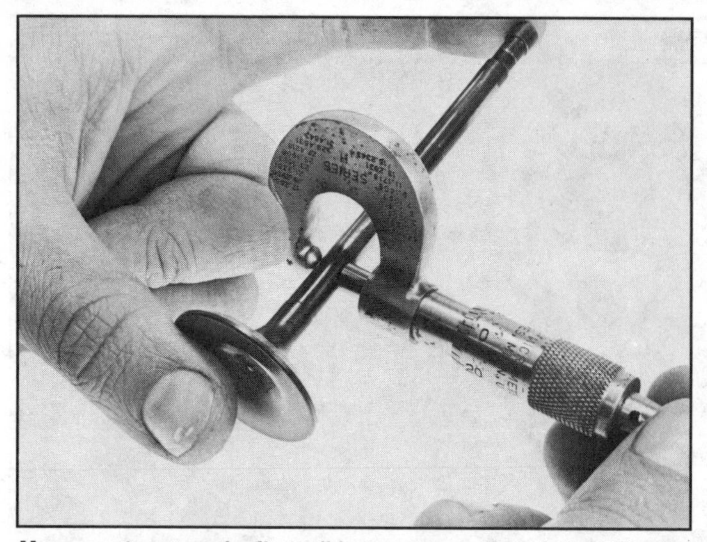
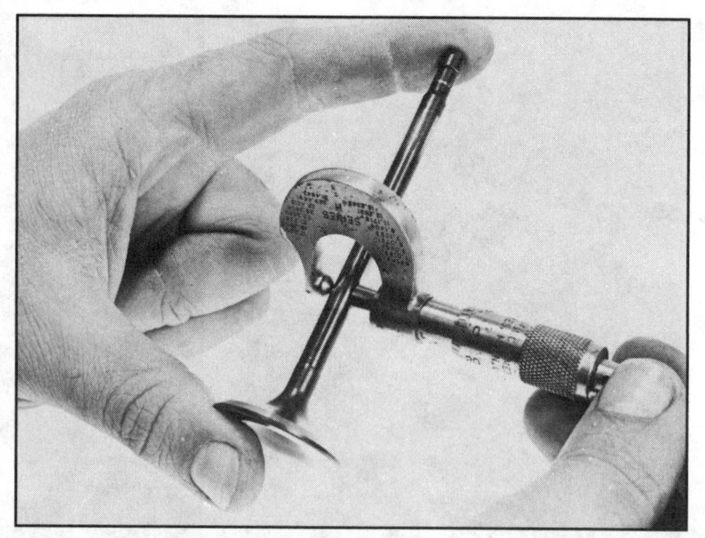

Measure stem wear by first miking unworn portion of stem. Then measure worn section and subtract from first measurement. See chart for maximum wear. Photos should show measurements taken at top of stem, not bottom. Top portion always wears more than bottom.

Lean mixtures and high operating temperatures of Hondas are tough on exhaust valves. Unless exhaust valves are practically new, replace them. Valve shown experienced normal service, maintenance and miles. Dark areas show where valve wasn't sealing. Cupped area at 7 o'clock is particularly bad. Grinding won't save this one.

Honda intake valves are allowed about 0.002-in. stem-to-guide clearance when new. New exhaust valves are allowed 0.003 in. (0.075mm). Wear limits are typically 0.003 in. for intakes, 0.005 in. (0.125mm) for exhausts. Whatever guide/stem combination you decide to run, the clearance should be below these limits. The tighter the clearance, the longer the guide/stem combination will last, but clearance shouldn't be so tight that the stem seizes in its guide.

Valve Grinding—Normal rebuilding practice is to grind valves to restore fresh sealing surfaces. When grinding limits are exceeded, the valve is replaced.

In practice, it's difficult to grind a Honda exhaust valve enough to remove all pitting without removing too much material. It's better to replace the hot-running exhaust valves and get maximum material for maximum life. Most Honda engine builders discard the old exhaust valves during disassembly. You may be able to salvage some exhaust valves from 1200s, the '75 1488 or any 1980-and-later engine, but even these "cooler"-running valves are best replaced.

Intake valves are usually ground. Because the incoming air/fuel mixture cools intake valves, they live much longer than exhausts.

Of course, each valve must retain the minimum *margin* after grinding. The distance from the outer edge of the valve face—angle portion of the valve head that closes against the valve seat in the cylinder head—to the valve-head surface is a valve's margin. Minimum margin is 0.030 in. (0.75mm) for intakes; 0.040 in. (1.0mm) for exhausts. If the margin drops below the minimum specification, replace the valve. Otherwise, the valve may burn.

Reconditioning a valve involves resurfacing its tip and *face*. This is a job for a machinist with a valve grinder.

The tip end of the valve stem needs attention first. The tip must be trued first because it pilots the valve, centering it in the grinder. An untrue tip will result in an untrue valve face.

As shown in the photo, the valve stem is clamped securely and the tip is passed squarely across the face of a spinning grinding wheel. A small amount of material—0.010 in. (0.25mm) or less—is ground away. This is only enough to remove small irregularities. Otherwise, the valve would be ground too short. Large irregularities, such as mushrooming, require discarding the valve. Finally, the tip is chamfered to remove the sharp edge or any burrs.

Valve faces are reconditioned by grinding away small amounts of material at the correct angle until all flaws are removed. This is also done in the valve grinder. Major components of a valve grinder include an electric motor to power the grinding wheel, coolant pump, and a chuck to hold the valve. The valve is installed stem-first in the chuck. Honda recommends grinding valve faces at 45°.

The valve is turned slowly while the spinning grinding wheel passes back and forth across the valve face. While the valve is being ground, coolant is pumped over it and the grinding wheel for cooling and to wash away grinding chips. Coolant is water with water-soluble oil added.

It is important that no more than a minimum amount of material is removed from the valve face to retain a sufficient margin. If the margin is thinned excessively, the valve will overheat and burn or warp easily—especially exhaust valves. A

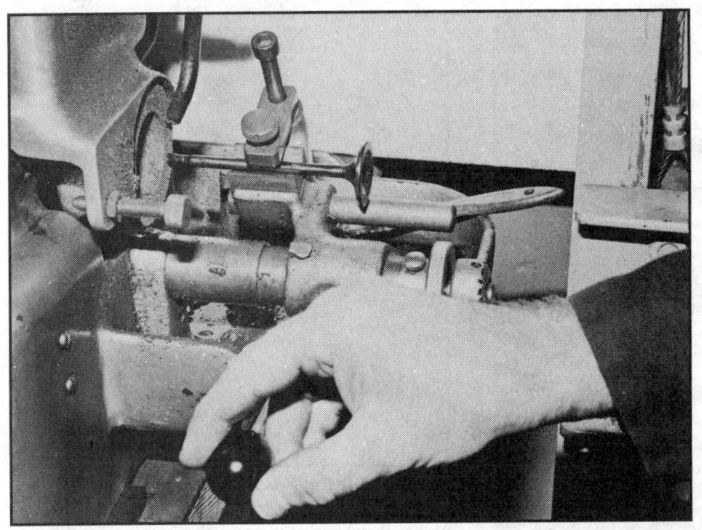

Half of valve grinder reconditions valve-stem tips. Other side grinds new 45° face on valve head. Enough material must remain for 0.030-in.-thick margin on intake valves, 0.040 in. on exhausts.

vernier caliper is best for measuring valve margin.

VALVE-SEAT RECONDITIONING

Valve seats are ground using special power tools and stones or cutters. If the valve seats have been ground before and there isn't enough seat material left to complete all necessary cuts, the old seats are removed and new ones installed. Both grinding and seat replacement are jobs for an expert with the proper tools.

Honda uses hardened-steel valve-seat inserts, press-fit into the head. If a seat is damaged or worn out, Honda recommends *replacing* the head. In fact, Honda doesn't even sell valve-seat insert replacements. However, independent Honda engine builders *do* replace seats. Where do they get the seats? Some save good seats out of junk heads and others buy new seats from aftermarket companies such as Manley Valve Division of Safeguard Engine Parts, Inc., York, PA.

To remove the old seat, Mike Paar of Hondaworks in Mesa, AZ, arc-welds a pull rod to the seat. He then chucks the rod into a slide hammer and pulls out the seat. Localized heating of the seat area in the head from the arc-welding operation aids in removal. After performing any necessary repair work to the valve-seat bore in the head—possibly welding and fly-cutting to undersize—the head is cleaned and then heated to 300F. The new seat is placed in the freezer. Seat installation requires positioning the seat over its bore and driving it into the head until flush with a driver and hammer. The new seat is then cut for three angles like the existing seats.

Not only does valve-seat grinding restore the seat, it has another very important benefit. Because the grinder *pilots* off a shaft installed in the valve guide, it makes the valve seat concentric with the guide. However, if you didn't do any guide work and the valve seats are OK, leave the seats alone. Guide work can upset the guide-to-seat relationship; the seats must be ground following guide work.

When correctly done, seats are ground at three different angles, 30°, 45° and 60°. The valve face contacts the 45° cut while the other two angles establish seat width and diameters. The 30° and 60° angles also improve gas flow into and out of the combustion chamber.

To restore valve seats, the machinist first makes the 45° cut with a dressed stone—shaped—at that angle. He then changes to a 30° stone and cuts the top of the seat to establish valve-seat OD. Seat OD is cut about 1/16-in. smaller than valve face OD. The 30° cut is called the *top cut.*

Next, the *bottom cut* is made with a 60° stone. This greater angle allows the stone to cut below the seat to establish seat ID and width. Seat width should be 0.0055 in. (0.140mm) with a 0.0079-in. (0.200mm) service limit.

Special equipment allows the machinist to check his valve-seat work quickly and accurately. He must if he's to stay in business. For instance, the valve-seat area is coated with Dykem or similar-type metal dye—blue or red—so the seat stands out from the bright metal. This makes it much easier to make measurements because one cut is easier to distinguish from another. Width and diameter measurements are made with dividers and a 6-in. machinist's scale. Also, a special dial indicator is used to measure valve-seat concentricity. With its mandrel centered in the guide and the indicator plunger positioned square against the valve seat, the dial indicator is rotated. If the indicator needle moves, the seat is not centered or concentric and must be reground.

Valve-seat inserts are *not* available from Honda; the factory recommends replacing the complete head if a seat is bad. However, aftermarket companies such as Manley Valve Division of Safeguard Engine Parts do. Photo by Ron Sessions.

Grinding valve seats restores guide-and-seat concentricity lost when new guides are installed. Cutting stone centers on *pilot shaft*, or *mandrel*, inserted into guide.

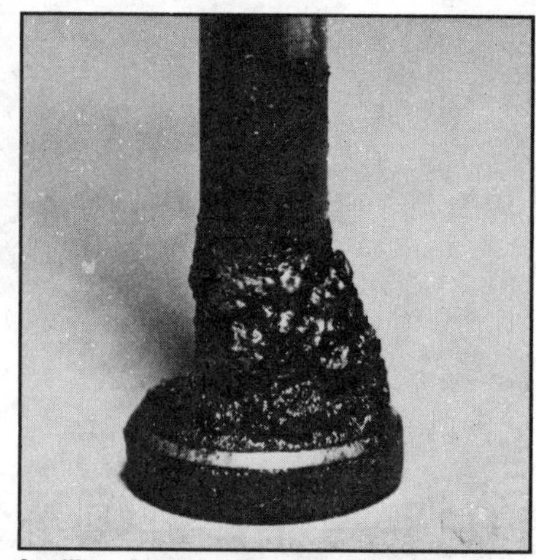

Auxiliary-intake valves don't wear out in normal use because they run cool and are very light. However, the rich auxiliary-circuit mixture tends to clog them. Remove residue with wire brush.

Lapping Valves—You may be familiar with the term *hand-lapping.* If so, you're probably wondering where it fits into the valve-reconditioning procedure. Honda does not recommend lapping conventional intake and exhaust valves. However, some independent rebuilders do, feeling that it helps ensure tight-sealing valves. Racers and other specialists hand-lap valves regularly. If your engine has a CVCC head, you are one of the "other specialists" as you'll see when it comes time to recondition the auxiliary valves.

With the exception of a new valve fitted on an old seat or the auxiliary valve, Honda says there's no reason to lap valves. If the valve job is done correctly, the angles will be correct and the valves will seal from the start. On the other hand, if the job is done incorrectly—the seat and guide may not be concentric—lapping won't cure the problem. It will, however, expose the problem, enabling you to fix it before the engine is assembled and compression is found to be way down. The seat-contact pattern will transfer to the valve seat and face—a real help when grinding. For instance, if this pattern is not consistent on the valve or seat, the guide and seat are not concentric; the seat and/or valve must be reground.

Lapping a valve to its seat is simple enough. Merely apply lapping paste to the valve face, insert the valve in *its* guide and attach a rubber suction-type lapping tool to the valve head. You then rotate the valve and tool back and forth while holding the valve lightly against its seat. Check your progress frequently. When the valve has a uniform gray band on its face corresponding to the desired seat-contact width, the lapping job is complete. Be sure to remove *all traces* of the abrasive lapping compound from the valve face and seat, or rapid valve and seat wear will result.

AUXILIARY-VALVE SERVICE

Working with CVCC auxiliary-intake valves is similar to working with conventional valves. The valve face and tip may need work, plus the seat could require "grinding." Everything is smaller, however, and there's the difference. These parts are so small that conventional valve-and-seat work is not possible. But, because almost all auxiliary-valve service is limited to replacement, there isn't much machining involved.

Start with the valve. It should be wire brushed to remove any carbon buildup. Inspect its tip and face. The tip should be in good condition because auxiliary-valve force doesn't exceed 25 lb!

The valve face should not have a concave wear pattern when viewed from the side. Rather, the face should be one continuous angle. Also, the valve-seat contact area should not exceed 0.028 in. (0.711mm).

Inspect the valve seat in the valve holder. It should not be pitted or worn concave. Auxiliary valves handle an extremely rich mixture, and therefore run cool. Wear comes from mechanical forces, not heat.

Now, slip the valve into the holder and wiggle it for a stem-to-guide clearance check. Maximum allowable clearance is 0.003 in. (0.075mm), measured with a dial indicator, as described on page 93.

Honda recommends that the auxiliary-valve seat be cut using three special hand-operated cutters. They measure 30°, 60° and 45°. Then the valve is lapped on its seat to reveal the sealing area. If the valve-contact area is centered on the seat, everything is OK. The valve is ready to go. If there are problems, such as an uncentered contact area, the valve and holder should be replaced as units.

The problem with following this procedure is finding the special cutters and reamers. Few rebuilders outside of a Honda dealer's service department have them.

Therefore, replace the auxiliary valve, lap it to the seat and leave it at that. Under normal conditions, this has several advantages. It cures any wear at the valve tip, stem and face. Thus, the tip should pose no problem

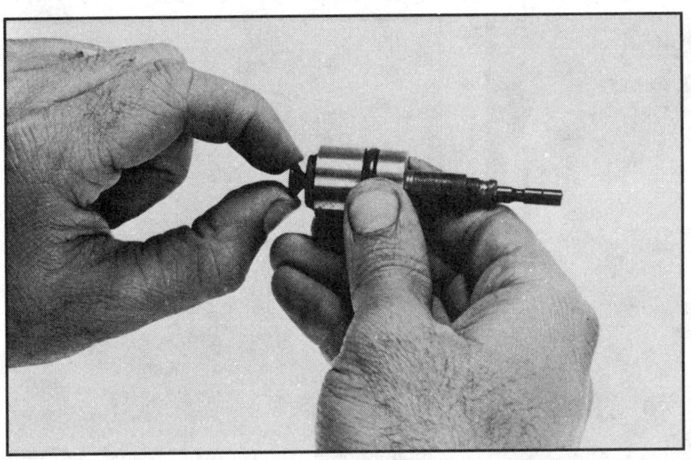
Check auxiliary-valve stem-to-guide clearance just as on a conventional valve. Maximum deflection at valve head is 0.002 in (0.05mm). Stem diameter should measure 0.215-in. (5.45mm) minimum.

Replace auxiliary valve if stem-to-guide clearance is excessive. New valve provides fresh sealing surface and valve-stem tip.

at all, stem-to-guide clearance is partially restored and the face is ready for a full life. In fact, replacing the valve—or valve holder—is the only way to restore auxiliary-valve stem-to-guide clearance.

The only two considerations remaining are guide and valve-seat condition. The new valve should cure excessive stem-to-guide clearance. Even if there's some slop left in the guide, it can't amount to much. It's such a small diameter that leakage and valve seating will barely be affected.

As for the valve seat, it's rarely pitted because the auxiliary valve runs so cool. You may see minor wear from valve-spring force—that's all. Lapping the new valve will restore the seat.

If yours is an extremely high-mileage engine, or on its second rebuild, you may need to replace both auxiliary valve and valve holder. About the only other way of handling these parts is taking the valve holder to a Honda dealer or Honda engine specialist with the correct tools and have him cut the seats for you. Otherwise, it's replacement time.

Lapping—Just how do you lap something that small? Here, we take a tip from motorcycle mechanics who have been dealing with small valves for a long time. Lightly chuck the valve holder in a vise, valve seat facing up. Dab fine lapping compound on the valve face and slip the valve in place. Push a length of tight-fitting rubber hose over the valve stem; it's hanging exposed under the valve holder. Work the hose back and forth between your palms. Lift the valve off its seat every few seconds to let fresh lapping compound get between valve and seat.

It shouldn't take long to lap these small valves. When an even, light-gray band appears all the way around the valve face and seat, the valve is fully lapped. Remove the valve from the holder and clean off all lapping compound. Use a rag dipped in solvent to clean the parts until *absolutely free* of grit.

ROCKER-ARM SERVICE

Rocker-arm service is limited to three areas: rocker-arm working surface (camshaft end), rocker-arm ID and rocker-arm shaft.

After solvent cleaning the rocker-arm assembly, position it upside down so the rocker-arm working surfaces are visible. These are the square shiny patches that rub against the camshaft. You'll see some scoring—that's normal. Run a fingernail across the working surface. If it catches in the scores, they're too deep. Also, these surfaces may be deeply grooved. This groove runs parallel with the camshaft and is cause for rocker-arm replacement.

Reposition the assembly so it stands on its pedestals. Grasp the end rocker arm and try to rock it 90° to its normal direction of movement. If you don't have much engine-rebuilding experience, compare the movement of each rocker against the others. If you have a feel for these things, rocker-shaft-to-rocker-arm clearance should be 0.003 in. (0.075mm) or less. That's not much. As with a valve in its guide, more than slight movement is too much.

This test tells you if there is excessive clearance. It doesn't tell you where excessive clearance may be coming from. It could be from rocker-arm ID or shaft wear. To get a quick check of what's at fault, push the rocker in question laterally on the shaft, against the spring. This exposes the shaft under the rocker arm so you can see if it's worn.

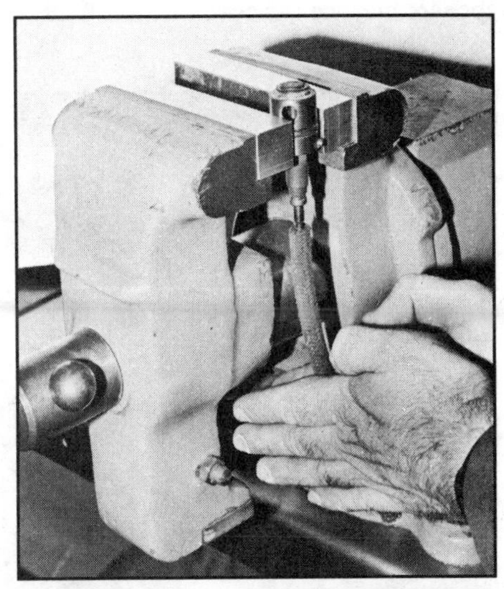
Lapping auxiliary valve can often restore sealing surfaces. It doesn't take long with a length of hose. Be sure to clean valve and holder with solvent when done.

Rocker-Arm Disassembly—If the rocker-arm assembly seems to be in good condition and the engine was clean inside, there's no need to remove the rockers. But, if you find a loose rocker, worn working surfaces or if you suspect gunk buildup inside the rocker-arm shafts, dismantle the assembly. You'll then be able to close-

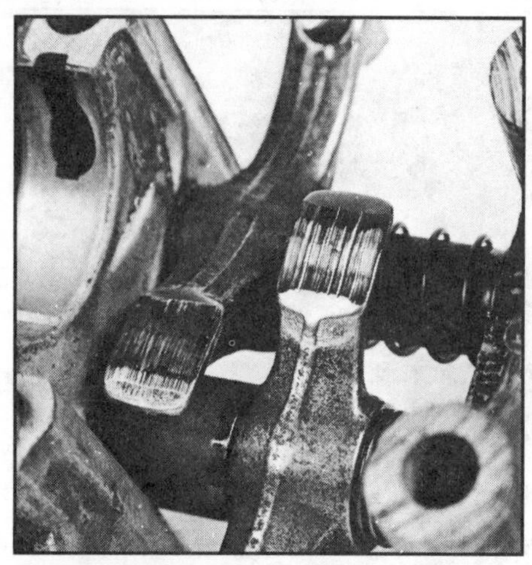

Cam lobes and rocker arms wear in together. Reusing these rockers would ruin a new cam in seconds. Both cam and worn rockers *must* be replaced if visible wear is present.

Normal rocker-arm wear looks like this. Moderately dark stripe across middle of working surface denotes small groove or flat spot. Rocker arms with small groove or flat spot must be replaced, also.

Disassemble rockers by squeezing and removing spring pin. Then slide end pedestal, rockers and springs off rocker-arm shaft.

Pulling rocker arms sideways exposes wear area on shaft. If oil and filter were changed frequently, and rocker-arm working surfaces look good, rocker assembly probably doesn't need to come apart. Try to wiggle arms on shaft to expose worn rocker-arm ID. If no wear is found, clean and store assembly.

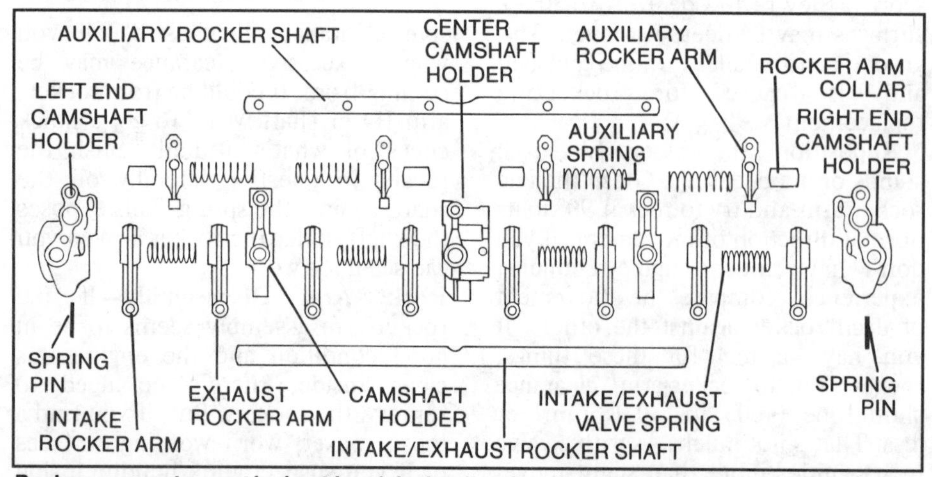

Rocker-arm parts *must* be kept in original order when disassembling. Inspect, clean and replace *one piece at a time* so order is maintained. Moly-lube shafts during assembly. CVCC-engine rocker assemblies shown. Courtesy of American Honda Motor Co., Inc.

ly inspect the rocker assembly, clean the rocker shafts and replace worn parts.

A word of caution before you take the assembly apart: In the typical CVCC assembly, there are 31 parts and almost all must go back in the exact order as they came off. So, do yourself a big favor and keep everything in order as it comes apart. Lay the pieces in order, on clean rags or towels, as you remove them. When you clean or inspect the parts, do one at a time, immediately returning each part to its place on the towel.

One particular component that can cause great grief on 1982—'83 engines is the number-8 rocker arm, or the last one at the flywheel end of the head. On previous engines, it was identical to other rockers. But, starting in 1982, it was changed to a two-piece design. It was intended to cure number-8 camshaft-lobe wear, so make sure this special one ends up in the right place.

It's easy enough to disassemble the rocker-arm assembly. At each end of the shaft is a solid pin or *spring pin*. A spring pin is split lengthwise in a sawtooth pattern, which allows you to squeeze it together. Examine the pin until you find the parting line. Then grasp the pin with dykes 90° to the parting line and pin axis. Clamp on the pin and lever it out of its hole. The pedestal should slide off the rocker shaft, as should the rocker arms, springs and remaining pedestals. Remove the pin at the other end to remove the last main-rocker-shaft pedestal. The auxiliary-rocker shaft is not pinned, so it will separate as soon as either main-shaft pin is pulled.

Now, mike the rocker shafts. Compare worn and unworn portions of the shaft. Use a telescoping, or snap, gage to measure rocker-arm ID. Combined wear of rocker and shaft cannot exceed 0.003 in. (0.075mm). Replacement is the only way to correct worn parts. Also, check the springs for obvious distortion, although you should have no problems with them.

If you are reusing the rocker shaft, clean it in clean solvent. Pay special attention to the oil holes. The rocker shaft doubles as an oil pipe, and the holes oil the rocker arms. If the holes clog, the shaft and rocker arms will wear rapidly. If you have one bad rocker, its oil hole may have been clogged.

Reassemble the rockers using plenty of oil or a light coat of moly lube. As a final touch, eyeball the valve adjusters. If the jam nut is rounded-off or screw threads are loose, replace them. Simply unscrew the old jam nut and adjusting screw and screw on the new ones.

VALVE-SPRING INSPECTION & INSTALLATION

Next up for inspection are the valve springs. The following steps must be performed to verify that the springs are usable. Those that aren't must be replaced; good valve springs are critical for good engine performance.

There certainly is a lot to consider when checking valve springs. The list includes *spring rate, squareness, free height, load at installed height, load at open height* and *load at solid height.* A *spring tester* must be used for making some checks. And, although it's a lot easier to use a spring tester, some checks can be made with a ruler. From these results, checks requiring a spring tester can be deduced.

SPRING TERMS

To understand what to look for when testing a spring, you should first understand some spring terms.

Spring Rate—Basic to a spring's operation is its *rate.* Spring rate is how much force a spring applies for a given amount of *deflection,* usually expressed in *pounds per inch* of compression for a coil-type valve spring. As a spring is compressed, its resisting force increases.

For example, a typical Honda intake-valve spring set—inner and outer combined—yields about 200-lb per in. A CVCC auxiliary-valve spring manages a little over 50-lb per in. It takes a higher rate spring to control the heavier valve. If the main intake valve weighed as little as the auxiliary valve, it too could use a lower rate.

Valve-spring *inertia* must be controlled by the valve spring. Inertia is the valve's resistance to change in movement. The more the valve weighs, the more inertia it has. Also, the *faster* the valve moves—resulting from engine speed and valve-train geometry—the greater the inertia. So, above a certain rpm, valves will *float.* To raise rpm above that, higher-rate springs are needed.

Valve float occurs when the inertia of the valve train exceeds the ability of the valve springs to close the valves—a spring ceases to hold the valve against the rocker, and the rocker against the cam lobe. When this occurs, compression and combustion pressures are lost out the intake or exhaust port, or worse, the piston contacts the valve. In the first instance, power is lost; in the second, valve-train or piston damage results.

Engineers try for the lowest-possible spring rate, even on high-rpm race engines. Only enough "spring" should be used to close the valves, not much more. It takes horsepower to turn the camshaft and raise the valves against their springs. Too-stiff springs mean wasted horsepower and unnecessary valve-train wear.

Consequently, there's little reserve strength in a valve spring. After exposure to high heat and millions of openings and closings, springs *fatigue,* or lose their load-producing capability. More precisely, the springs will *sag*—much like a chassis spring. When this happens, the valves float at lower and lower rpm.

Free Height—Free height is simply the unloaded height, or *length,* of a spring—how high, or *tall* it is when sitting on the bench. The force or load exerted by a spring is directly proportional to how much it is compressed. Consequently, a shorter, fatigued spring has less free height—exerts less force because it's compressed less.

Because you probably don't have a spring tester to determine spring force—or spring load at different heights—use free length to judge valve-spring condition.

Check valve-spring squareness with carpenter's square. Rotate spring so maximum lean shows against square. Replace springs that have more than 1/16 in. of lean at top.

Load at Installed Height—Load at installed height is the force an installed valve spring exerts when the valve is closed. It is also a specification used for checking valve springs. An average figure for Hondas is 19 lb at 1.437 in. of compressed height.

Load at installed height is important because, like free height, it gives a good indication of spring condition. For normal service, the minimum load is 5% below the spring rating, or 18 lb in our example. If you want the engine to be capable of revving to its design limit, use specified load at installed height, not the minimum limit.

Load at Open Height—Similar to load at installed height, load at open height occurs when the valve is fully open—or when the valve spring is compressed more than its installed height by the amount of maximum valve opening.

Typically, load at open height is 83.4 lb at 1.094 in.—or 64.4 additional pounds for 0.343-in. additional compression. Again, this figure can be 5% lower than rated specifications for light-duty use.

Don't waste money on new springs if your car will only be used for going to the store and back, and you find a few "5% springs." On the other hand, don't think high-rpm use means racing—if you use the lower gears for passing or hill climbing, then your engine needs springs that are right on spec.

VALVE SPRING

Engine	Year	FREE HEIGHT EXHAUST INNER in. (mm)	FREE HEIGHT EXHAUST OUTER in. (mm)	FREE HEIGHT INTAKE INNER in. (mm)	FREE HEIGHT INTAKE OUTER in. (mm)	FREE HEIGHT AUX. in. (mm)	INSTALLED HEIGHT EXHAUST INNER Height in. (mm)	INSTALLED HEIGHT EXHAUST INNER Load lb (kg)	INSTALLED HEIGHT EXHAUST OUTER Height in. (mm)	INSTALLED HEIGHT EXHAUST OUTER Load lb (kg)	INSTALLED HEIGHT INTAKE INNER Height in. (mm)	INSTALLED HEIGHT INTAKE INNER Load lb (kg)	INSTALLED HEIGHT INTAKE OUTER Height in. (mm)	INSTALLED HEIGHT INTAKE OUTER Load lb (kg)
1170	73	1.6535 (42.0)	1.5728 (39.95)	1.6535 (42.0)	1.5728 (39.95)	—	N/A	N/A	N/A	N/A	N/A	N/A	N/A	N/A
	74-77	1.6535 (42.0)	1.5728 (39.95)	1.6535 (42.0)	1.5728 (39.95)	—	N/A	N/A	N/A	N/A	N/A	N/A	N/A	N/A
	78-79	1.9882 (50.5)	2.1181 (53.8)	1.6654 (42.3)	1.6649 (42.29)	—	N/A	N/A	N/A	N/A	N/A	N/A	N/A	N/A
1335	80	1.988 (50.5)	2.118 (53.8)	1.665 (42.3)	1.665 (42.3)	1.146 (29.1)	1.331 (33.8)	27.5 (12.5)	1.437 (36.5)	51.3 (23.3)	1.401 (35.6)	8.82 (4.0)	1.488 (37.8)	28.22 (12.8)
	81	1.988 (50.5)	2.118 (53.8)	1.665 (42.3)	1.665 (42.3)	1.17 (29.7)	1.331 (33.8)	27.5 (12.5)	1.437 (36.5)	51.3 (23.3)	1.401 (35.6)	8.82 (4.0)	1.488 (37.8)	28.22 (12.8)
	82	1.988 (50.5)	2.118 (53.8)	1.665 (42.3)	1.665 (42.3)	1.17 (29.7)	1.331 (33.8)	27.5 (12.5)	1.437 (36.5)	51.3 (23.3)	1.401 (35.6)	8.82 (4.0)	1.488 (37.8)	28.22 (12.8)
	83	1.988 (50.5)	2.118 (53.8)	1.665 (42.3)	1.665 (42.3)	1.17 (29.7)	1.331 (33.8)	27.5 (12.5)	1.437 (36.5)	51.3 (23.3)	1.401 (35.6)	8.82 (4.0)	1.488 (37.8)	28.22 (12.8)
1488	75	2.047 (52.0)	2.138 (54.3)	1.583 (40.2)	1.573 (39.95)	1.146 (29.1)	1.331 (33.8)	23.1 (10.5)	1.432 (36.1)	51.3-62.3 (23.3-28.3)	1.358 (34.5)	12.1 (5.5)	1.437 (36.5)	11.0 (5.0)
	76-77	1.988-2.047 (50.5-52.0)	2.118 (53.8)	1.583 (40.2)	1.573 (39.95)	1.146 (29.1)	1.358 (34.5)	25.8-31.5 (11.7-14.3)	1.437 (36.5)	54.8-60.0 (24.4-27.2)	1.358 (34.5)	20.1-23.2 (9.1-10.5)	1.437 (36.5)	18.5-19.0 (8.4-8.6)
	78-79	1.988 (50.5)	2.118 (53.8)	1.665 (42.3)	1.665 (42.3)	1.146 (29.1)	1.331 (33.8)	27.5 (12.5)	1.437 (36.5)	51.3 (23.3)	1.40 (35.6)	8.82 (4.0)	1.488 (37.8)	28.22 (12.8)
	80	1.988 (50.5)	2.118 (53.8)	1.665 (42.3)	1.665 (42.3)	1.122 (28.5)	1.331 (33.8)	27.5 (12.5)	1.437 (36.5)	51.3 (23.3)	1.40 (35.6)	8.82 (4.0)	1.488 (37.8)	28.22 (12.8)
	81-82	1.988 (50.5)	2.118 (53.8)	1.665 (42.3)	1.665 (42.3)	1.17 (29.7)	1.331 (33.8)	27.5 (12.5)	1.437 (36.5)	51.3 (23.3)	1.40 (35.6)	8.82 (4.0)	1.488 (37.8)	28.22 (12.8)
	83	1.988 (50.5)	2.118 (53.8)	1.665 (42.3)	1.665 (42.3)	N/A	1.331 (33.8)	27.5 (12.5)	1.437 (36.5)	51.3 (23.3)	1.40 (35.6)	8.82 (4.0)	1.488 (37.8)	28.22 (12.8)
1600	76-77	1.988-2.047 (50.5-52.0)	2.118 (53.8)	1.583 (40.2)	1.573 (39.95)	1.146 (29.1)	1.358 (34.5)	25.8-31.5 (11.7-14.3)	1.437 (36.5)	53.8-60.0 (24.4-27.2)	1.358 (34.5)	20.1-23.2 (9.1-10.5)	1.437 (36.5)	18.5-19.0 (8.4-8.6)
	78	1.988 (50.5)	2.118 (53.8)	1.665 (42.3)	1.665 (42.3)	1.146 (29.1)	1.331 (33.8)	27.5 (12.5)	1.437 (36.5)	51.3 (23.3)	1.401 (35.6)	8.82 (4.0)	1.488 (37.8)	28.22 (12.8)
1751	79-80	1.988 (50.5)	2.118 (53.8)	1.665 (42.3)	1.665 (42.3)	1.339 (34.0)	1.358 (34.5)	25.8-31.5 (11.7-14.3)	1.437 (36.5)	51.4-62.4 (23.3-28.3)	1.402 (35.6)	7.9-9.7 (3.6-4.4)	1.488 (37.8)	25.4-31.1 (11.5-14.1)
	81	1.988 (50.5)	2.118 (53.8)	1.665 (42.3)	1.665 (42.3)	1.339 (34.0)	1.402 (35.6)	7.9-9.7 (3.6-4.4)	1.488 (37.8)	25.4-31.1 (11.5-14.1)	1.402 (35.6)	7.9-9.7 (3.6-4.4)	1.488 (37.8)	25.4-31.1 (11.5-14.1)
	82-83	1.67 (42.3)	1.665 (42.29)	1.67 (42.3)	1.665 (42.29)	1.17 (29.7)	1.40 (35.6)	7.9-9.7 (3.6-4.4)	1.49 (37.8)	25.3-31.0 (11.5-14.1)	1.40 (35.6)	7.9-9.7 (3.6-4.4)	1.49 (37.8)	25.3-31.0 (11.5-14.1)

N/A Not Available
— Does not apply
AUX. = Auxiliary

Solid Height—If a valve spring is compressed until each of its adjacent coils touch and it cannot be compressed any farther, the spring is at its *solid height*—it is *coil bound.*

In the real world, a valve spring that *goes solid* or *coil binds,* creates great havoc with its valve train. A not-too-uncommon occurrence, coil binding is caused by a cam with more lift than the original cam, or by a spring *shimmed* excessively to bring it within specifications, page 104.

Squareness—How straight a valve spring stands on a flat surface is called *squareness.* If a spring *leans* too much—1/16 in. or more measured from the top coil to a vertical surface—it should be replaced.

SPRING TESTS

Now that you've been bombarded with basic valve-spring terms, apply this information to put your valve springs through several tests. Check for squareness, free height, installed load, open load and solid height.

Unless you are among the fortunate few, you probably don't have a spring tester. An industrial type fetches $500 or so—this puts it out of the occasional-user league.

Fortunately, there is a valve-spring tester that is within the average do-it-yourselfer's range. Although not as convenient to use, all you'll need, in addition to the tester, is a vise and a vernier caliper. Clamp the spring and tester in the vise, compress the spring to its rated installed or open height—measured with the vernier caliper—and read spring load directly.

If your local auto-parts dealer or speed shop doesn't sell such a tester, it is available from: Goodsen, 4500 West 6th Street, Winona, MN 55987; or C-2 Sales & Service, Box 70, Selma, OR 97538.

SPECIFICATIONS

		OPEN HEIGHT									
		EXHAUST				INTAKE					
AUXILIARY		INNER		OUTER		INNER		OUTER		AUXILIARY	
Height in. (mm)	Load lb (kg)	Height in. (mm)	Load lb (kg)	Height in. (mm)	Load lb (kg)	Height in. (mm)	Load lb (kg)	Height in. (mm)	Load lb (kg)	Height in. (mm)	Load lb (kg)
—	—	N/A	N/A	N/A	N/A	N/A	N/A	N/A	N/A	—	—
—	—	N/A	N/A	N/A	N/A	N/A	N/A	N/A	N/A	—	—
—	—	N/A	N/A	N/A	N/A	N/A	N/A	N/A	N/A	—	—
0.906 (23)	14.5-16.7 (6.6-7.6)	1.024 (26.0)	38.5-49.5 (17.5-22.5)	1.108 (28.0)	79.2-90.2 (36.0-41.0)	1.008 (25.6)	30.87 (14.0)	1.094 (27.8)	108.03 (49.0)	0.787 (20)	21.2-25.5 (9.6-11.6)
0.906 (23)	14.5-16.7 (6.6-7.6)	1.024 (26.0)	38.5-49.5 (17.5-22.5)	1.108 (28.0)	79.2-90.2 (36.0-41.0)	1.008 (25.6)	30.87 (14.0)	1.094 (27.8)	108.03 (49.0)	0.787 (20)	21.2-25.5 (9.6-11.6)
0.98 (25)	13.9-17.0 (6.3-7.7)	1.024 (26.0)	38.5-49.5 (17.5-22.5)	1.108 (28.0)	79.2-90.2 (36.0-41.0)	1.008 (25.6)	30.87 (14.0)	1.094 (27.8)	108.03 (49.0)	0.87 (22)	25.8-31.5 (11.7-14.3)
0.98 (25)	13.9-17.0 (6.3-7.7)	1.024 (26.0)	38.5-49.5 (17.5-22.5)	1.108 (28.0)	79.2-90.2 (36.0-41.0)	1.008 (25.6)	30.87 (14.0)	1.094 (27.8)	108.03 (49.0)	0.87 (22)	25.8-31.5 (11.7-14.3)
0.906 (23)	14.5-16.7 (6.6-7.6)	0.988 (25.0)	38.5-49.5 (17.5-22.5)	1.108 (28.0)	7.92-90.2 (36.0-41.0)	1.024 (26.0)	59.0-68.0 (26.77-30.83)	1.094 (27.8)	72.4-83.4 (32.9-37.9)	0.787 (20)	21.1-25.5 (9.6-11.6)
0.906 (23)	14.5-16.7 (6.6-7.6)	1.024 (26.0)	38.5-49.5 (17.5-22.5)	1.108 (28.0)	79.2-90.2 (36.0-41.0)	1.024 (26.0)	59.0-68.0 (26.77-30.83)	1.094 (27.8)	72.4-83.4 (32.9-37.9)	0.787 (20)	21.2-25.5 (9.6-11.6)
0.906 (23)	14.5-16.7 (6.6-7.6)	1.024 (26.0)	38.5-49.5 (17.5-22.5)	1.108 (28.0)	79.2-90.2 (36.0-41.0)	1.008 (25.6)	30.87 (14.0)	1.094 (27.8)	108.03 (49.0)	0.787 (20)	21.2-25.5 (9.6-11.6)
0.984 (25)	13.4-17.4 (6.1-7.9)	1.024 (26.0)	38.5-49.5 (17.5-22.5)	1.108 (28.0)	79.2-90.2 (36.0-41.0)	1.008 (25.6)	30.87 (14.0)	1.094 (27.8)	108.03 (49.0)	0.866 (22)	26.0-31.3 (11.8-14.2)
0.984 (25)	13.9-17.0 (6.3-7.7)	1.024 (26.0)	38.5-49.5 (17.5-22.5)	1.108 (28.0)	79.2-90.2 (36.0-41.0)	1.008 (25.6)	30.87 (14.0)	1.094 (27.8)	108.03 (49.0)	0.866 (22)	25.8-31.5 (11.7-14.3)
N/A	N/A	1.024 (26.0)	38.5-49.5 (17.5-22.5)	1.108 (28.0)	79.2-90.2 (36.0-41.0)	1.008 (25.6)	30.87 (14.0)	1.094 (27.8)	108.03 (49.0)	N/A	N/A
0.906 (23)	14.5-16.7 (6.6-7.6)	1.024 (26.0)	38.5-49.5 (17.5-22.5)	1.108 (28.0)	79.2-90.2 (36.0-41.0)	1.024 (26.0)	59.0-68.0 (26.77-30.83)	1.094 (27.8)	72.4-83.4 (32.9-37.9)	0.787 (20)	21.1-25.5 (9.6-11.6)
0.906 (23)	14.5-16.7 (6.6-7.6)	1.024 (26.0)	38.5-49.5 (17.5-22.5)	1.108 (28.0)	79.2-90.2 (36.0-41.0)	1.008 (25.6)	30.87 (14.0)	1.094 (27.8)	108.03 (49)	0.787 (20)	21.1-25.5 (9.6-11.6)
0.984 (25)	24.5-28.4 (11.1-12.9)	1.024 (26.0)	38.5-49.6 (17.5-22.5)	1.102 (28.0)	79.4-90.4 (36.0-41.0)	1.008 (25.6)	27.8-34.0 (12.6-15.4)	1.094 (27.8)	99.2-116.8 (45.0-53.0)	0.866 (22)	32.6-37.9 (14.8-17.2)
0.984 (25)	13.9-17.0 (6.3-7.7)	1.008 (25.6)	27.8-34.0 (12.6-15.4)	1.094 (27.8)	99.2-116.8 (45-53)	1.008 (25.6)	27.8-34.0 (12.6-15.4)	1.094 (27.8)	99.2-116.8 (45.0-53.0)	0.866 (22)	29.8-36.4 (13.5-16.5)
0.98 (25)	13.9-17.0 (6.3-7.7)	1.03 (26.1)	26.7-32.8 (12.1-14.9)	1.11 (28.3)	93.5-114.2 (42.5-51.9)	1.03 (26.1)	26.7-32.8 (12.1-14.9)	1.11 (28.3)	93.5-114.2 (42.5-51.9)	0.87 (22)	25.7-31.5 (11.7-14.3)

Squareness of a spring is the easiest check, so let's start there. The equipment needed is a carpenter's framing square or any right-angle measuring device. Stand the spring upright in the corner of the square. Turn the spring until you determine its maximum *lean*—distance between the top coil and the vertical leg of the square. Maximum allowable lean is 1/16 in. Replace the spring if it leans too much.

Free height is a fairly accurate indication of valve-spring condition. Although you won't be able to measure spring loads, you can measure *free height* with a 6-in. machinist's rule. If free height is within spec, the installed and open loads should also be OK.

Find the specifications for your springs in the accompanying table, then measure free height. Chances are springs that are short—have *sagged*—by more than 0.100 in. (2.54mm) of their free height, have fatigued. They should be replaced or checked in a spring tester for installed and open loads.

A good rule of thumb is to replace *any* spring that's 1/8-in. (3.175mm) or more short. If the spring is only 1/16-in. (1.59mm) short, it can be used in an engine destined for light service. But don't install such a spring in an engine that will be operated in the upper-rpm range.

You may have weeded out a spring or two and replaced them. But just because the replacement springs are new doesn't mean you shouldn't check them. Put them in your lineup and perform the remaining tests on them. That brings up another point—spring order. You should've kept the springs in order, unless you reground the valve seats or valves. If they got mixed up during cleaning or whatever, put them back in line and

This is why most Honda camshafts are replaced—extreme wear on cylinder-4 lobes. Even with conscientious maintenance, Honda cams can wear out like this by 75,000 miles. Wear occurs sooner with longer oil-change intervals.

Check oil-pump and distributor-drive gears for wear. This pattern is excellent because wear is confined to center of teeth and is not excessive. Gear was replaced anyway to match new camshaft and companion gear.

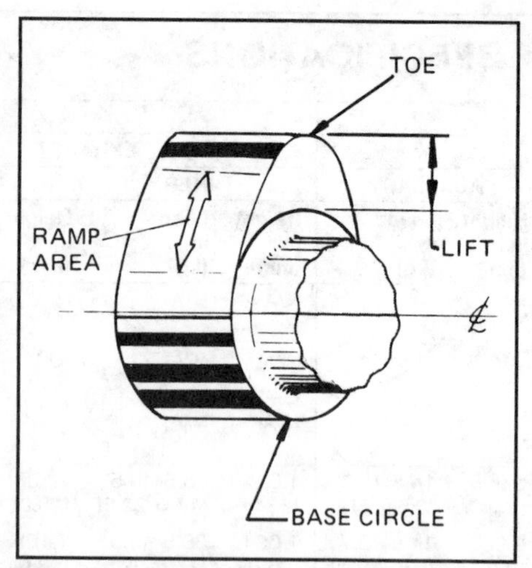

Lobe lift is distance from center of lobe to toe less base-circle radius. Drawing by Tom Monroe.

Keepers are often overlooked during inspection. They don't often wear out, but check them just in case. Keeper at left has narrow, shiny band with concave wear pattern, midway down—replace it. Right keeper has wider band with no appreciable wear.

To aid exhaust-valve life, use 1975 1488 thrust washers under spring seats. Unlike more common, ratchet-type rotaters, these turn valves through coil-spring *windup* only.

maintain this new order from now on. Spring inspection from this point on requires a spring tester. If you don't have one, gather up your springs and take them to your engine machinist.

Springs that fail the free-height test by 1/16 in. or less can be *shimmed* back to original height. Valve-spring shims are available at auto-parts outlets. You must have a spring tester to see if a spring will work when shimmed, and how thick a shim it needs.

If you shim a spring to restore its installed and open loads, check for coil bind! Because the spring will be compressed more than its specified open height, its reserve travel is reduced when the valve is open—distance between coils will be reduced. This is especially true of high-performance engines fitted with high-lift camshafts. There must be 0.010 in. (0.254mm) between coils when the spring is compressed to its open height.

It's better to replace a spring than to shim it. For example, a new spring has its entire life ahead of it, the old spring is already on its way out—that's why it needs shimming. Also, a shimmed spring is compressed more. Thus, the increased compression subjects the already-fatigued spring to greater stress than it was designed for. This fatigues the spring at an increasing rate.

Unless you are patching an engine to run a few-thousand more miles—with knurled valve guides, pistons and such—replace valve springs before you shim them to obtain installed and open loads. If you do shim a spring, wire the shim to the spring so they don't get separated. Otherwise, you'll have to recheck the springs.

If you want, you can install 1975 1488 exhaust-valve thrust washers under the exhaust-valve spring seats. This very thin shim will not noticeably affect spring performance. It will, however, turn the exhaust valve for longer life.

CAMSHAFT

If there's one Honda engine part known for wearing out, it's the camshaft. It's a fact that Honda engine cam wear is much higher than the norm—so much that the camshaft is a normal replacement item. I'll give the full-blown inspection procedure for checking Honda cams, but I don't think you'll need it.

For most Honda rebuilds, a visual inspection will verify a worn-out cam. Go straight to the lobes for cylinder 4. Roll the camshaft in your hands as you watch light reflect off the lobes. If you see sharp definitions in the reflected light, the cam is worn. Run your finger over each lobe; feel for sharp edges. On the typically worn lobe, there will be at least two of these ridges, just off the very top, or *toe,* of the lobe. If you aren't sure what you are feeling for, compare what you feel on number-4's lobes to other lobes on the cam. A good lobe will have one smooth curve all the way around the lobe. There should be no sharp edges.

Worn cam lobes are easy to spot. They usually look like someone filed two perfectly flat surfaces onto the lobe.

As you've guessed, number-4's lobes are the main troublemakers. For some reason, probably poor oiling, they wear rapidly—a problem that transcends all model years of

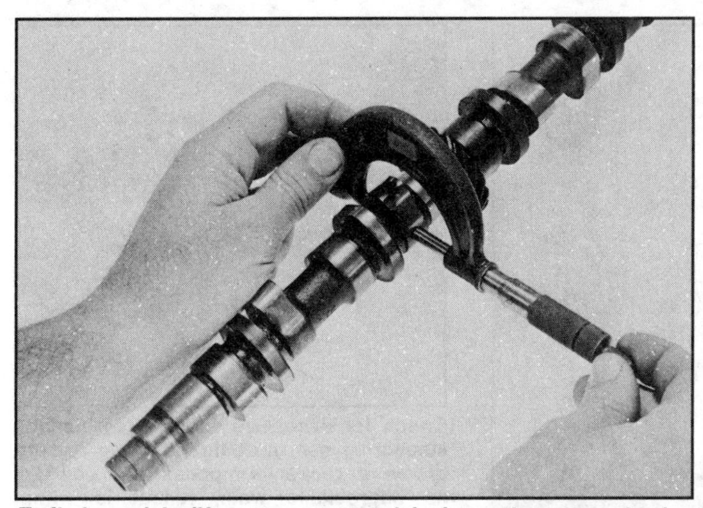
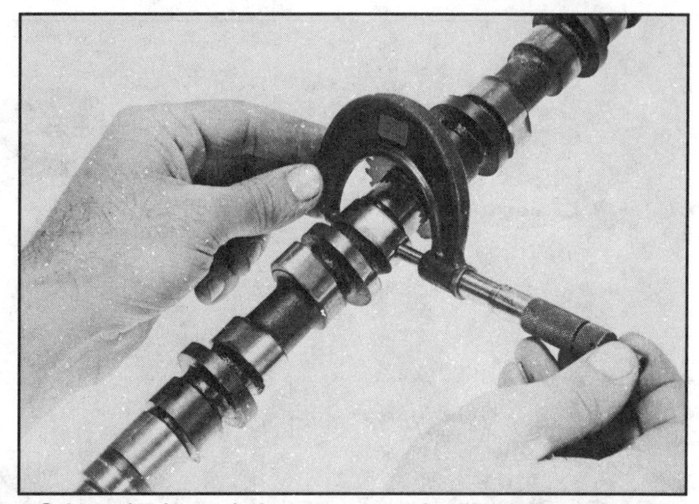

To find cam-lobe lift, measure across lobe base, then across heel and toe. Subtracting base-circle measurement from heel-to-toe measurement gives *approximate* lobe lift. This method is not extremely accurate, but is good for comparing lobes for wear.

CVCC engines. Because it takes only one bad lobe to ruin an entire cam, this means a new camshaft.

Service has less to do with cam wear than you might think. Even with frequent oil and filter changes, it's not uncommon for cylinder-4's lobes to *flatten* before 75,000 miles. Of course, running dirty oil practically guarantees a bad lobe or two.

So, if your cam has flat lobes, throw it away and get a new one. Along with the new camshaft, get a new oil-pump drive-shaft gear—the one that meshes with the camshaft. The camshaft (drive gear) and the oil-pump- drive-shaft (driven gear) wear into a matched set, so replacing one gear means replacing both. If you use the old oil-pump gear on a new cam, it will quickly ruin the drive gear.

If the lobes pass visual inspection, check the oil-pump and distributor-drive gears. Normal wear should look like a light polishing *centered* on the gear teeth. When the wear pattern extends the length of teeth, can be felt with a fingernail or there's visual concavity, the gear is worn out. Oil-pump-drive-gear wear was a minor problem on 1973–'74 1200s, but hasn't been much of a factor since.

If the camshaft passes visual inspection, there are other checks to perform.

Lobe Lift—If you checked valve lift with a dial indicator during the diagnostic session, there's no need to check *lobe lift.* Refer back to your notes to see which lobes were worn, if any. Remember, a dial indicator is the *only* accurate way to check valve lift. If you didn't measure valve lift before, check the lobes now. You'll need a vernier caliper or micrometer for making measurements.

Measuring cam lobes is a comparison check, one lobe to another. Unlike the check in the diagnosis chapter, the purpose here is to compare relative lobe *wear,* not *lift.* Because cam-lobe wear accelerates *once the surface hardening is worn through,* you should have little trouble pinpointing a bad lobe.

Take the first measurement across the lobe, halfway between its heel and toe; this is the *minor dimension.* Then measure the major dimension, from the tip of the toe to the heel, as shown. Both measurements should be made at the points of maximum visible wear and the same distance for the same edge of the lobe. Now, subtract the minor dimension from the major dimension to get *approximate* lobe lift.

You can't depend on these measurements to reflect lobe lift accurately, particularly with a high-performance cam. This is because you'll be measuring across the ends of the *opening* and *closing ramps*—transition sections of the lobe between the toe and the base circle. The reading you get will be greater than base-circle diameter. As a result, lobe lift will appear to be less than actual.

Measure all lobes, including the auxiliaries, recording your results as you go. Compare your findings by placing all lobe-lift figures for the intake lobes in one column, exhaust figures in another and auxiliary figures in a third. All lobes should be within 0.005 in. (0.125mm) of others of that type. If there is *one* bad lobe, the cam and oil-pump drive gear must be replaced.

Oil starvation due to clogged oil screen ruined this 1237 cam. Raised portion at right shows original lobe profile.

Camshaft-Bearing Journals—Camshafts are more durable than cranks in one respect; their journals seldom wear out. There should be less than 0.002-in. (0.05mm) variation between journal diameters when measured with a micrometer.

Runout—To check cam runout, support it with two V-blocks, one under each end journal. Or, support it between centers, such as in a lathe. Indicate the center bearing journals as you did with the crankshaft. With a dial indicator set up at the center journal, rotate the cam. Runout should not exceed 0.004 in. (0.10mm). The final runout check is to install the cam in the head and spin it in well-oiled bearings. Unless the cam shows play or binds as you rotate it, it's OK.

End Play—End play is measured with a dial indicator while the cam is installed in the head. End play on a new

Cam runout should be checked in lathe, or on precision V-blocks with a dial indicator. But, laying cam in head, rotating it and checking runout at center journal with dial indicator gives "quick-and-dirty" indication.

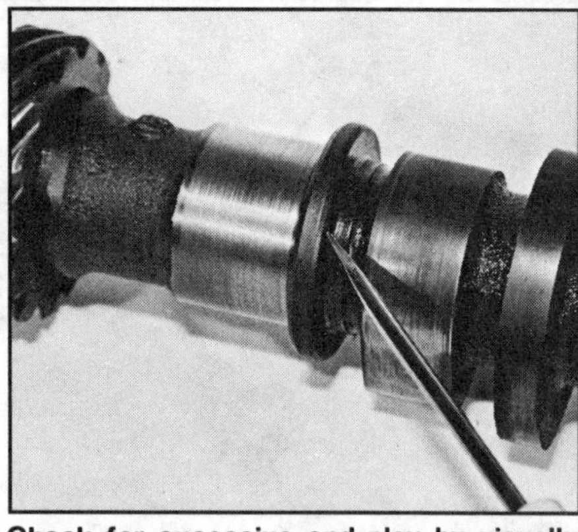

Check for excessive end play by visually inspecting camshaft thrust *fence* and its groove in rocker-arm pedestal. If you see no appreciable wear, runout is within specs. If you suspect excessive end play, check with dial indicator, page 110. Usually, camshaft is replaced because of lobe wear, automatically restoring fence to new thickness. Excessive end play is frequent problem on 1200s, but rare on CVCC engines. If thrust groove in rocker pedestal is worn out, fix is *head replacement*, or installing *new pedestal and align-honing cam bores.*

engine should be 0.002—0.006 in. (0.05—0.15mm). The wear limit for used cams is 0.020 in. (0.5mm). Thrust is taken by a *fence* just forward of the rear cam journal—flywheel end—and a groove in the rear rocker-arm-assembly pedestal. Check the cam fence and rocker-arm pedestal for wear and replace the cam or pedestal as necessary if end play is excessive.

CYLINDER-HEAD ASSEMBLY

Items you'll need for assembling the cylinder head include a valve-spring compressor, ruler and *possibly* an assortment of valve-spring shims. Don't buy the shims until you determine that they are needed—chances are they won't be. It's also a good idea to have a small tube of anti-seize compound handy. Use it as a thread lubricant for head bolts, manifold studs and sparkplugs.

Installed Height—It's time to check spring *installed height*—height of an installed spring with its valve closed. Directly related to load at installed height, correct installed height ensures that the springs are compressed enough to exert correct closing force to prevent valve float.

Don't be confused about the possibility of having to shim the valve springs again. The shimming you may have done earlier was to compensate for sag or to restore installed and open loads. This checking and possible shimming is to ensure that the distance between the spring seat and its retainer is at the specified installed spring height.

For instance, if a valve stem extends too far out of its guide—if the valve seat or valve face needed considerable grinding—the spring will not be compressed to its installed height. The spring must be shimmed to compensate for the difference in installed height. Therefore, you may need to add more shims to springs that already have them; one set to correct for installed and open load, and possibly one for installed height.

Check Installed Height—Installed height is checked with the valve installed in *its* guide *without* the spring. Start by placing the *spring seat* and any *previous* shims or thrust washers—not the ones used for correcting installed and open loads—over the guide.

Insert the valve into its guide. Now, install the retainer with its keepers. With the valve against its seat, pull the retainer up snugly against the keepers. While lifting firmly on the retainer, measure the distance from the spring seat to the underside of the retainer. This check must be accurate so make sure the lighting is good enough to see what you're doing. One of two methods can be used for

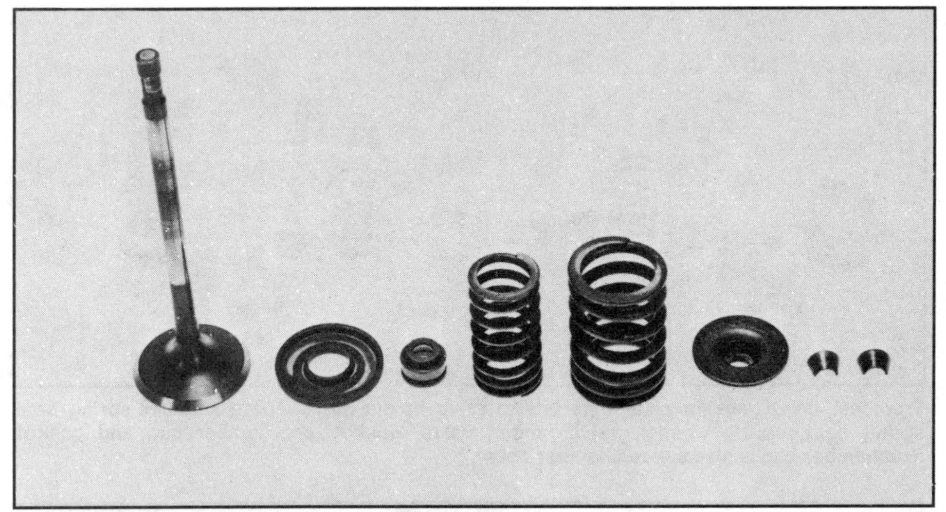

Parts needed to assemble one conventional valve are: valve, spring seat, seal, inner spring, outer spring, retainer and keepers.

Insert valve into head, then lay spring seat over valve guide.

After oiling valve stem and guide, gently push seal over stem and guide. Seal will seat on guide with a small "pop."

Slide inner and outer springs over valve stem.

measuring installed height.

If you have a telescoping gage, use it and a 1—2-in. micrometer for checking gage length. Or, check installed height by making a gage from a piece of coat hanger or other heavy wire. Snip out a straight section, then grind it to the spring's installed height. Stand the gage between the retainer and the spring seat. If there's a gap, use a feeler gage between the gage and spring retainer to determine needed shim thickness.

If you use a telescoping gage and micrometer, subtract specified installed height from your measurement to find the shim needed—if any. Ignore anything less than 0.015 in. (0.381mm). Compensate for differences over 0.015 in. with a shim of the same or *less* thickness. Don't over-shim a spring.

Because I told you not to buy shims until you knew what was needed, I'll assume that you don't have any. So, record the thickness needed on a narrow piece of paper and slip it under the valve spring that needs the shim. Also, write it down on your shopping list. After you've checked all the springs, go buy the shims.

Install the shim(s) under each correct spring seat. With the valve in its guide, the valve-stem seal, if used, can go on. 1975 1488s and all 1200s do not use stem seals on the exhaust valves, but 1488s can be retrofitted with them. Oil the valve stem and, as gently as possible, push the seal over the stem and over the valve guide.

Retainer-to-Guide Clearance—The bottom of the valve-spring retainer should clear the valve-guide top and oil seal by at least 0.060 in. (1.5mm). Check by opening the valve—while holding the retainer and keepers in place—to its full-open position. Use a machinist's 6-in. rule against the retainer to check valve opening. Eyeball the retainer-to-guide clearance. If it appears to be remotely close, measure it with a feeler gage.

If there's insufficient retainer-to-guide clearance, the top of the guide must be machined shorter. Have a machine shop do this.

Cylinder-Head Assembly—After purchasing the shims, you should have everything needed to assemble the head. If you have everything laid out, head assembly will go quickly and with little effort. Don't worry if more than one shim goes under some of the springs. You could have shims for free height, installed height or no shims at all.

Organize all parts in front of you. Have the heads with the valves, seals, shims and spring seats already installed, and their springs in order. Remember, springs, shims and valves must be installed exactly as they were

Position retainer with a slight wiggle to seat springs, attach compressor and slip in keepers. Compress valve springs only enough to install keepers. Over-compressing springs can damage them. Release compressor and install seven remaining valves.

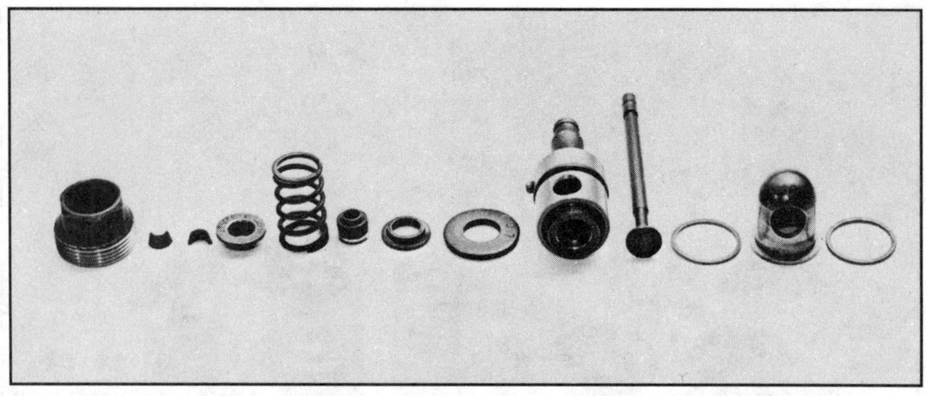

From left, CVCC auxiliary-valve parts are: valve-holder nut, keepers, retainer, spring, seal, spring seat, spring washer, valve holder, valve, gasket, prechamber cup and gasket. Prechamber cup is also called chamber collar.

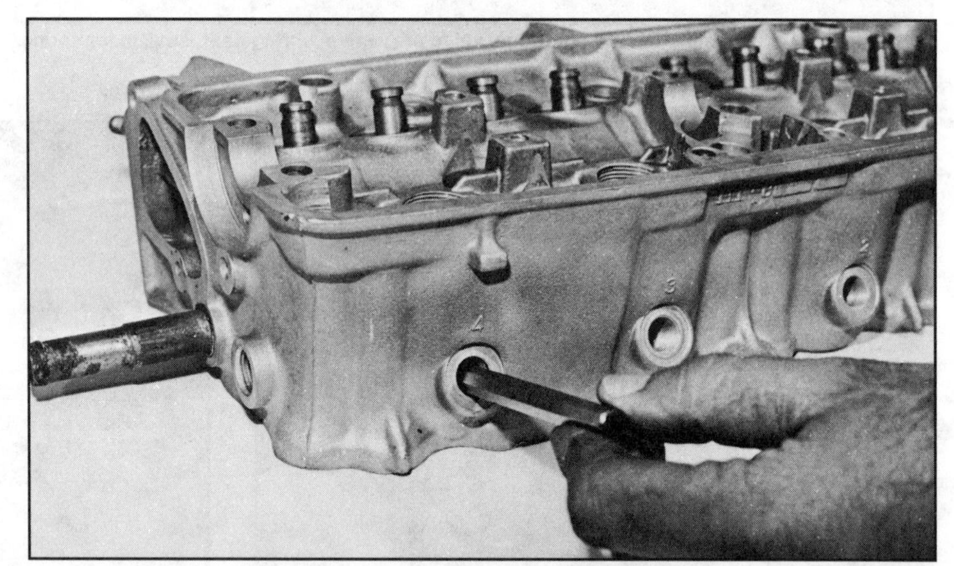

CVCC Auxiliary-valve assembly begins with prechamber cup. Using *new gaskets*, lay a gasket into head, then cup, then other gasket. Align oval cup hole toward main combustion chamber, and round hole toward sparkplug hole. If cup has multiple holes on one side, position them toward main combustion chamber. Insert drift through sparkplug hole to keep prechamber cup from rotating.

machined or checked.

You'll need a valve-spring compressor and an oil can. Remember to compress a valve spring only enough to install the keepers, no more. Also, if a spring has closer-spaced coils at one end, put the coil-heavy end at the seat, not the retainer. After the spring and retainer are compressed, install the keepers. A dab of grease on the keeper groove will hold the keepers in place. This will allow you to use both hands to release the compressor.

CVCC Auxiliary-Valve Assembly— With a CVCC head, you need to assemble and install the auxiliary valves. Oil the auxiliary-valve stem and slip it into the valve holder. Lay the spring washer on the valve holder with the side marked LOW facing down. Some washers are marked UP on their top sides. Set the spring seat on the spring washer, then slip the oiled valve-stem seal onto the valve holder. Set on the valve spring and retainer, compress the spring by hand or with a compressor and install the keepers. Finish by stretching a well-oiled O-ring into its groove in the valve holder.

The chamber collar must be installed in the head before the auxiliary-valve assembly. Select two new collar gaskets from the gasket set. Lay one of these in the head against the shoulder near the bottom of the hole for the auxiliary valve. Now, lay in the chamber collar. Most collars have two holes in their sides; one is round, the other oval. Other collars have one large round hole and multiple small holes. The round hole is for the sparkplug and should be aligned with the sparkplug hole. The smaller oval or multiple small holes are the passage(s) that feed the rich auxiliary chamber mixture into the main combustion chamber. Therefore, the hole(s) should face the hole in the roof of the combustion chamber.

Honda has a special tool for keeping the chamber collar aligned during auxiliary-valve installation, but it isn't necessary. All you need is a tapered punch that slides through the sparkplug hole, into the chamber collar and through the other side into the combustion chamber.

Drop in the chamber collar and insert the punch. Place the other chamber-collar gasket on top of the collar, then slide the auxiliary-valve assembly into place. Wiggle the auxiliary-valve assembly as you push lightly downward on it. This helps seat the valve assembly against the

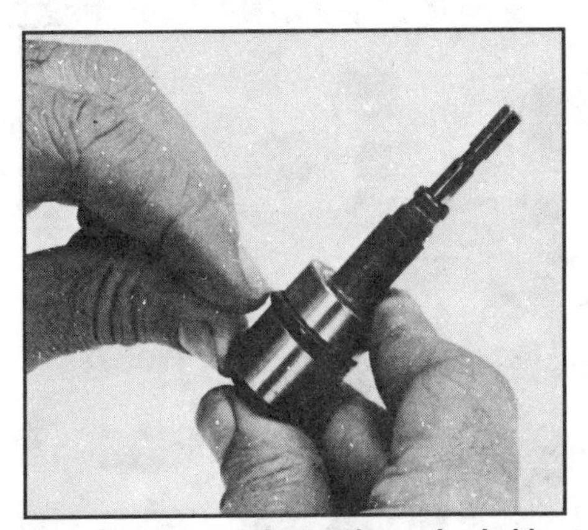

New O-rings *must* be used on valve holder. Reusing old O-rings leads to excessive oil consumption.

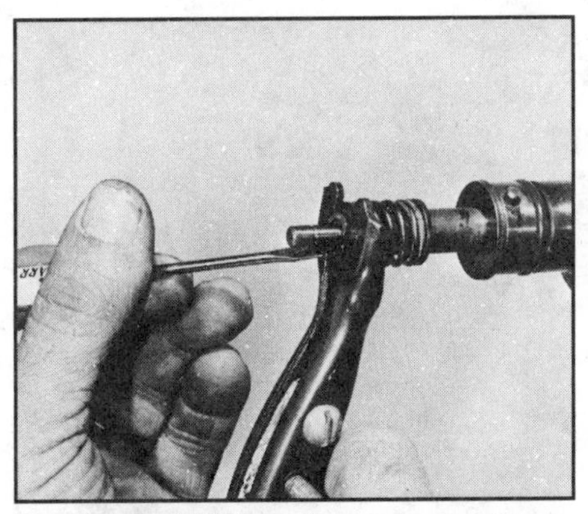

Remaining CVCC auxiliary-valve parts are most easily assembled with valve-spring compressor. First slip in valve, then spring washer, spring seat, seal, spring, retainer and keepers. It's also possible to assemble auxiliary valve with its holder in head.

Moly lube is almost same color as camshaft, so it doesn't show well in photo. Give each cam lobe an even coating, including auxiliary-valve lobes. *Do not lube cam journals,* yet.

Lube O-ring and slip complete CVCC auxiliary-valve assembly into head. Tang on valve holder fits through slot in head.

Pass CVCC valve-holder nut over assembly and thread into head. Torque nut to 50 ft-lb on pre-'80 1488s and 58 ft-lb on all other CVCC engines. If nut is improperly torqued, gaskets and O-ring will move, resulting in high oil consumption. Excess oil in auxiliary valve clogs it, requiring teardown and cleaning.

chamber collar without distorting the O-ring. When the auxiliary-valve assembly is down as far as it will go, thread the valve-holder nut into the head and tighten to 54 ft-lb. After torquing, remove the punch from the sparkplug hole. Install the three remaining assemblies the same way.

Install Camshaft—First, the camshaft must be well lubed. Although regular motor oil works well on bearing journals, it won't work on the cam lobes. A thin, residual coating of oil can't tolerate the extremely high pressures between the cam lobes and rocker arms during initial engine start-up. There isn't enough oil until the engine is *running* for a couple of seconds at fast idle. Between the time you first crank the engine and when full-volume oil is flowing to the cam is critical. Improperly lubed cam lobes will wear, setting up a pattern that will quickly destroy the cam—perhaps in less than an hour of engine operation.

Cam-lobe assembly lubricant must have excellent high-pressure qualities and be thick enough so it won't run off the cam. Any of the special cam lubes containing *Molybdenum Disulphide* (MoS_2)—*moly* will do. These compounds are specifically designed for camshaft lubrication during engine rebuilds. They are thick and grease-like, so they can't run off between the time the cam is installed and the engine started. Moly gives top-notch, high-pressure lubrication during initial start-up. Once oil flows, the moly is washed away and eventually is deposited in the oil filter and sump bottom.

Small cans or tubes of moly lube are available at parts houses. You don't need very much—a little goes a long, long way. Coat each cam lobe until it is completely covered with a thin, black film. Don't goop it on. Do not smear some on the journals, if you'll be Plastigaging them.

After moly-lubing the cam, oil the camshaft bearings in the head, then lay the cam in them. Pretty simple.

Rocker-Arm Assembly—Set the rocker assembly on its pedestals so you can reach the valve adjusters. Loosen the jam nuts and back the valve-adjusting screws all the way out to keep the rocker arms from pushing against the cam lobes during rocker-

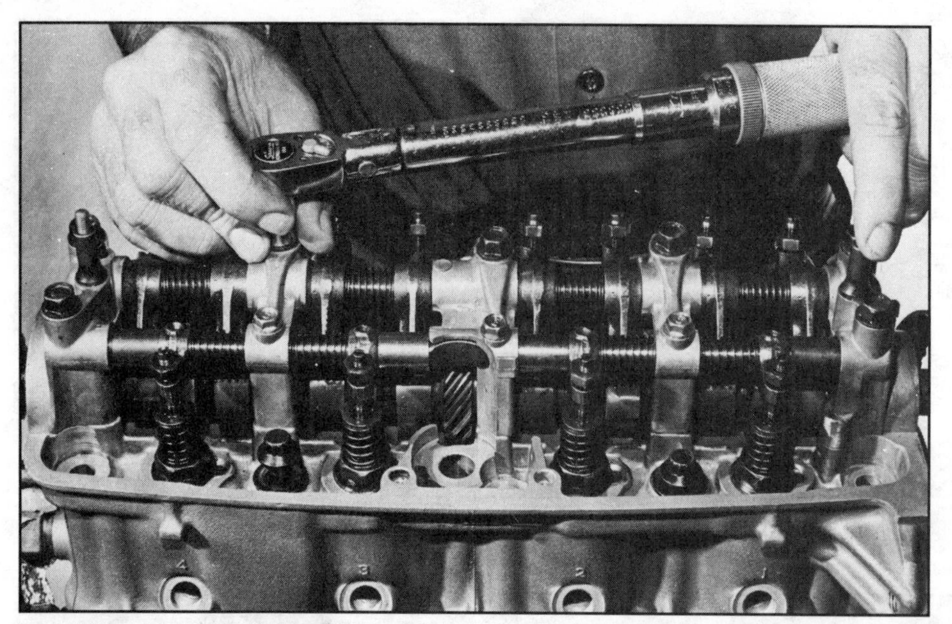

Lay strips of *fresh* Plastigage on cam journals, then position rocker-arm assembly. Start *all* bolts finger tight. Torque rocker-pedestal bolts with 12mm heads 16 ft-lb; bolts with 10mm heads 7 ft-lb. Start in center and work toward ends. Don't forget two screws on 1200-engine rocker assemblies.

Reverse torque sequence and remove rocker-arm assembly. Check Plastigage against scale on package. Clearance should be 0.002—0.006 in.

arm installation. This promotes more accurate hold-down-bolt torque and prevents valve-train binding. Besides, a valve job will totally change the previous valve adjustment.

By grinding the valves and seats, the valves will sit deeper in the combustion chambers. That is, more of the valve stem will protrude from the rocker-arm side of the head. This closes the old valve adjustment. If the valve seats were ground a lot, this change could be troublesome during rocker-arm installation. So, because you have to readjust the valves anyway, loosen the adjustment now to avoid any binding.

With all the adjusters loose, flip the assembly upside down and coat the rocker-arm rubbing surfaces with moly. Spread some oil in the upper cam-bearing bores, too.

Lay the rocker assembly on the head. Oil the bolt threads, thread them in and torque them. Torque small rocker-assembly bolts 10 ft-lb, the larger ones 17 ft-lb, except for 1200s. Torque all 1200 rocker-arm-shaft bolts 13—16 ft-lb.

For now, leave the valve adjustment loose. They are easier to adjust during engine assembly.

INTAKE & EXHAUST MANIFOLDS

Working on the intake and exhaust manifolds may not seem like part of the typical *valve job,* but it is a logical extension of head work, so I've included it here.

Intake Manifold—Manifold service is neatly divided into two categories: cleaning and replacement. After stripping the intake manifold of any emission-control equipment—EGR valve, air-suction valve, etc.—clean it. Soaking the intake manifold in carburetor cleaner or strong solvent will remove all but baked-on oil. Use a wire brush or blunt screwdriver to get in the corners. A nylon toothbrush does an excellent job without scratching, but because of its size, takes longer.

Clean as much of the intake runners as you can. And on CVCC engines, pay close attention to the auxiliary-circuit runners. These small runners can get blocked with oil crud. To solve this problem, rod out the runners as best you can with stiff wire. Then, flush the runners with plenty of solvent while working the wire around inside. It's best to finish this job with a dip through the cold tank to ensure you removed *all* the grit. Have the machine shop do it or, if you do a lot of car work, buy a five-gallon can of carb cleaner.

Because the manifold won't completely fit into a five-gallon can of carb cleaner, dip one end in the solution overnight. Remove the manifold and insert the other end for several hours. After it comes sparkling clean, rinse it off with solvent. Be very careful with

Reassemble rocker arms to head for final installation. If you didn't do so already, check end play with dial indicator. Checking end play with rockers loaded against cam lobes makes moving cam more difficult. Move cam toward timing-belt end by striking opposite end with a lead mallet. Mount dial indicator and knock camshaft rearward to read end play directly. Preferred range is 0.002—0.006 in. (0.05—0.15mm). Service limit is 0.020 in. (0.50mm).

carb cleaner. It's highly corrosive. Wear eye protection and don't splash it on yourself.

On 1200cc engines, and late-model CVCC engines with coolant-heated intake manifolds, give the intake manifold a good solvent cleaning and rod out its coolant passages. Use a rat-tail file and blunt screwdriver to remove as many mineral deposits as possible. Aluminum engines are prone to mineral-deposit buildup from the use of antifreeze formulated for cast-iron engines.

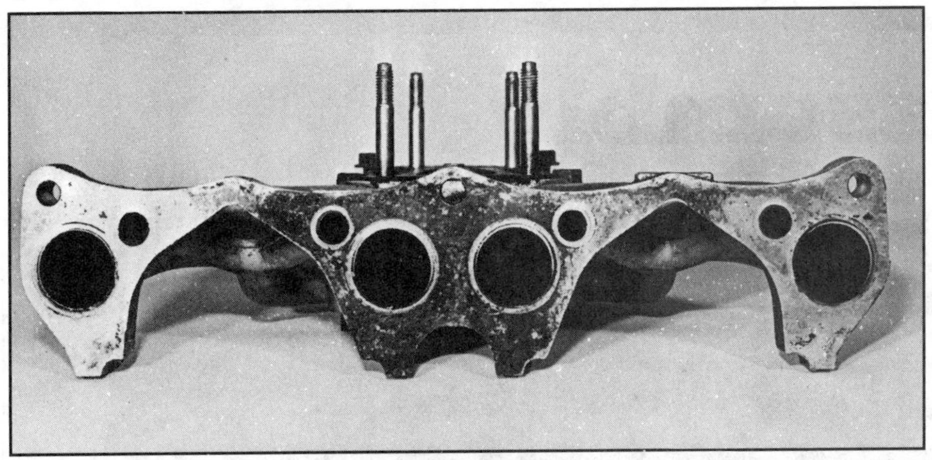

Don't forget to clean intake manifold. Small CVCC runners near main runners are prone to clogging. This is especially true of '75 1488 because its CVCC auxiliary runners have a 90° bend. Use pipe cleaners, gun-bore brushes and coat hangers to *decarbonize*—remove carbon from—auxiliary runners.

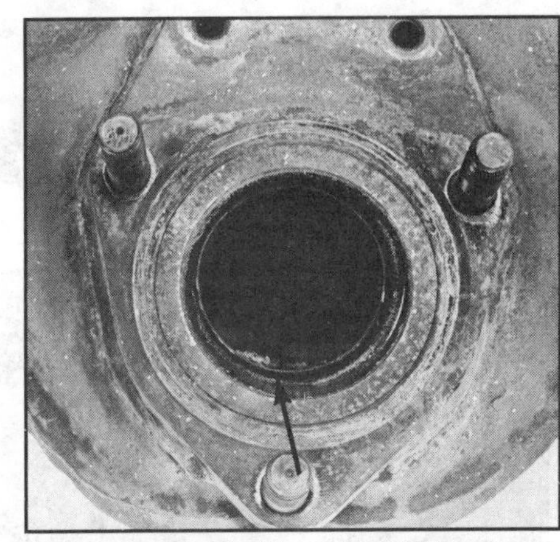

Look for cracks inside early CVCC exhaust manifold. Don't be fooled by sheet-metal ring just inside opening (arrow). Although it looks like a crack, it's supposed to be there.

Exhaust Manifold—Unlike the intake manifold, you don't have to worry about degreasing the exhaust manifold. Because it operates in excess of 1800F, oil and grease gets burned off. In most climates, an exhaust manifold gets a rusty scale buildup on its exterior. This leaves a rusty mess on whatever touches the manifold—wire-brush it off.

More importantly, check the exhaust manifold and pipe—still in the chassis—for cracks. On all but the 1200s and late-model CVCC engines, Honda exhaust pipes are double-walled. This double-walled construction is used to route some of the exhaust gases around the outside of the inner pipe. This keeps the inner pipe very hot, helping to burn hydrocarbon exhaust pollutants—kind of a poor-man's catalytic converter. That's why the exhaust manifold has that bulky, two-piece metal shroud around it and weighs so much.

Carefully inspect the outside of the bolted or welded shroud seam for cracks. A crack here allows exhaust gases to leak. You'll hear it as hot gases "spit-spit-spit" under the hood.

Baffling Question—Look inside the exhaust manifold through its outlet. The interior baffling is partially visible. If the baffling starts to crumble, pieces can block the outlet. This creates a massive flow restriction and will keep top speed below 40 mph or so. The manifold will have to be replaced. To get you by until you can get a new manifold, gut the old manifold with tin snips and needle-nose pliers. Pull out the baffle pieces through the exhaust runners and pipe-flange holes. Of course, the engine won't run quite as Honda intended until you replace the manifold.

Crusty exhaust residue in mating area of early CVCC exhaust manifold is caused by loose gasket. Scraping off grit, using a new manifold sealing ring and torquing nuts to specs solved problem. Broken piece of sheet-metal shroud hurts cold-weather driveability slightly, but doesn't cause exhaust leak.

Pipe Inspection—Also, on early CVCC models, inspect the exhaust-header pipe in the chassis. This is the *A-pipe* in Honda parlance. If the A-pipe is cracked around its throat, exhaust gases will shoot into the hot-air system and be sucked into the carburetor through the hot-air hose connected to the air filter. Trouble is, the exhaust gases make an absolute mess out of the carburetor, CVCC intake runners and auxiliary valves. That's why they get so dirty. And, the hot, oxygen-starved gases murder engine performance.

You must replace a cracked exhaust manifold or A-pipe. These parts are very expensive, but it's smarter to pay for them now than pay for cleaning up the carburetor, intake manifold and auxiliary valve, *as well as* paying for the exhaust parts later.

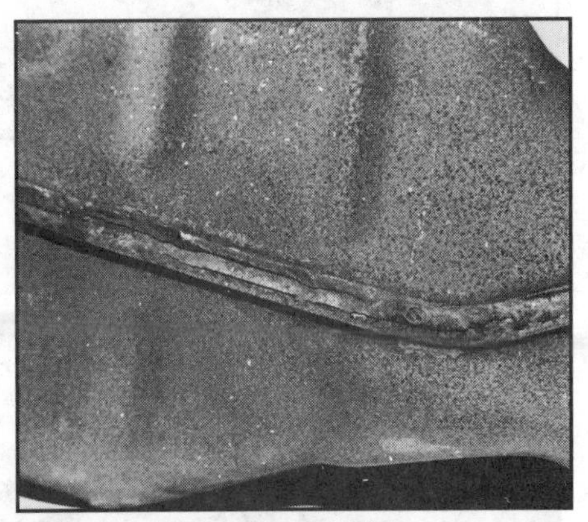

Examine early CVCC exhaust-manifold seams for cracks. If manifold is cracked, it is usually disintegrated internally as well. Replacement is cure.

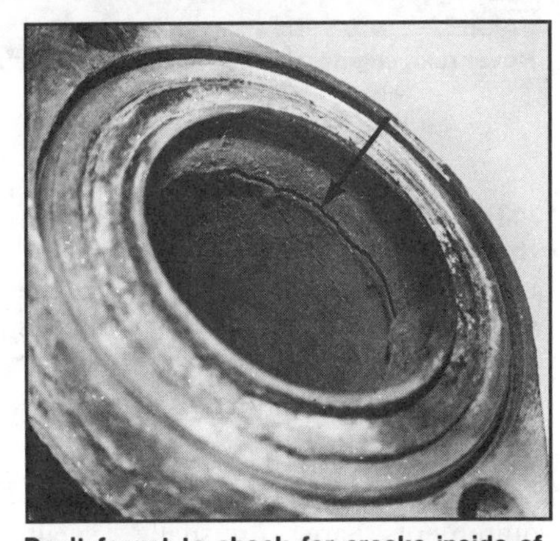

Don't forget to check for cracks inside of A, or header, pipe. This little crack (arrow) can destroy the A-pipe, carburetor, intake manifold and auxiliary valves because hot exhaust gases are drawn into induction system.

Engine Assembly 7

Never rush engine assembly by skipping steps. All bearing and sliding surfaces should be liberally coated with engine oil.

From here, it's all downhill. All that's left to do is bolt your engine back together and install it in the chassis. Most of the running around is done and the parts are clean. It's time to put your engine together.

THINGS TO HAVE

Sealer—A Honda engine's sealer requirements are unusually small. You'll need room-temperature vulcanizing—RTV—silicone sealer and a non-hardening sealer. RTV is a liquid that cures, or vulcanizes, when exposed to the atmosphere. It does a great job, and can actually replace gaskets in certain applications. About the only bad thing about RTV is it doesn't get tacky enough to hold gaskets in place when assembling parts. With Honda engines, however, this isn't a problem. In fact, with so many preformed rubber gaskets and circular oil seals, there are few applications that require sealer.

Non-hardening sealer will seal metal-to-metal applications, such as core plugs. It's also great as a lube for installing O-rings, as on the water pipe. For conventional paper gaskets, RTV is preferred only because it is not as messy.

Oil & Grease—In addition to the sticky stuff, you'll need lubricants. First on the list is a quart or more of the engine oil you'll use in your new engine. Some people are partial to one particular brand of motor oil—that's OK. The important part is that oil for your Honda should carry at least an *SE designation.* If you use an *SF* oil, that's even better.

Under no circumstances should you break-in your engine on non-detergent oil, and then switch to a modern, detergent oil. The rules of 30—40 years ago don't hold true; today's rings, cylinder walls and other internal engine parts break-in quickly. There's no need for some trick engine oil for break-in. Use SE or SF for break-in and forever after. To lubricate parts as you assemble them, fill your squirt can with fresh motor oil of the same type you'll use in the engine.

In addition to motor oil, get some *moly* grease. You probably used some of this for cylinder-head assembly. A small tube of this cam-saving grease will suffice.

Another special lubricant you'll need is a small tube of anti-seize compound. This thread lubricant is great for head bolts or studs. Iron-block engines can get by without anti-seize if you're cutting corners, but it's a must on aluminum blocks. High-temperature applications like the exhaust manifold need anti-seize, too. In fact, anything that threads into aluminum, like sparkplugs, can use a dab of anti-seize. It helps prevent ruining the aluminum threads upon fastener removal.

Tools—Following is a list of specialized tools you'll need for engine assembly. Round them up if you don't have them at hand.

Torque Wrench—First on the list is a torque wrench. I've mentioned this tool before, so I won't go into a full description again. You must realize that there's no way you are going to assemble an engine correctly without one. Regardless of how experienced or talented a mechanic may be, there's no human way to "feel" the correct amount of force necessary to torque a nut or bolt to the *exact specification* required.

Try to get a torque wrench with a relatively low range. You are interested in torquing bolts from 10—75 ft-lb. A 30—150-ft-lb wrench is typical, but a little high for your needs. You may want an *inch-pound*—in-lb—torque wrench. If so, get a large one; at least 800 in-lb or more. It is possible to use a ft-lb torque wrench to arrive at in-lb values—just multiply them by 12. For example, 20 ft-lb equals 240 in-lb. To convert in-lb to ft-lb, divide by 12; 540 in-lb equals 45 ft-lb. Remember that torque wrenches are less accurate at low and high extremes of the scale.

Ring Compressor—To install a piston, the rings must be squeezed together so they will go into the bore. You'll need a ring compressor for this chore. Plier- and cylinder-type compressors are equally popular. The plier-type compresses a C-shaped sheet-metal band around the rings. The cylinder-type wraps completely around the piston and tightens with a ratchet mechanism. Some mechanics prefer the plier-type because it doesn't bind like the cylinder-type, plus it's quicker. Use whatever tool is available.

Plastigage—You *must* check bearing clearances on Honda engines. It's too easy to end up with the wrong bearings, or mix them up during installation. Also, if the crankshaft was ground and rehardened, checking the clearances with *Plastigage* will reveal any machining problems. Plastigage is a precision-thickness wax, and is sold in foot-long lengths in auto-parts stores. Purchase the green variety, 0.001—0.003 in. You don't need much; one strip is plenty.

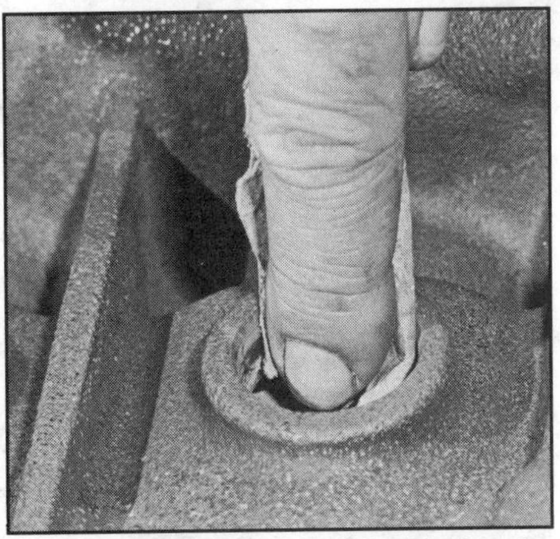

Prep core-plug bore by removing old sealer with sandpaper.

Honda core plugs are 30mm. Brass and steel plugs are available. Brass plugs are more expensive, but won't rust like steel.

Make sure the strip you purchase is *fresh*. Plastigage that's been laying around for a year or two will lose its *plasticity*. Then, it will crumble like dried-out wax and give inaccurate readings. Resulting readings will indicate more-than-actual clearance.

Ring Expander—A tool that isn't mandatory, but one that makes life easier and potentially much less frustrating, is a ring expander. It's actually a specialized pair of pliers that holds and spreads—expands—a piston ring so it can be slipped over a piston and into its groove.

The advantage of a ring expander on Honda-size pistons is the reduced risk of breaking a ring. It is all too easy to break a ring by hand. A ring expander is also more pleasant to use than gouging your thumb tips against ring tension.

Soap—Finally, your hands must be clean when assembling the engine. Have some abrasive soap and waterless hand cleaner nearby to keep your hands absolutely clean. Otherwise, dirt on your hands will end up being circulated throughout the engine's lubrication system.

The shop and tools must be clean, also. It makes no sense to practically sterilize all the engine parts, then assemble them in a pigpen. At least, clean the bench, wiping up all oil and grit. Also, wipe the tools with a solvent-soaked, clean rag. Check all sockets for grit hidden in the hex.

Paper Towels—Rags may be OK for the bench and tools, but only paper towels should be used to wipe anything on or in the engine. Period.

BLOCK PREPPING

Bring the cylinder block out from under the bench and uncover it. If you have an engine stand, mount the block on it. Otherwise, place the block on the bench. Inspect the block for cleanliness.

Core Plugs—The 1170, 1237 and 1335cc aluminum-block engines have no core plugs in the cylinder block. On all other engines, the core plugs should be replaced, *now*. It's cheap and easy, especially compared to the pain of changing them with the engine in the car. Don't be fooled by good-looking core plugs. They rot from the inside out, so by the time they look bad they're on the verge of leaking.

Core plugs can be bought separately or in a kit. Get the kit if you can; it's easier and cheaper to buy the plugs in a set. Wipe the core-plug bore in the block with abrasive paper, removing all old sealant and rust. Coat both the bore and core-plug outer edge with non-hardening sealer. Place the plug into its bore and drive it in. To evenly distribute the hammer force, use a punch or socket that just fits the core plug without being larger than its OD. The hardest part is starting the core plug. If the plug cocks in its bore, tap it sideways, remove it and start over

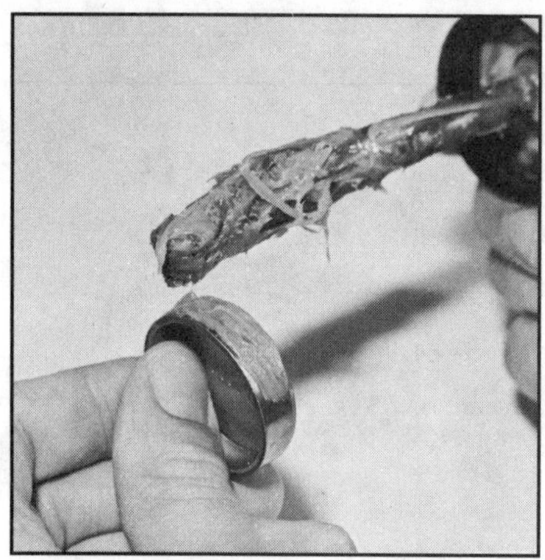

Coating periphery of plug with non-hardening sealer helps installation and sealing.

Use a 15/16-in. or 30mm socket and short extension to drive in core plugs.

Bearing bores must be perfectly dry for bearing installation. Wipe bores with paper towels. Shop towels or cotton rags leave non-oil-soluble lint.

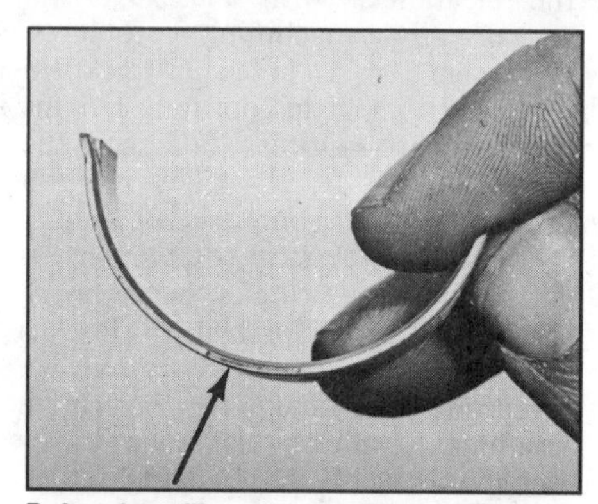

Before installing them, check Honda bearings for proper color code. Color code is on bearing edge (arrow).

again. It must go straight in.

The plug is fully seated when the outer lip is flush with the block surface. Wipe off excess sealer and install the next plug. There are seven core plugs in a cast-iron block; three on the water-pump side, two on the oil-filter side and one at each end. There are no replaceable core plugs in the head; CVCC or crossflow.

Oil-Gallery Plugs—If you opened the oil-gallery plugs for cleaning, replace them now. Use non-hardening sealer on the plug threads, then screw them in until good and tight. On cast-iron blocks, one plug is in the flywheel area; another in the timing-belt area. Oil galleries in aluminum-block engines are in the main-bearing-cage assembly. It's the cast section running the length of the assembly. Look at each end of this section for an Allen-type socket-head plug.

CRANKSHAFT INSTALLATION

It's time to *lay the crank*. You need the crankshaft, main bearings, caps, bolts and Plastigage. Place the block upside down so the main-bearing bore is up.

Start by installing the main bearings. Wipe the bores with a paper towel so they are perfectly clean and dry. Do the same to the other half of the bore in the main-bearing caps. Install the first main-bearing insert in the block. Make sure you have the correct-size or -color insert for that number bearing. Both top and bottom bearing halves are grooved, so you don't have to worry about the top and bottom bearing halves being different.

Before installing a bearing, wipe its back side clean. There should be no dirt or lubricant between bearings and their bores. Align the bearing tang with the notch in the block. Use your thumbs, one at each end of the bearing, to push it into the block. Install the other bearing-insert half in the main-bearing cap. Do all five main bearings.

Alternating between block and cap helps keep the bearings paired, especially with Honda OEM bearings. Because Honda packages their bearings two to a plastic bag—most of the time—this means there will be only two bearings available for installation at any time. As a result, this reduces the likelihood of slipping a blue bearing in the block and a red bearing in the cap.

Because the crankshaft and bearings are going to be Plastigaged, *do not* lubricate the bearing inserts just yet. Get the crankshaft out of storage and unbag it. With paper towels, wipe off the protective oil from the bearing journals. Gently lay the crankshaft on its bearings. You'll find the crankshaft easier to work with if you rotate it so the counterweights are level with the sides of the block and not sticking up and out of the crankcase.

Slip two thrust-washer halves into their recesses at the number-4 main bearing. The flat side goes against the block; grooved side faces the crank. Coat the washers with motor oil to keep them in their recesses until the main caps are bolted on.

Carefully open the Plastigage package and snip off a piece long enough to stretch the length of a main-bearing

Use both thumbs to push bearing inserts into place. Fully grooved insert halves are the same. Tang in bearing must align with recess in bearing bore. Don't lube bearings or journals until after crank has been Plastigaged.

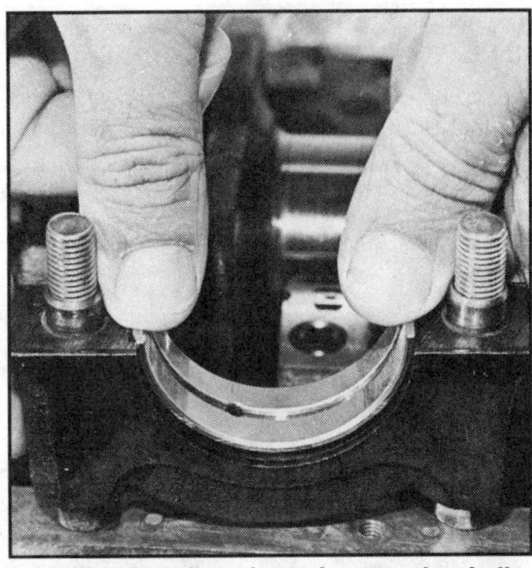

Installing bearings in main caps is similar to those in block. Wipe bore and back of bearing dry with paper towel, then install with thumbs. Honda bearings are fully grooved.

Oil the bearing-cap bolts. A very light coat of oil is plenty.

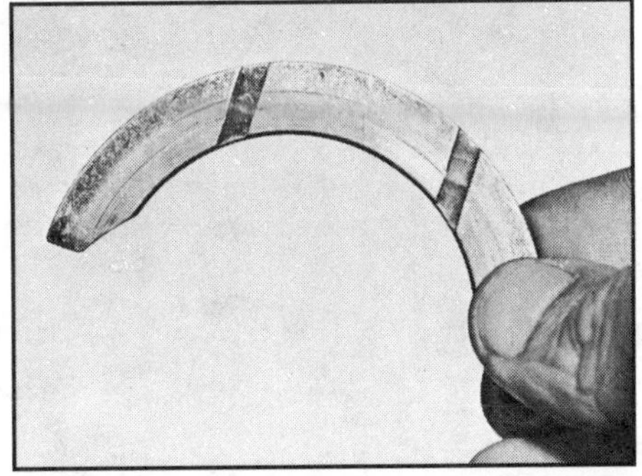

Don't forget two upper and two lower thrust washers. Remember that 1200s use *no* thrust washers against bearing cage. This side, with two grooves, faces crank. Back side is plain and butts against block.

Install each upper thrust washer by placing it against crank in the lower position, then rotate it into block. *Do not rotate the crank!* Oiling block and crank beforehand keeps thrust washers from sliding around. If any oil gets on crank journals, wipe it off before applying Plastigage.

journal. If there's oil on the journals or bearings in the caps, wipe it off. Lay the wax onto the journal, avoiding any oil holes. Continue until all main-bearing journals have a Plastigage strip. Double-check that all wax is in place, then carefully set all main-bearing caps on their journals, and thread in the bolts.

When installing main-bearing caps, keep several points in mind. First, *don't oil the bearings*—oil dissolves Plastigage. If any gets on your carefully laid strips, they will yield false readings and you'll have to repeat the process. Remember to keep the caps in order and properly positioned when installing them. On the aluminum-block engines, you have no choice because of the cap design. But on cast-iron blocks, each individual cap must go in its own register and in the correct position. Use the cast-in numerals as reference. As for right and left orientation, the bearing tangs are your best guide. When correctly installed, the bearing tangs of the upper and lower bearing halves will be on the same side.

As you work with the main caps, avoid rotating the crank in any direction. If you do bump the crank, it will tear and smear the Plastigage, ruining your readings. You'll have to remove the main caps and start over.

When torquing the main-bearing caps, tighten them in progressive steps, such as 10, 20, 30 and 35 ft-lb. Lightly oil the bolt threads to help prevent binding. On aluminum-block

Snip off fresh piece of Plastigage (arrow) and lay it full length of each main-bearing journal. Plastigaging *must* be done at room temperature. Too cold and Plastigage will indicate excess clearance; too hot and it will indicate too little clearance.

Install main-bearing caps using cast-in arrows and numerals for order and orientation. Torque main caps to 29 ft-lb on aluminum-block engines; 33 ft-lb on 1488s, 1600s; and 48 ft-lb on 1751s.

Remove each main cap or aluminum-block main-bearing-cage assembly to check oil clearance. The 0.002-in. oil clearance shown is at high end of Honda's 0.001–0.002-in. specification. If clearance is out of spec, use chart on page 71 to select Honda color-coded bearing required. Lift out crank, oil bearings and reinstall.

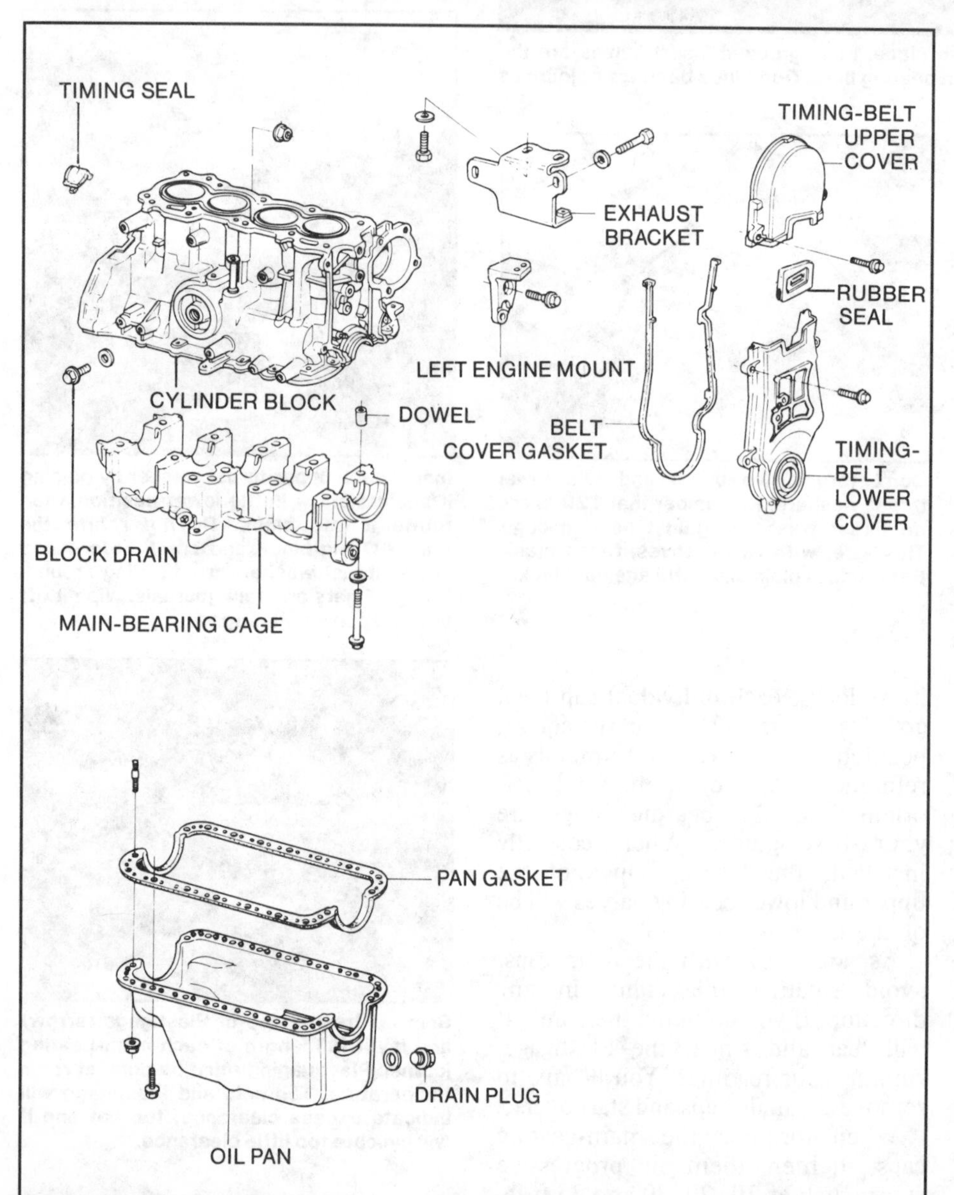

Exploded view of engine block, bearing cage and related components; 1335 shown, 1200 similar. Courtesy of American Honda Motor Co., Inc.

engines, torque to 27–31 ft-lb; 1488s and 1600s, 30–35 ft-lb; 1751s, 44–51 ft-lb.

Once all caps are final torqued, remove them. Again, avoid bumping the crankshaft. Carefully rock the caps off their dowels and lay them in order on the bench. The squished Plastigage will stick to the journal or bearing. Either is OK. Use the scale on the Plastigage sleeve to measure the wax.

Oil Clearance—Standard clearance for cast-iron blocks is 0.001–0.002 in. (0.025–0.050mm) with a 0.003-in. (0.075mm) wear limit. Aluminum blocks are slightly less at 0.0009–0.0017 in. (0.023–0.043mm), and have a 0.0028-in. (0.071mm) wear limit. As you can see, these limits leave a lot of leeway considering the 0.0002-in. (0.005mm) difference in Honda bearings.

What it boils down to is this: An oil clearance of 0.0005–0.0035 in. (0.013– 0.089mm) will work. That's a wide margin, but 0.002 in. (0.05mm) clearance is best.

The larger oil clearance—0.0035 in.—will work *for a while.* The oil pump is capable of maintaining oil pressure provided everything is OK. But the moment the oil screen becomes a little clogged, the oil pump can't supply the required large volume of oil—the bearing journals will score. Anything over 0.003 in. (0.075mm) is undesirable for this reason. Also, oil clearance increases with engine operating time. Therefore, the engine will last longer and have a much larger safety reserve at the tighter oil clearance.

At the other end of the spectrum, 0.0005 in. is too tight. At 0.0005 in. oil clearance, there isn't much room for oil and the journals run the risk of scoring. If the engine gets hot from extended operation—you drive in stop-and-go traffic in very hot summer weather—the crankshaft and block will expand and choke off this already tight clearance. Don't set oil clearance below 0.001 in. (0.025mm).

That leaves the 0.001—0.003-in. range. A racer will set oil clearance at 0.003 in. to allow for heat expansion and maximum oil flow. Because oil absorbs considerable engine heat, it's best to get maximum flow without causing an oil-pressure loss. But for the street, the engine will enjoy faster lubrication-system priming and have a larger safety reserve at 0.002 in.

If you didn't grind and reharden the crank, you can shuffle Honda OEM bearings to get oil clearance *exactly* at 0.002 in. For example, the crankshaft and block markings may call for a brown bearing, but Plastigaging indicates a 0.0015-in. oil clearance. That's a completely acceptable clearance. But if you like to tinker, you could switch to a red bearing and get oil clearance closer to 0.002 in.

Of course, a ground crankshaft requires aftermarket bearings. You are stuck with whatever oil clearance the journal size and compatible aftermarket bearings give. If the crank was correctly ground, you'll get the desired clearance of 0.001—0.002 in. But, if oil clearance is tight, try polishing the crank. A light polish job is good for 0.0005 in. or so; heavy polishing can remove 0.001 in. Remember, if you take your crank to the machinist and say, "Polish off 0.0004 in., please," he's not going to give any guarantees. On the other hand, if oil clearance is excessive, take your genuine complaint to the crank grinder with your findings. See if different bearings, crankshaft or grind will help.

Final Crankshaft Installation— Once you've arrived at the proper oil clearance, it's time to install the crankshaft—for good. First, wipe off any Plastigage residue with paper towels.

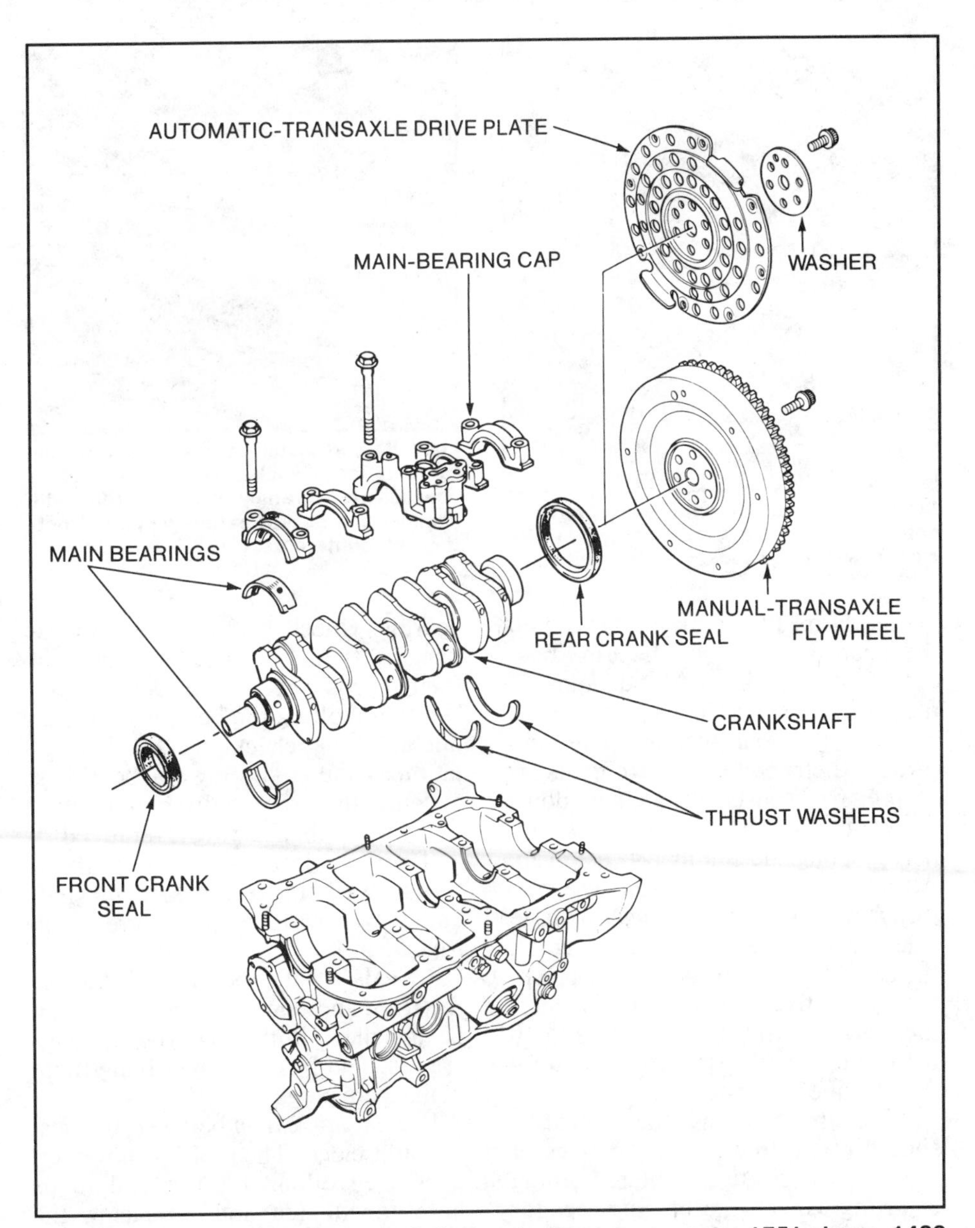

Exploded view of engine block, crankshaft and related components; 1751 shown, 1488 and 1600 similar. Courtesy of American Honda Motor Co., Inc.

After oiling main journals and torquing main-cap bolts, crankshaft should rotate freely without binding.

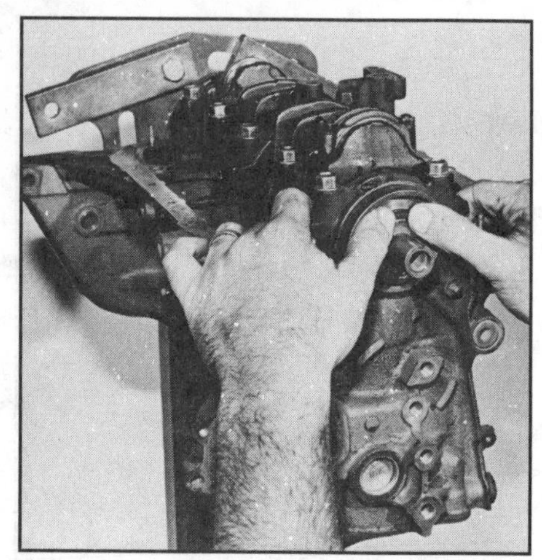
Lube front-seal ID with motor oil and square it in its bore. Carefully tap it in with a deep-well, large socket or pipe section.

Lube ID and install rear seal as you did front. If large socket or pipe section is unavailable, carefully drive in seal with hammer and socket extension until it bottoms in counterbore.

Lightly oil the crank journals, then the bearing inserts in the cap. Take care *not* to oil the rod journals now—they haven't been Plastigaged! Position the main-bearing caps and torque their bolts in increments as before. On iron-block engines, don't forget the other two thrust-washer halves alongside number-4 main cap. Aluminum-block engines use *no* thrust washers on the bearing-cage side. The cage itself takes crank thrust. To even stress on the block, torque the five caps together, in increments of 10 ft-lb. Bring all caps to 10 ft-lb, then to 20 ft-lb and so on until final torque is reached.

After final torquing, the crankshaft should rotate freely in its bearings. If it binds or won't turn at all, something is wrong. Check the bearing caps for proper orientation and position. If you didn't Plastigage the bearings, you may have installed the incorrect-size bearing. Check bearing size by Plastigaging them, or check the back of their shells for a size designation. When unbolting the bearings caps, do one at a time, checking crankshaft rotation as you go. If only one bearing is at fault, the crankshaft should rotate freely when that bearing is removed.

If you replaced the crankshaft or thrust washers, recheck crankshaft end play now. The procedure is on page 55.

Oil Seals—With the crank installed, the front and rear crankshaft seals can be installed. Both ring-type lip seals are installed by carefully driving them in like core plugs.

Use *no lubrication* between the seal OD and block. The seal ID, however, *does* need oiling. Fit the seal to its bore, open side and lip facing the block. The flat, closed side of the seal with ID and size numbers molded into it faces away from the block. With *light* hammer taps, drive in each seal. The trick is to start the seal straight, and drive it in square. If it gets cocked in the bore, lever it out, and buy a *new* seal.

To drive the seal straight in, tap it alternately at 12, 3, 6 and 9 o'clock. Keep moving around the seal with the hammer. Don't get anxious; the seal will eventually start in. Once the seal is started, place a socket between the hammer and seal to distribute the blows. The seal is fully installed when its outer edge is flush with the edge of the block. Additionally, you'll feel solid resistance through the socket when the seal bottoms against the block.

If the block is on an engine stand, you probably won't have access for installing the rear seal. That's OK. There is no need to install the rear seal until *before* you install the flywheel or driveplate. Therefore, wait until the block is off the stand to install the rear seal. Don't forget it!

PISTON-RING CHECKING & FITTING

You should have assembled the pistons and connecting rods by now. If you haven't, refer back to page 80 and get them together.

With the connecting rods and pistons assembled, prepare to install the rings. As with the bearings, today's rings are the result of millions of dollars of research and development. Its rare that will you have any trouble with rings, but you should still check them.

End Gap—The most important ring check is end gap. It must not be so

RING END GAP
in.(mm)

Engine	Year	1st Compression	2nd Compression	Limit	Oil	Limit
1170	73	0.008-0.016 (0.20-0.40)	0.008-0.016 (0.20-0.40)	0.023 (0.60)	0.008-0.035 (0.20-0.90)	0.043 (1.1)
1237	74-79	0.010-0.016 (0.25-0.40)	0.010-0.016 (0.25-0.40)	0.023 (0.60)	0.012-0.039 (0.30-0.90)	0.043 (1.1)
1335	80-83	0.006-0.014 (0.15-0.35)	0.006-0.014 (0.15-0.35)	0.023 (0.60)	0.012-0.035 (0.30-0.90)	0.043 (1.1)
1488, 1600	75-79 76-78	0.008-0.016 (0.20-0.40)	0.008-0.016 (0.20-0.40)	0.023 (0.60)	0.008-0.035 (0.20-0.90)	0.043 (1.1)
1488	80-83	0.006-0.014 (0.15-0.35)	0.006-0.014 (0.15-0.35)	0.022 (0.55)	0.012-0.035 (0.30-0.90)	0.043 (1.1)
1751	79-83	0.006-0.014 (0.15-0.35)	0.006-0.014 (0.15-0.35)	0.023 (0.60)	0.012-0.035 (0.30-0.90)	0.043 (1.1)

small that the ring ends touch or so large that power is lost. If ring ends touch, the ring may break and score the bore. If end gap is excessive, combustion-chamber pressure will be lost into the crankcase.

Checking End Gap—Measuring ring end gap is easy. Remember the bore-taper measuring method computed from end gap, page 65? This is the same check, only simpler. Select a top ring—also known as the *first compression ring*—from the ring pack and place it in its bore. Each ring set should be kept with the bore it's checked in. Use the top of a piston to push the ring down squarely in the bore.

If you didn't rebore the block, push the ring down to the bottom of the bore. Bore diameter will be least there because of little or no bore wear. Consequently, the end gap will be smallest. Use your feeler gages to measure end gap.

All 1979-and-earlier iron-block engines use a 0.008—0.016-in. (0.2—0.4mm) end gap. The service limit for these engines is 0.023 in. (0.60mm). From 1980 on, both cast-iron and aluminum engines use a 0.006—0.014-in. (0.15—0.35mm) end gap with a 0.022-in. (0.55mm) service limit. Earlier aluminum-block engines differ. The 1170cc engine uses the same ring end gap as all other pre-'80 engines. All 1237cc engines use a slightly wider 0.010—0.016-in. (0.25—0.40mm) gap with a 0.023-in. (0.60mm) service limit.

The above tolerances are pretty wide. If possible, keep end gap around 0.010—0.012 in. (0.25—0.30mm). This narrower gap will keep power losses to a minimum. A wider gap gives more overheating insurance, so there are advantages there, too. For a street engine, small differences in end gap are not critical.

The above figures are for the top two rings—the compression rings. The bottom ring is the oil-control ring. Its end gap is less critical. Anything between 0.012 and 0.035 in. (0.30—0.90mm) is fine. Honda specifies a service limit of 0.043 in. (1.1mm).

Measure the gap on all rings for each cylinder. Besides verifying that the rings are correct, checking end gap also double-checks any cylinder boring you might have had done.

Setting End Gap—If ring gaps are excessive, you probably have the wrong ring set. Exchange it for the correct one. If the gaps are too small, they can be increased by filing. You must have a file fine enough to do the job without cutting gouges in the ring end. What you need is a small *smooth, double-cut* flat file. Secure the file in a vise or clamp it to the edge of the workbench with Vise-Grips or a C-clamp.

By clamping the file, you'll have much better control. Pass the ring over the file, making all cuts from the outside of the ring to the inside. File one end only. Filing from the outside in keeps a large burr from forming and prevents moly rings from chipping off at their outside edges. It takes little filing to get the ring to size, so don't file too much without checking your progress. After filing a little, deburr the ring end and recheck gap in its bore.

If you are lucky enough to be working in a machine shop, or just like buying tools, you should be aware of a special rotary file for gapping rings. It doesn't do a better job than you can do with a fine file; it just does it faster and easier. You'll still need a fine file for deburring the ring ends.

Finish the job by *breaking* the sharp edges formed by filing. The outside edge of the ring probably doesn't have a sharp edge, but the inner edge and sides will need a *light* cleaning up with a *very fine* file or 400-grit sandpaper. Go lightly. Just give the inner edge a pass or two with your fine file or sandpaper. When your finger cannot feel any roughness at the ring-end edges, it's done.

Check end gap one ring set at a time, then move on to the next. Remember to keep each set of rings with the bore it was measured in, or the gaps will not be as checked when installed. The paper pack the rings come in is an excellent container for keeping rings organized; simply mark the cylinder number on each package.

Also, don't confuse the top and second rings in the set. They look alike, but are different. By working with only one ring at a time, you can avoid mixing them up. If you do get mixed up, check the marking near the end of the ring. These marks are there to designate the top of the ring and are usually different for the top and second ring.

If you bought pistons with rings already fitted, don't be lazy. Remove at least the top ring from each piston and check its end gap.

If end gap is insufficient, clamp file in vise and file ring end. Hold ring square to file and file only in direction shown. Recheck end gap frequently. After filing, deburr ring ends with light stroke of file on four corners of filed end.

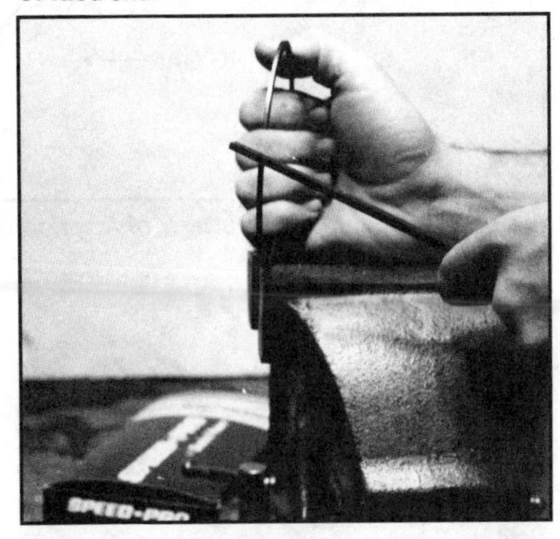

INSTALL PISTON RINGS

Now that the rings are gapped, clean them thoroughly before putting them on the pistons. I've already mentioned the ring expander. If you were thinking of getting one, now is the time.

Steady the rod/piston assembly in a vise. Except for holding the rod between your knees while sitting, this is about the only workable way of holding the piston motionless so you can install the rings. When you clamp the rod in the vise, make sure there are *soft jaws*—brass or aluminum—or two pieces of wood on each side of the rod. Never clamp a rod directly in a vise. You'll cut a sharp groove in the rod beam, creating a starting point for a stress-induced crack. Using a rag between the vise jaws and rod does little good. Position the assembly so the piston skirts sit on top of the jaws. This will steady the piston so it

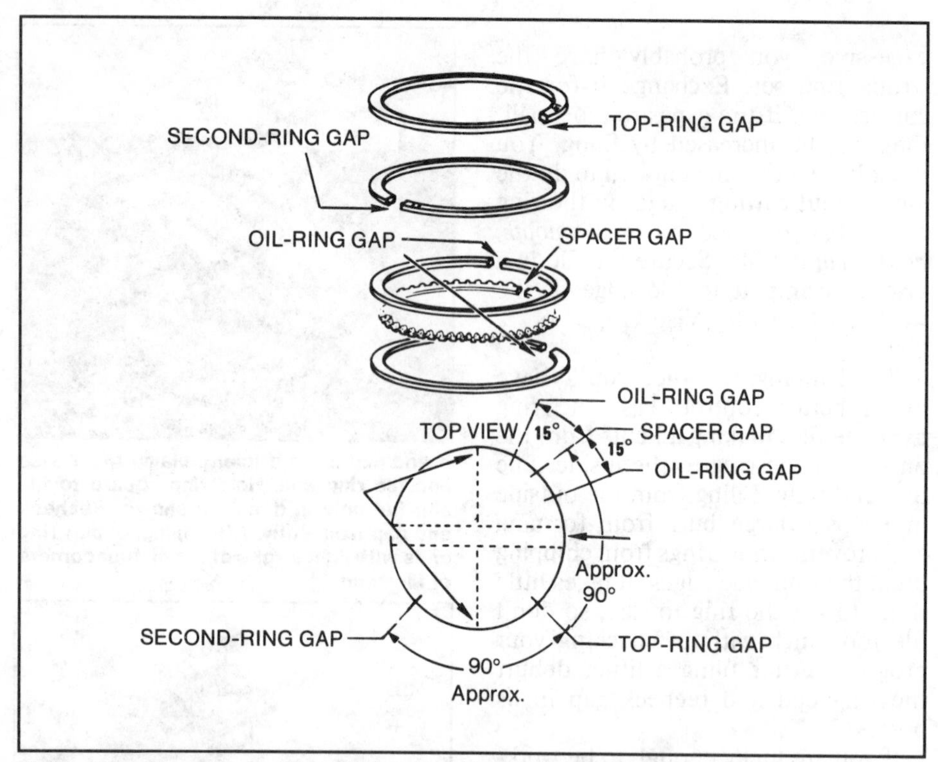

Piston-ring alignment. Courtesy of American Honda Motor Co., Inc.

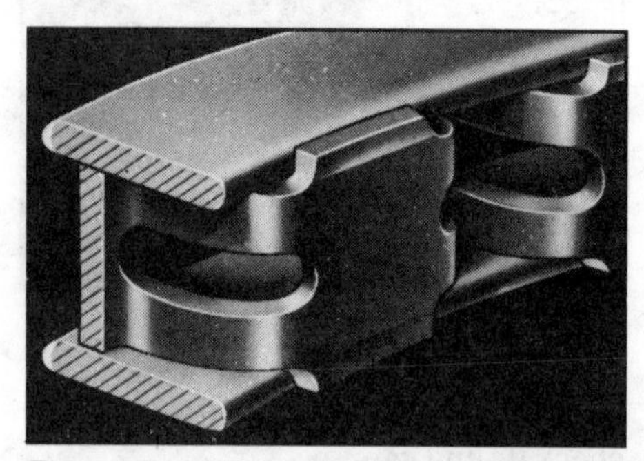

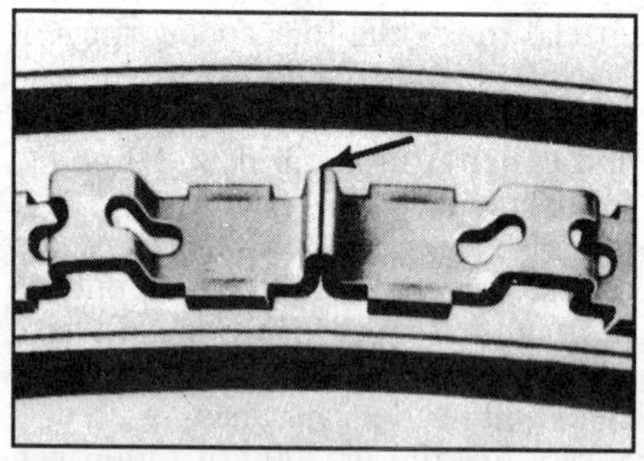

Three-piece oil rings have two scraper rails installed over expander/spacer. It is essential that the expander/spacer ends butt (arrow), not overlap. Otherwise, you'll end up with a scored cylinder wall. Drawing courtesy of Sealed Power Corporation.

doesn't wobble back and forth. Don't clamp the piston!

If you don't have a vise, buying one would be money well spent, even long after your Honda is back on the street. If you don't want to buy one, secure the rod to a bench with a C-clamp or have a friend hold the rod/piston assembly while you install the rings. Believe me, using a vise is *much* easier.

Install Oil Rings—Rings are always installed over the top of the piston from the bottom up, so the oil ring goes on first. As you know, the oil ring is a three-piece assembly: two side rails, or scrapers, and an expander/spacer. Each piece must be installed separately. Study the ring-placement illustration so you'll get the end gaps in the right place before starting. All oil-ring components are light and handle easily, so you won't need the ring expander for these.

However, three-piece oil-ring components are *not* installed from the bottom up. Because the middle section, the expander/spacer, has tabs that lock behind the top and bottom oil scrapers, the expander/spacer is installed first. Then, you install the *top* scraper, and finally, the *bottom* scraper.

Select the oil-ring expander/spacer or the thin-gage, wavy ring. Fit it in the bottom ring groove with its ends positioned according to the above illustration.

Next on is the top rail, or scraper. This thin ring fits between the expander/spacer and the groove. Because the rail is so flexible, you can fit one end in the groove and hold it

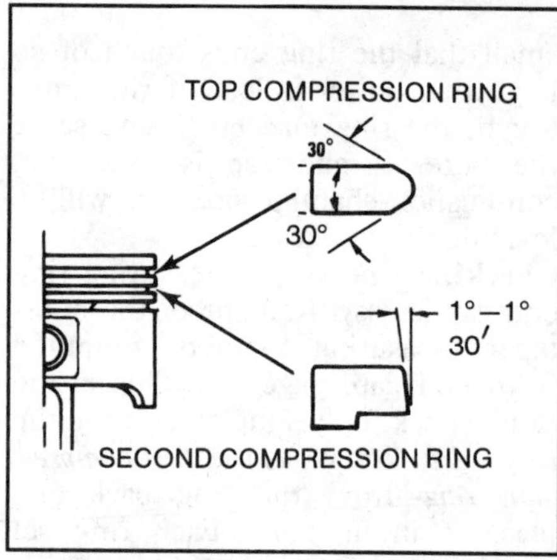

Compression-ring identification. Courtesy of American Honda Motor Co., Inc.

there with your thumb. Then, insert the rest of the scraper by running your finger around its edge.

The object here is to keep from scratching the piston with the end of the scraper. Hold the end of the ring away from the piston with your thumbnail. Sometimes, you'll get your thumbnail caught between the piston and ring, but that's better than scratching the piston. Install the bottom rail, or scraper, in the same manner.

Once the bottom scraper is in place, check that the ends of the expander/spacer don't overlap. Remove the scraper and start over if the expander/spacer ends overlap. If you don't, severe bore damage will result.

Install Compression Rings—Although they appear to be simpler than the oil ring, compression rings have hidden design features that you must be aware of for proper installation. For example, an oil ring has no top or bottom; compression rings do. The top and bottom are a result of the *twist* built into a compression ring. Twist is the angle the ring sits in its groove. Although twist is virtually invisible, it's there to help the ring seal to its grooves.

So you'll know which side is up, the top side of a compression ring is marked with some sort of dot or indentation near the gap. Honda top rings are marked 1R and second rings 2R. The mark must face up when the ring is installed, or twist will be going the wrong way. That could result in excessive blowby, power loss, increased oil contamination and a host of other bad things. You'll find an in-

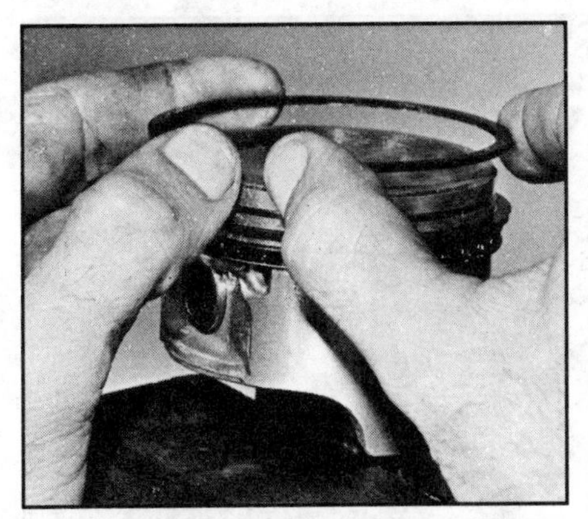

If you don't have a ring expander, install rings by hand. Spread rings with your *thumbs only*—levering against fingers may break ring. Spread ring only enough to clear piston and keep ring ends from gouging piston. I didn't do it here, but it's a good idea to tape your thumbs to protect them from sharp ring ends.

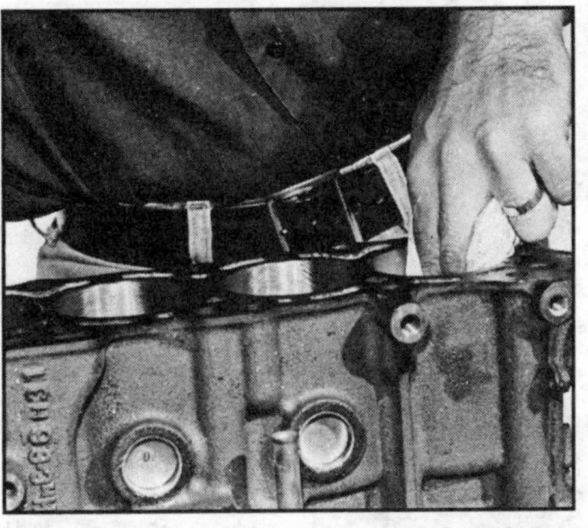

Bores must be perfectly clean for piston installation. Wipe them with paper towel to remove dust clinging to oiled cylinder walls. Towel must come out clean before bore is ready for piston.

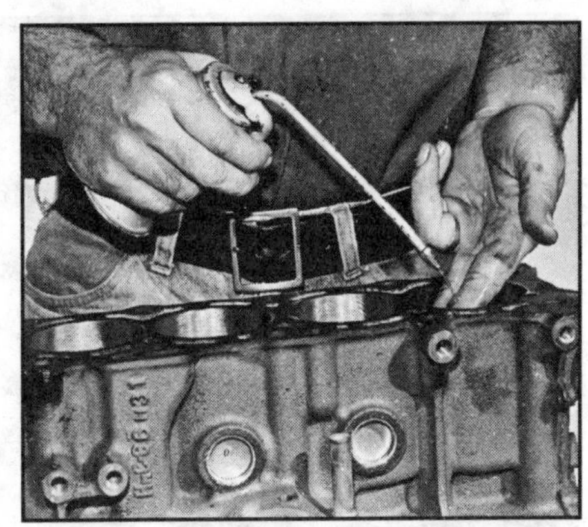

Liberally oil cylinder walls. Clean fingers are best for spreading oil evenly in bore. Do all four bores, then wipe off your hand. No matter what you do, installing pistons is a real oil bath.

struction sheet with your rings. *Read the instructions carefully* and follow them to the letter, even if they differ from these instructions.

Spreading, or expanding, compression rings is the difficult part of installing them. They are very rigid and, worse yet, brittle. If you spread or twist one a little too much, you'll have a broken ring on your hands. Most mechanics can tell you about a ring that broke during installation for seemingly no reason at all. Most of these horror stories resulted from installing rings by hand.

Using a ring expander reduces the risk of ring and piston damage to practically zero. This is especially true when you use the type that cradles the entire ring. Even the inexpensive style, which only grasps the ring by the ends, is much better than doing it by hand. The rule is to not expand rings more than necessary. Spread the ring only enough to clear the piston, move it to its groove, then relax it into position.

Be gentle when installing compression rings. Use both hands to guide the ring over the piston. Don't cock the ring on the piston because this makes it necessary to expand the ring farther, increasing the chance of breaking it.

Install the bottom, or second, compression ring first. When the bottom compression ring is installed, install the top one. If you get the top ring on first, remove it and install the bottom one. There's no way you'll get one compression ring over another without breaking it.

Work with one piston at a time, installing all of its rings. When you are done, check the marks to make sure the compression rings are installed top-side up. Then, position the gaps according to the drawing. Set the finished piston aside and start with another. The rings will move around somewhat as you handle the rod/piston assemblies, so recheck ring-gap spacing immediately before installing each piston in its cylinder.

If you are installing compression rings by hand, there are some tricks you should know about. First, wrap your thumbs with a couple of layers of masking tape. This will protect them somewhat from the sharp ring ends and allow you more control over the ring. Take the second compression ring and hold the ends with the tip of your thumbs. Spread the ring and guide it over the piston with the ends *slightly lower* than the rest of the ring. Gently spread the ring *with your thumbs only.* Do not allow your fingers to press against the outer edge of the ring; they will act as a fulcrum and the ring will break before you know what's happening. Instead, use your fingers to gently guide the ring downward. Pass the ring over the piston to its groove, then relax your thumbs.

Install the next compression ring after your thumbs revive. After you've installed all the rings, you're ready to fit the pistons and rods to the block.

PISTON & CONNECTING-ROD INSTALLATION

To install the pistons and rods without damage, you'll need several tools: a ring compressor, crank-journal protectors that slip over the rod bolts, an oil squirt can, a large can half full of motor oil and a hammer. You'll use the hammer handle to push the pistons down their bores. The large can is for dipping the pistons in oil before installation. With Honda-size pistons, a 1-qt motor-oil can or coffee can with the top removed does nicely. It's the easiest way to oil the pistons thoroughly before installation.

Get Organized—Piston installation should go smoothly if everything is laid out beforehand. All your tools must be clean and within reach. The block must be positioned so you can work the piston in from the top and guide the rod in from the crankcase with your other hand.

If you have an engine stand, turn the block so the cylinders point straight up. This will let the rod hang without touching the bore. This makes it easier to guide. If you are working on a bench, turn the block on its side with the deck facing you. In this position, you can reach around through the bottom of the crankcase to guide the rod through the cylinder.

Clean the Bores—Wipe the protective oil coating from the bores with paper towels. The towel should come out clean. Use your squirt can to put a fresh coat of oil on the cylinder walls. Spread the oil around with your *clean* fingers. If you feel any grit, now's the time to clean it out. Inspect the pistons, rods and your tools for dirt, and clean them as necessary with fresh oil. Remove the rod bearings from their boxes.

Arrange the rod/piston assemblies in order next to the block so you

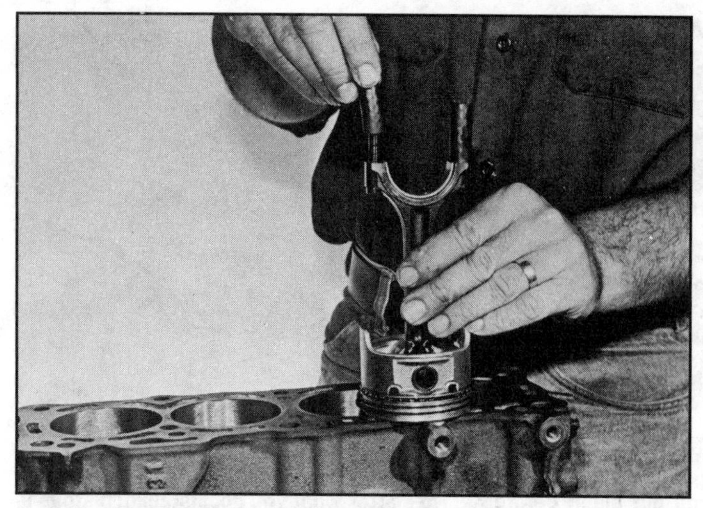
Install rod bearings as you did main bearings—dry back side, tangs aligned. Then fit journal protectors over rod bolts.

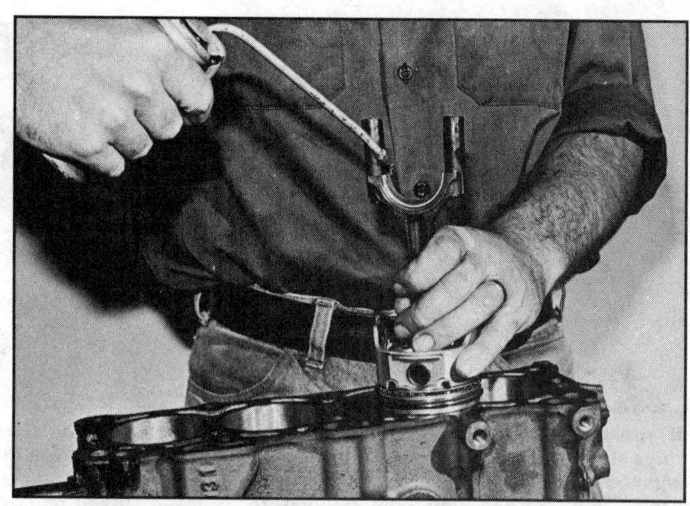
If you're Plastigaging rod bearings, don't oil the bearings and crank journals just yet. Oil softens Plastigage and takes up clearance. Otherwise, oil them liberally.

Double-check bearing color or undersize before installing it in cap.

won't have to walk around searching for the right piston.

Position Crankshaft—Go through your collection of parts until you find the crankshaft pulley, bolt, washer and Woodruff key. After making sure the bolt threads are clean, temporarily install the pulley.

You'll now be able to use a socket on the pulley bolt to turn the crankshaft. This is necessary so you can turn each connecting-rod crank journal to BDC. That way, you'll have room to reach the rod as you guide it onto its journal. As you start out, you may not need the bolt, but as more pistons are installed, drag will increase until you'll want a socket and breaker bar to turn the crank.

Ready Rod/Piston—To install a rod/piston, first turn its crank journal to BDC. Remove the cap from the connecting rod. Wipe the rod and cap bores clean and dry with a paper towel, then wipe the back of the bearing. Any dirt or oil *behind* bearings causes insufficient oil clearance at the crank. Press the bearing inserts into place in the rod and cap. Don't oil the bearings or journals if you're checking oil clearance with Plastigage. Slip journal protectors over the rod bolts—*this is a must.* If you don't have the ready-made protectors, use two short lengths of 5/16-in. fuel hose—4-in.-long hose works well.

Again, don't lubricate the bearings or journals, just yet. You're going to Plastigage the rod bearings later.

Next, coat the rings and pistons with oil. You can do this with a squirt can. However, a large can half filled with oil is much easier. Turn the rod/piston assembly upside-down and dunk the piston in over the rings and piston pin. Move the rod back and forth to work oil in between the pin and pin bore. Lift the piston out and let the excess run off into the can. Don't worry about the mess. You can clean it up after installing the rods and pistons.

Compress Rings—I've already mentioned the two most common ring compressors—plier and cylinder types. The plier type is preferred because it's a bit easier to use. Some mechanics claim it does a better job of compressing the rings. You compress the compressor band and rings by squeezing the plier handles. The ratchet locks, keeping the rings compressed. Release the compressor by tripping the ratchet latch.

Cylinder-type compressors are usually tightened with a large, square key. They, too, have a ratchet released by tripping a latch. Both types of compressors have a limited range of travel, particularly the plier type.

The plier-type ring compressor is limited by the metal band that fits around the piston. Several come in the tool kit. A band for 2-3/4- to 3-1/8-in. pistons covers all Hondas.

You may have noticed that the ring compressor has a top and bottom; an arrow on the cylinder or band indicates bottom. The ratchet or handle attachment is closer to the bottom. Make sure the handle or constricting bands are at bottom when compressing the rings. This orientation keeps the rings tightly compressed as they enter the bore. If the compressor is inverted, the rings may pop out from under it before entering the cylinder. They may do it anyway.

Check that the ring end gaps are correctly positioned. Then, fit the relaxed, cleaned and *oiled* compressor over the piston so the bottom is toward the connecting rod. Begin to tighten the compressor after centering it on the piston; the piston should just peek out the top of the compressor. As the compressor tightens around the piston, wiggle it back and forth on the piston so the rings will relax as you compress them. You'll feel the compressor relax as the rings compress. Tighten the compressor until it is firm against the piston.

Double-check the position number of the rod/piston assembly before placing the rod in the cylinder. Also, make sure it's correctly aligned in the block, once installed. The dot on the piston should be on the same side of the rod as the oil hole. And, both the oil hole and piston dot or triangle face the intake-manifold side of the block.

If you are using aftermarket pistons, the piston marks may be different. The instruction sheet that came with the pistons will show which way the piston fits on the rod. Regardless of which pistons are used, make sure the rod oil hole faces the intake manifold.

Install Rod/Piston—Guide the rod into the cylinder with your free hand. Don't let the rod bang against the bore on its way down. The piston skirts will slip easily into the bore until the ring compressor butts against the deck.

Force the compressor down so it butts squarely against the block deck. Do this by *lightly* tapping around the upper end of the compressor with a hammer handle. With a cylinder-type compressor, you can actually watch the overlap portion of its band align as you tap it down.

Use your free hand to guide the rod into the crankcase. Sometimes, the piston will turn on its way in and the big end of the rod will hang up on a crank counterweight or the block. If so, turn the rod/piston assembly so it lines up with its journal.

Check the ring compressor to make sure it is still tight around the piston. Then use the hammer handle to *tap* the piston into its bore. If the piston *goes solid*—doesn't want to go any farther—**STOP!** A piston ring has popped out of the ring compressor and has snagged just before entering the bore.

If you try to force the piston, you'll break a ring or ring land. If you break the ring and continue installation, the broken ends of the ring will gouge the cylinder. You won't notice it until too late, and then you'll really be angry.

If in doubt, prevent a potential disaster by stopping and pulling the rod/piston assembly from its bore. Remove the compressor and check the rings. If everything is OK, refit the ring compressor and start over.

When the last ring enters the bore, the ring compressor will relax and slip off the piston. Set the compressor aside. Push the piston gently down its bore with one hand while guiding the rod big end with your other hand. Fit the rod to the journal. Remove the two journal protectors.

Fill spotlessly clean can with enough 30W oil to immerse piston and pin. Dunk piston and work rod back and forth. After excess oil drains off, install ring compressor.

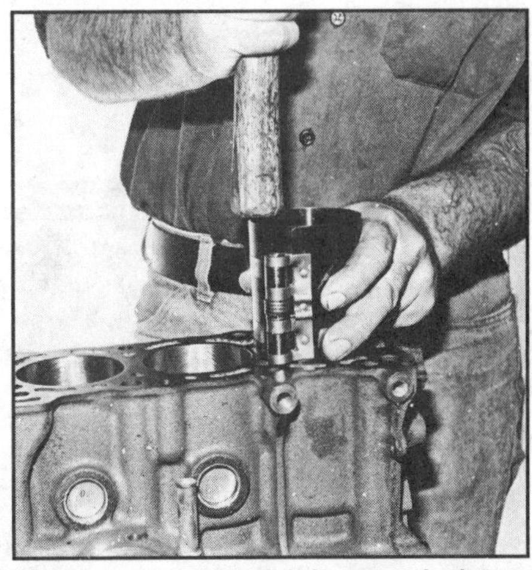

Clamp compressor tightly around piston, then position rod/piston assembly over its bore. Make sure marks on piston are properly oriented. Tapping edge of compressor with hammer handle aligns compressor bands; keeps rings from popping out at deck.

Tap piston into cylinder while guiding big end of rod past crank. If you feel solid resistance, *STOP!* Ring has popped out from under compressor and will damage ring and piston if forced. Instead, pull assembly from bore, refit compressor and try again.

Install Connecting-Rod Cap—Because you are going to check rod-bearing oil clearance with Plastigage, don't oil the rod-bearing half in the cap. Wipe any oil from rod-journal 1 and lay a strip of Plastigage on it. Place the rod cap on the rod bolts only after you make doubly sure it's the correct cap and is aligned properly. Remember, both bearing tangs must be on the *same* side of the rod. Also, check the rod-size code. The numeral is marked on both halves of the rod big end. When the rod cap is correctly fitted, the numeral will be complete and legible. If the cap is backward, the size-code numeral will appear incomplete—one-half on both sides of the rod.

As for keeping the right cap on the right rod, the best method is to never remove more than one cap at a time. Then, there's only one rod and cap combination to go together. If the caps get mixed up, use your punch marks or stamped numbers to reorganize them.

Run the rod nuts on finger tight. Then, torque them *in stages* to 20 ft-lb; 22—25 ft-lb for the 1751. First, bring both bolts to 10 ft-lb, then torque each in 5-ft-lb stages, alternat-

Fit cap to rod so bearing tangs are on the same side of rod. Bearing-selection number and your own numbers should align. Torque is 20 ft-lb on all engines except 1751; torque 1751 nuts to 22-25 ft-lb.

After Plastigaging, clean off wax residue, oil cap bearing and install. Because Honda torque specs are so low, I use locking compound on bolt threads. Threads must be oil-free for maximum effectiveness. Don't turn crank until each rod bearing has been lubricated.

Filling oil pump with motor oil now helps pump priming on initial start-up. Inverted, the pump can be filled. Make sure pan is installed before uprighting engine.

On aluminum engines, don't forget to mount oil-pump bypass block with a new O-ring.

ing between them until final torque is reached. Fully torquing one bolt with the other finger tight will distort the bearing.

Remove the rod cap. Again, go evenly when loosening the rod nuts. Break *both* bolts loose before completely removing one.

Read Plastigage—Measure the squished Plastigage with the scale on the Plastigage sleeve. Normal oil clearance is 0.0008—0.0015 in. (0.020— 0.038mm), with a service limit of 0.0028 in. (0.07mm). Optimum rod-bearing oil clearance is 0.001—0.0015 in. (0.025—0.038mm), so aim for that range, if possible. Anything within specifications will work, although 0.0025-in. (0.063mm)-or-greater clearance may result in shorter bearing life. Just as with main-bearing oil clearance, a wide range will work just fine—but there is an optimum point for perfectionists.

Oil the rod journal, install the rod cap and torque to specifications. You're ready to Plastigage the next rod bearing. Install the rod/piston assembly for cylinder 4. If clearance is OK, oil number-4 journal and install bearing and cap. Turn the crank 180° and check and install numbers 2 and 3.

If you replaced any rods or the crankshaft, recheck rod side clearance now. See the procedure on page 54.

OIL PUMP & PAN

When the pump was assembled after inspection, you should've packed it with petroleum jelly or oil. This ensures that it will prime immediately when the engine is first started. If the pump was assembled dry, fill it with oil through the pickup.

Once the pump has been lubed, installation is simple. On cast-iron blocks, bolt the pump to the center main-bearing cap. Don't forget the bolt that passes through the pump under the pickup screen. Use no gasket or sealer between the pump and main-bearing cap.

Oil-pump installation on aluminum-block engines is similar, but there's no hidden bolt under the pickup screen. Don't forget to mount the bypass block—the tube with the 90° bend that links the pump to the bearing-cage assembly. Fit a new O-ring to the bottom of the bypass block before installation.

Tighten all bolts as evenly as possible, then torque them 10 ft-lb. Always use a *new* oil-pump pickup screen. Old screens are next to impossible to get clean. On cast-iron engines, the screen snaps *into* the pickup. Lightly tap the screen edge to seat it. On aluminum engines, fit a new oil screen *over* the pickup.

Bottom-End-Inspection—Before you bolt the oil pan into place, make one last inspection of the bottom end. First, retorque all main- and rod-cap bolts and nuts, respectively. If you get slight movement from some of them, don't worry. What you felt was a bearing cap that had relaxed, or *normalized*—conformed to the torque-induced strain applied to it. By retorquing, you minimize any danger from a loose rod nut. The part will not loosen again because built-up strain was compensated for by retorquing.

Torque for oil-pump bolts is scant 9 ft-lb. Low torque is best measured with an inch-pound torque wrench. So, 9 ft-lb X 12 in. per ft = 108 in-lb.

Once pump bolts are torqued, install oil-pump screen. Tapping too hard will distort screen.

Later 1751 oil-pump screen uses metal cover and knob (arrow) for screen alignment.

If you have a fastener that turns, but never tightens, or feels rubbery, remove it and inspect the installation. Either the bolt is bad or the part misaligned. Replace the bolt or remove the part and inspect it and the threads on the bolt and nut or in the hole. If all is OK, reinstall the part, make sure it's firmly seated, and install the bolts or nuts and retorque them.

Also, check the rods to make sure they are not cocked on the crankshaft. Merely grasp the big end of each rod in your finger tips and attempt to move it back and forth on the journal. You may find a rod that *pops* loose; it could have been cocked on its journal. This cocking could harm the journal or bearing had it not been *popped* loose. After taking a last look at the bottom end for another 100,000 miles, bolt on the pan.

Oil Pan—Although Hondas use a neoprene, one-piece pan gasket, you should still apply RTV to both sides of it. Start by laying a thin bead of RTV along the perimeter of the block. Pay special attention to the four corners where the main-bearing caps meet the pan rails. A little extra dab of RTV goes in these corners. Lay the pan gasket on the block, fitting it over the pan studs. Gently press the gasket into the corners until it is aligned with its bolt holes. Lay another thin bead of RTV on the pan gasket, again putting an extra dab in the four corners. Finally, fit the pan to the block and start all pan bolts and nuts.

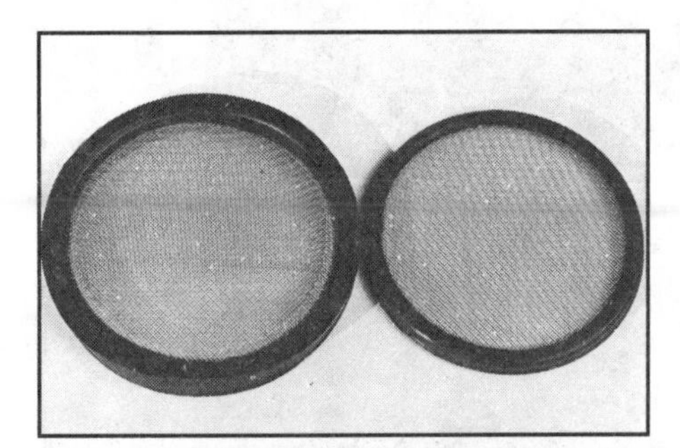

The 1200 oil screen at left fits *over* pump pickup. Screen at right fits all earlier iron-block oil pumps and snaps *into* pump pickup. *Always* use a new screen.

Because the pan can be distorted if tightened unevenly or overtightened, follow the oil-pan torque sequence shown on page 126. Tighten the nuts and bolts in at least three steps—that is, go through the sequence three times. Final torque on pan bolts is 13 ft-lb. *The nuts get only 8 ft-lb.* That isn't much, so just snug the bolts on each pass.

If you start at one end of the pan and cinch all bolts down, you'll likely warp it. It might not be much of a warp—just enough to leak oil. Think of the pan as a carpet. Unless you lay it evenly from the center out, you'll end up with a wrinkle somewhere.

Before leaving the pan, remove the drain plug and replace the washer. There should be a new drain-plug washer in the gasket set. Reinstall the plug and torque to 29—36 ft-lb.

Paint Block—Now's the best time to paint a cast-iron block. An even coat of gloss-black paint will restore the factory-fresh look. Mask the deck, exposed crankshaft areas, machined surfaces and all threaded holes first.

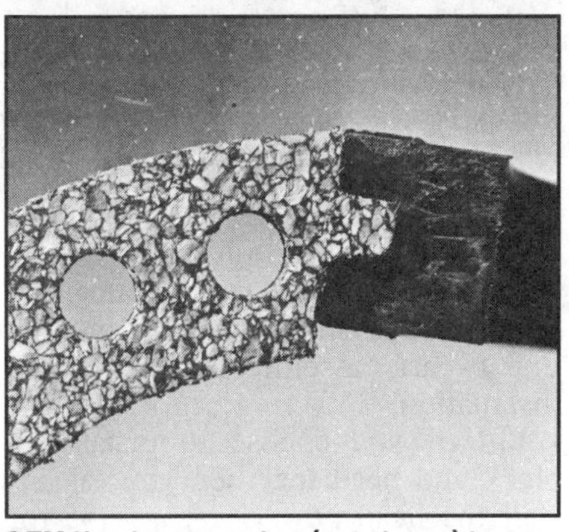

OEM Honda pan gasket (not shown) is one-piece. Apply RTV to four corners where pan rails meet main-bearing caps. This Fel-Pro gasket is four-piece, with interlocking tabs at all four corners.

TIMING BELT AND SPROCKETS

If you are installing the engine without the head, you can't completely install the camshaft-timing components

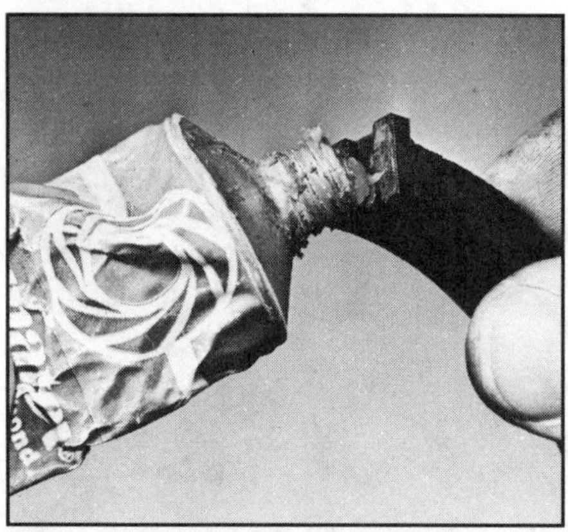

With multi-piece pan gaskets, dab RTV at corners of two end pieces . . .

. . . then press them into position. Make sure tangs and cutouts mate.

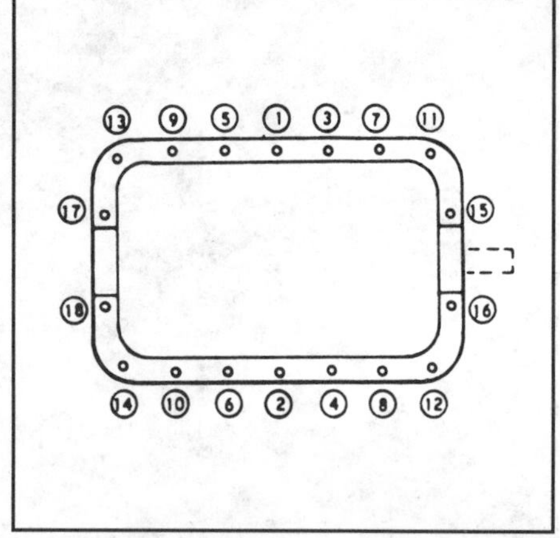

Torque pan bolts and nuts using this tightening sequence. Courtesy of American Honda Motor Co., Inc.

Most common oil-pan-installation mistake is overtorquing bolts and nuts. Overtorquing distorts pan rails around bolts or nuts and may cause oil leaks. Follow tightening sequence shown. Torque pan bolts to 13 ft-lb; nuts to 8 ft-lb. Retorque bolts and double-check because pan gasket relaxes.

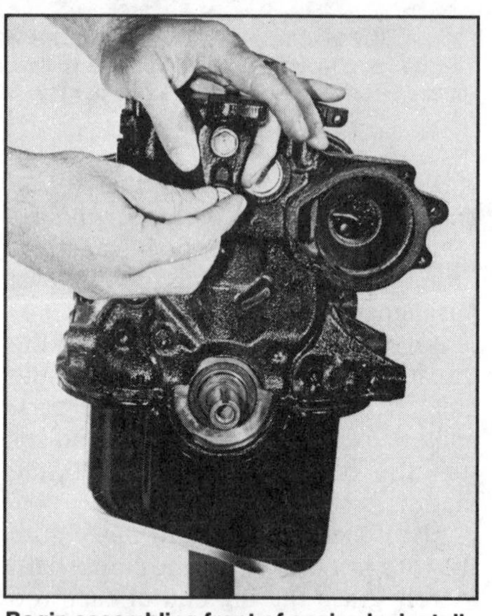

Begin assembling front of engine by installing left engine mount.

yet. This is no disadvantage, however, as the lower timing-belt parts can be installed now and the upper parts assembled during head installation.

But, if you choose to install the block and head together, you can install the head now. See page 145 for head-gasket installation, then pages 128—136 for final head assembly.

For now, everything that fits under the timing-belt lower cover can be installed. Start by bolting on the left engine mount at the top of the timing-belt area.

Next, install the timing-belt tensioner with its coil spring and two long-shank bolts. The coil spring hooks onto the tensioner via a small V-shaped cutout in the tensioner arm. The other end of the spring hooks onto a small column cast into the block. The two long-shank bolts clamp the tensioner to the block and provide for belt-tension adjustment. Inspect the two bolt holes in the tensioner arm. One is round, the other slotted. The round hole is for the pivot bolt; the slotted hole is for the adjusting bolt. It doesn't matter which bolt you place in which hole.

The coil spring passes between the tensioner and block. It's easier to hook one end of the spring onto the V-groove cut into the tensioner and hold it there while bolting the tensioner to the block. Start both bolts and run them in until they just touch the tensioner. Clamping force isn't required yet. Now, hook the free end of the spring onto its column.

Gather up the crankshaft timing sprocket and fences. The outer fence has holes; the inner fence is usually solid. Slide the inner fence over the crank with its curved lip toward the block. Then, slide on the sprocket.

Thread on the timing belt. Install a *new* timing belt, unless the old one is relatively new. If reusing the old belt, it must turn counterclockwise—the same direction as it did when originally installed. If you didn't mark the direction of rotation before removal, inspect the belt teeth. One edge will be worn more than the other. This is the driven side. When correctly mounted, the driven side will be the bottom half of the tooth when looking at the right side of the belt.

After looping the belt over the crankshaft sprocket, pass the right side of the belt to the left of the tensioner. The back side of the belt should rest against the tensioner drum while the left side is straight. For now, let the belt flop over to the left. You can't finish installation until the head is installed. Slip the outer,

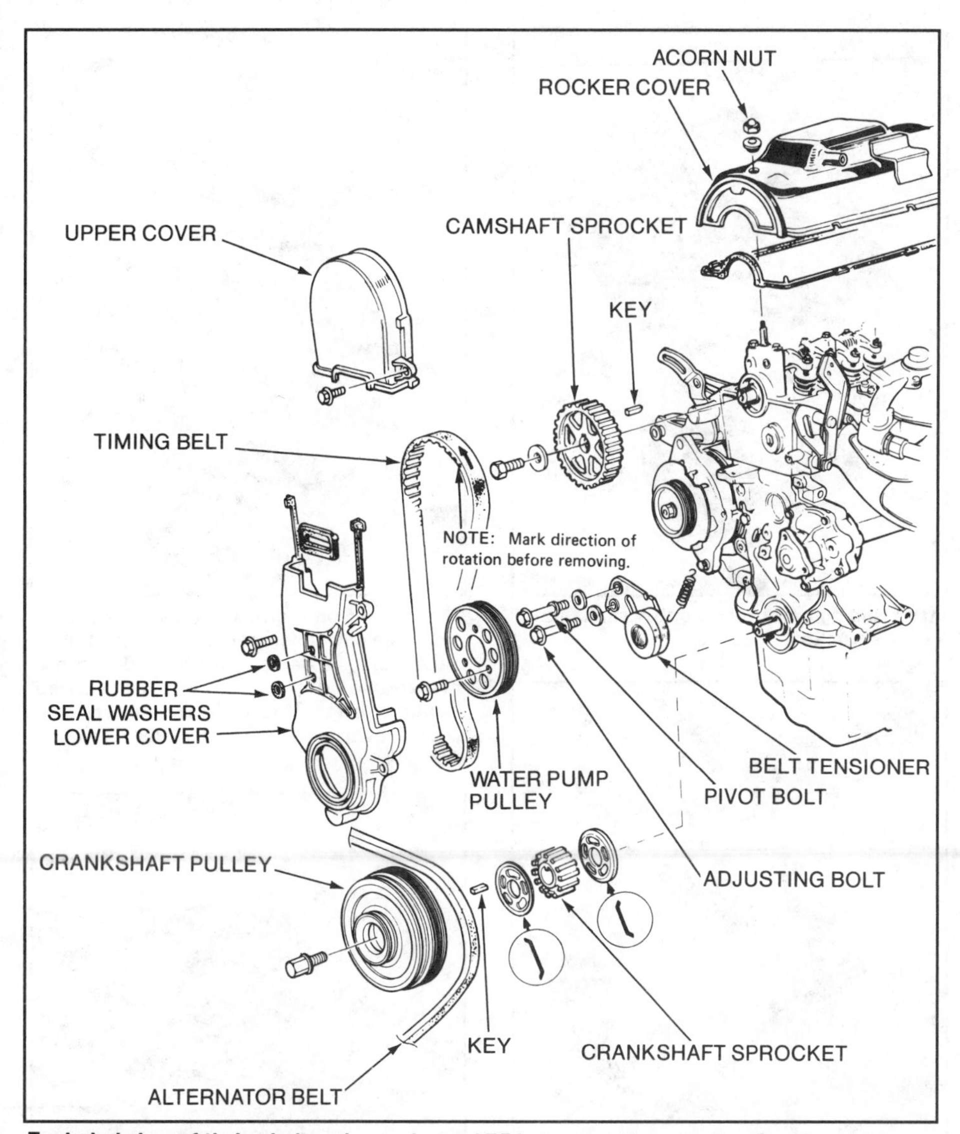

Exploded view of timing belt and sprockets; 1751 shown, others similar. Courtesy of American Honda Motor Co., Inc.

Belt tensioner is next: Leave bolts finger tight for now. Don't forget large washers between bolts and tensioner.

Hook angled eye of tensioner spring into V-notch in tensioner (arrow), and rounded eye onto post on block. Spring fits between tensioner and block.

holed fence into place against the crank sprocket.

Timing-Belt Lower Cover—The front of the engine is ready for the timing cover. In the gasket set, find the cover gasket(s), left engine-mount gasket and the two tensioner-bolt seals.

Finding a one-piece lower-cover gasket can be puzzling until you realize that the gasket is longer than the cover. When installed, it rests against the head. For now, it extends loosely above the block's deck. Remember that some engines use a two-piece lower-cover gasket. Once you've found the correct gasket(s), lay it into the recess in the timing cover. Also, set the left engine-mount gasket in the cover. This is the square-cut neoprene gasket at the cover's top center. Bolt the cover to the engine, taking special care to fit the cover around the water-pump bracket.

Finish the installation by stretching the two seals over the tensioner bolts. The seals merely fit between the bolt heads and timing cover, so don't try to force them into the cover.

Water Pump—Go through the gasket set and find the D-shaped, formed O-ring for the water pump. Pry out the old O-ring and lay in the new one. Also, find the new rubber-faced bracket that attaches to the water pump. Exchange it for the old bracket under the water-pump mounting bolts. You can now bolt the water pump to the block. Don't forget the air-pump lower bracket on 1975—'79 1237cc engines.

Drive Pulleys—Bolt on the water-pump and crankshaft pulleys. The water-pump pulley is a straightforward

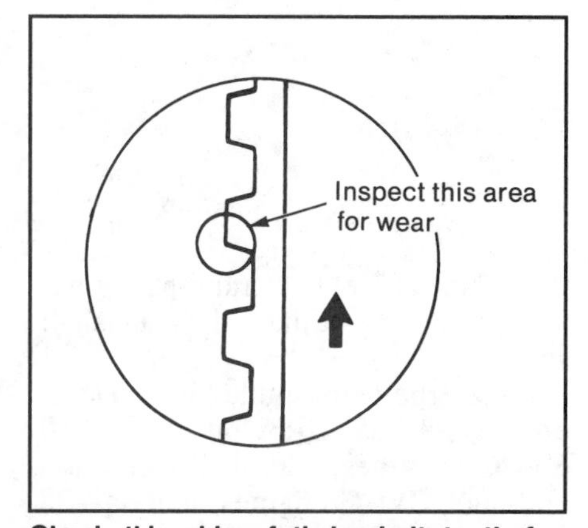

Check this side of timing-belt teeth for wear. Replace belt if used for more than 30,000 miles. Courtesy of American Honda Motor Co., Inc.

Holding three crank-sprocket pieces together shows how curved edges of fences form pulley groove.

Slide outer fence onto crankshaft sprocket.

After sliding inner fence onto crank, install crank sprocket.

Fit new neoprene gasket into timing-belt lower cover and bolt cover to block. Note how ends of gasket extend above lower cover. They will seal timing-belt upper cover.

Thread belt onto crank sprocket with back side against tensioner. If reusing belt, it must turn in same direction as before. Arrow should point to left as viewed. Honda crankshaft rotation is counterclockwise—reverse of most other engine designs.

With a little prying, install two rubber grommets over tensioner-pivot and adjusting bolts.

three-bolt affair. The crank pulley has a separate key like the camshaft sprocket.

Install the crank-pulley bolt. Tighten to 58—65 ft-lb on 1975—'80 CVCC engines, 80 ft-lb on '81-and-later CVCC engines, and 34—38 ft-lb on 1200s. To keep the crankshaft from rotating, thread two flywheel or driveplate bolts into adjacent holes in the crank flange. Position a hammer handle between the bolts to give the needed leverage to hold the crankshaft.

CYLINDER HEAD

I like to use the short block to hold the head for assembly. This is the same arrangement used for head disassembly. Use the old head gasket or what's left of it to protect the head and block head-gasket surfaces. On iron-block engines, snug two diagonally opposite head bolts to hold the head in place. On aluminum-block engines, the block studs will hold it in place. Of course, if you want to install the engine with the head on, skip ahead to page 145 for tips on bolting the head down, then return to this page for cylinder-head assembly.

Before you set the head on the short block, rotate the crankshaft so cylinder 1 is at TDC. Now rotate the crank *clockwise* a quarter turn to move the pistons about midpoint in their cylinders. This procedure will ready the engine for timing the camshaft later on.

After positioning the crank and setting on the head, you're ready to complete cylinder-head assembly.

Oil Seal—Tap in the camshaft oil seal

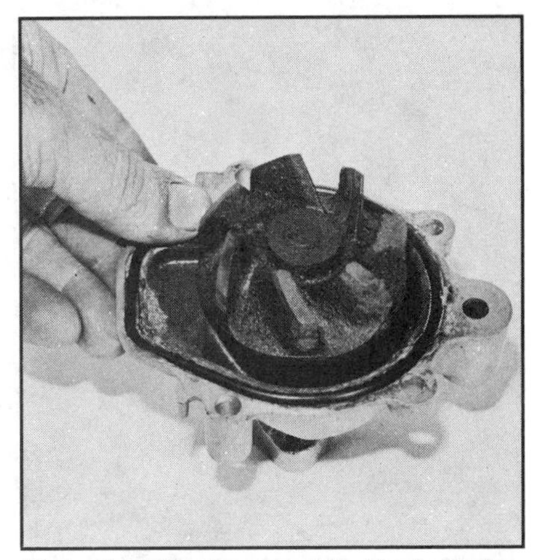

Discard old water-pump-to-block O-ring and install new one.

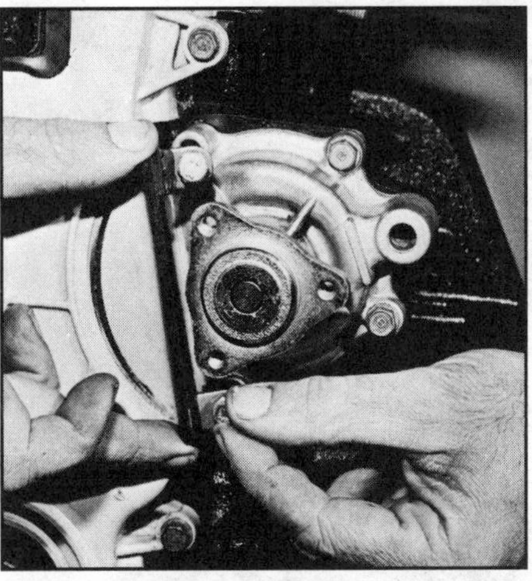

Install neoprene seal for side of timing cover. Two water-pump bolts pass through it. Torque water-pump bolts 7 ft-lb. Bolt pump pulley in place after torquing pump bolts.

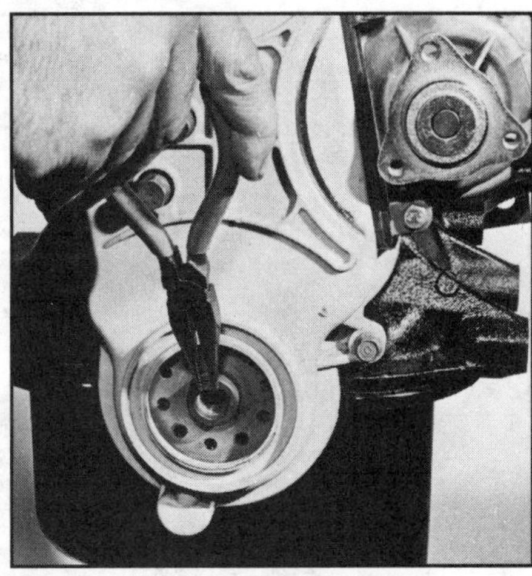

Unless you have extraordinarily small fingers, fitting woodruff key to crank is best done with needle-nose pliers.

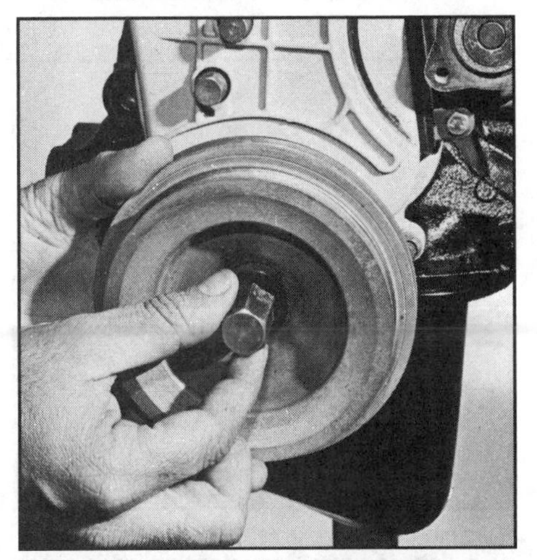

Slide crankshaft pulley onto crank snout and into engagement with key. Then, install large washer and pulley bolt. Tighten to 35 ft-lb on aluminum blocks; 60 ft-lb on all 1980-and-earlier iron blocks. Tighten bolt to 80 ft-lb on later iron blocks.

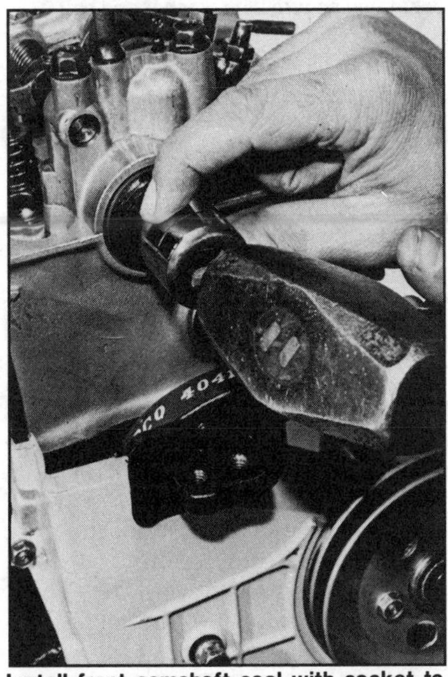

Install front camshaft seal with socket to distribute hammer blows. You'll feel seal *go solid* when it bottoms in bore.

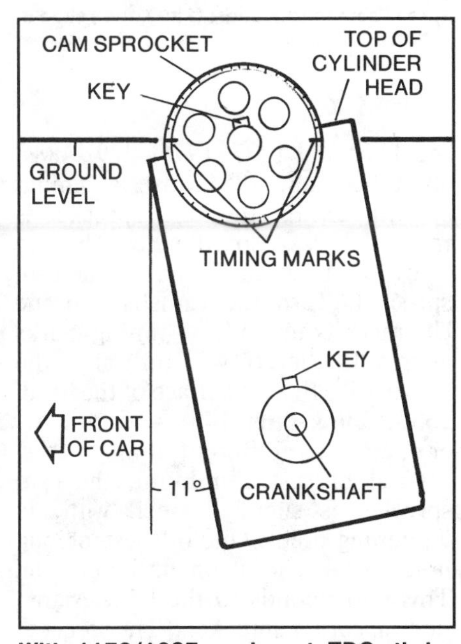

With 1170/1237 engine at TDC, timing marks are parallel with ground, not top of cylinder head. Engine tilts 11° as installed. Drawing by Paul Fitzgerald.

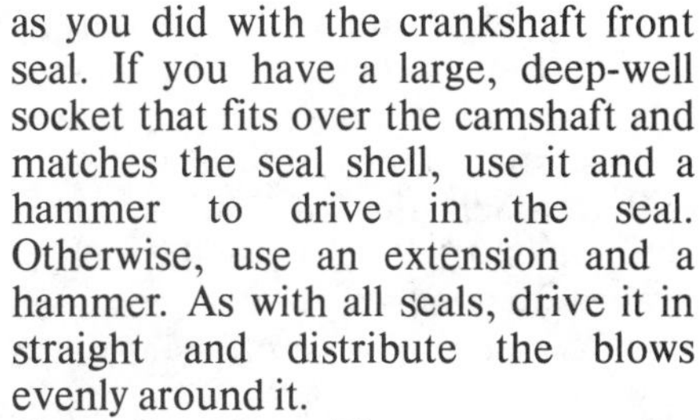

as you did with the crankshaft front seal. If you have a large, deep-well socket that fits over the camshaft and matches the seal shell, use it and a hammer to drive in the seal. Otherwise, use an extension and a hammer. As with all seals, drive it in straight and distribute the blows evenly around it.

Cam Sprocket—The cam sprocket uses a loose-fitting, square key. Fit the key into the slot in the cam, slide on the cam sprocket and bolt it down. Make sure the key doesn't slip out the back side of the sprocket during assembly. Also, position the cam's keyway at 12 o'clock when installing the sprocket. This will aid cam timing, especially on engine sprockets without the UP designation.

Next, you must establish valve, or camshaft timing—the relationship between camshaft and crankshaft events. If this is not done, the valves and pistons will touch during valve adjustment.

Camshaft Timing—Because of differences in camshaft sprockets and timing procedures, I give specific timing instructions for each engine.

1170 & 1237—The camshaft sprocket is indexed with two marks on its outer rim. Rotate the camshaft to the cylinder-1 TDC position until the key is facing up and the two marks are parallel with the ground—*not the top of the head.* As shown in the drawing, the timing mark on the water-pump side of the sprocket will be about 1/2 tooth above the top of the head. If you can't see the key behind the bolt's large washer, loosen the bolt several turns and pull back the washer until you see it.

1488, 1600 & 1751—1975 1488s use the same camshaft-timing marks as

Install camshaft woodruff key, then slide on sprocket and secure with washer and bolt.

If no arrow is cast into cylinder head, align timing marks with head's rocker-cover surface. And if UP cast into cam sprocket is missing, turn cam so keyway faces up.

Turning cam sprocket to TDC cylinder 1. A few degrees more counterclockwise rotation and cam will be timed. UP mark must be up and two lines aligned with arrow cast in head. Cast arrow is visible in shadow to left of sprocket (arrow).

the 1170 and 1237. 1976–'79 1488s, all 1600s and 1751s use a similar system, except the cam sprocket is marked UP. Also, the head has an arrow, cast just to the left of the cam sprocket. Turn the camshaft so the UP mark is up and the timing marks align with the arrow. Don't align the marks with the top surface of the head.

1335, '80-&-later 1488—These newer engines use different timing marks. Instead of cast aluminum, the cam sprocket is stamped steel, with six lightening holes. One of these oblong holes has a cutout on its inner side. This corresponds to the UP designation on earlier sprockets. The sprocket rim is marked with two timing marks, as are other sprockets.

On 1335cc engines, line up the two marks with the head's upper surface and the cutout in the cam-sprocket lightening hole at the top. On 1488s, there's an arrow cast into the head. Align the marks on the arrow with the cutout at the top.

Connect Crank & Camshafts—Now that you know what the cam sprocket should look like with cylinder 1 at TDC, turn it there *by hand.* This is easier said than done, but *do not wrench on the camshaft-sprocket bolt.* You'll either loosen it or overtorque it. If you use a large round drift and are very careful what you lever against, you can pry the sprocket around using the lightening holes. There's a risk of gouging the head or rocker assembly with this method, so use it only if you can't manage by hand. Sometimes, it's nice to have a pet orangutan.

Remember that the valves open and close as you rotate the camshaft. If you feel firm resistance and can no longer turn the cam sprocket by hand, *stop*—it's possible two pistons are at TDC and the valves have butted against those pistons. If you put a socket and breaker bar on that sprocket, you could *bend a valve.* That's why the crankshaft must be positioned so the pistons are down their bores, away from TDC. Then, there's is no danger of the valves contacting the pistons.

Once the cam is positioned, move the crankshaft counterclockwise one-quarter turn to TDC for cylinder 1. Go slowly, just in case you bring a piston and valve together. If you feel resistance, stop. Try rotating the crank in the opposite direction. If that doesn't work, remove the head from the block and retime the crankshaft. This time, bring the crank to cylinder-1 TDC, then refit the head, which has already been turned to TDC, to the block.

With both crank and camshaft at cylinder-1 TDC, you can install the timing belt. Slip the belt onto the camshaft sprocket. For valve adjustment, there's no need to adjust the tensioner, so leave it loose.

Valve Adjustment—Because each engine's valve-adjusting procedure is different, I'll cover them individually.

Valves are numbered 1–8 starting at the cam sprocket. CVCC-engine auxiliary intake valves are numbered 1–4 starting at the cam sprocket.

Before adjusting any valves, rotate the crankshaft 360° to load the rockers against the valves. If you don't do this, clearance may be off by as much as 0.006 in.

1170 & 1237—With cylinder 1 at TDC, adjust valves 1, 2, 3 and 6. Then, rotate the crankshaft counterclockwise 360° and adjust valves 4, 5, 7 and 8. Finally, rotate the engine another 360° to bring cylinder 1 back to TDC. Valve clearance is 0.006 in. (0.15mm).

1975 1488—With cylinder 1 at TDC, adjust valves 1, 2, 3, 5 and auxiliary valves 1 and 2. Rotate the crankshaft counterclockwise 360° and adjust valves 4, 6, 7, 8 and auxiliary valves 3 and 4. Then rotate the crankshaft counterclockwise 360° to bring it back to cylinder-1 TDC. Valve clearance is 0.006 in. (0.15mm).

'76—'79 1488, 1600—With cylinder 1 at TDC, adjust valves 1, 2, 4, 6 and auxiliary valves 1 and 2. Rotate the crankshaft counterclockwise 360° and adjust valves 3, 5, 7, 8 and auxiliary valves 3 and 4. Rotate the crank another 360° counterclockwise to return cylinder 1 to TDC. Again, valve clearance is 0.006 in. (0.15mm).

1335, 1751 and '80 & later 1488—With cylinder 1 at TDC, adjust all valves on cylinder 1. Rotate the crankshaft 180° counterclockwise. The two camshaft marks will again align with the head. Adjust cylinder 3's valves. Rotate the crankshaft another 180° and adjust the valves at cylinder 2. Another 180° rotation and cylinder 4's valves can be adjusted. After doing these valves, rotate the crankshaft another 180° to bring it back to cylinder-1 TDC. Intake-valve clearance, for both main and auxiliary, is 0.006 in. (0.15mm). Exhaust-valve clearance varies: 0.011 in. (0.28mm) for the 1751; 0.008 in. (0.20mm) for 1335 and 1488cc engines.

Valve-Adjusting Tips—Once you know what valves to adjust at which crankshaft position, and their valve-clearance specifications, the rest is easy. Clearance is measured between the valve-stem tip and rocker arm with feeler gages. Clearance is adjusted by loosening the jam nut on the rocker arm, turning the slot-head adjusting screw and retightening the jam nut. A box-end wrench is best for loosening and tightening jam nuts. Open-end wrenches may slip off, or round the jam-nut corners.

Let's say you need to adjust a valve to 0.006 in. Select the 0.006-in. (0.15mm) feeler gage from your set. Loosen the jam nut. Use a 10mm box end on 1170, 1237 and '75 1488s; 12mm on all others. Because you backed off the adjusting screws previously, run the adjusting screw in until it stops. The adjuster is now touching the valve—*clearance is zero.* Back off the adjuster a turn or two while trying to insert the 0.006-in. feeler gage between the valve and adjuster. Once the gage is in, slip it back and forth while tightening the adjuster. When you feel a light drag on the gage, the valve is adjusted. Double-check the adjustment by fitting the next larger and smaller gages—0.007 and 0.005 in. (0.175 and 0.125mm) in this case. The larger gage should just barely start and pass through the gap with considerable effort. The small gage should slip through with no drag.

After tightening valve-adjuster locknut, double-check valve adjustment. It usually changes slightly after tightening.

Once adjusted, tighten the jam nut while holding the adjuster in position. *Recheck the adjustment after tightening the jam nut.* If the adjustment changed, repeat the procedure until you get the right valve clearance. The clearance might change because the adjuster slips while tightening the jam nut. Or, the adjuster may stretch, loosening the adjustment. If the adjuster is stretching, chances are you're overtightening the jam nut. However, the adjuster will stretch a little regardless, so reduce valve clearance slightly below specifications, then tighten the jam nut. The too-close valve clearance will increase when the adjuster stretches. But recheck it just to be sure.

If you have a rocker cover full of stretched adjusters, it's time to replace them. Buy new adjusters and jam nuts from Honda.

Auxiliary valves are adjusted the same as the main valves, except everything is smaller and needs less force.

FINAL HEAD ASSEMBLY—1170, 1237cc

From this point on, I'll differentiate assembly procedures between 1200 and CVCC heads. After finishing head assembly, instructions are again combined beginning on page 136.

Manifolds—Mounting the crossflow-head manifolds is the first final cylinder-head assembly step. Slip four exhaust-manifold gaskets over the studs. Don't bother with gasket sealer. Exhaust parts run too hot for normal sealers to do any good. Slide the exhaust manifold into place followed by the hot-air baffle, if so equipped. Start all eight nuts, remembering that the two taller nuts go on the studs at the water-pump end of the head. Most 1200s have a hose bracket attached here, too. Torque all exhaust-manifold-to-head nuts to 13—17 ft-lb. Start at the *center* of the manifold and work out to each end. Bolt on the air-pump upper bracket.

The intake manifold requires a light bead of RTV around both sides of each gasket. After sealing, slip the two gaskets over the studs and install the manifold. Torque the intake-manifold nuts 13—17 ft-lb.

Fuel-pump spacer on 1200s uses two gaskets; one against head and one against pump. When cylinder 1 is at TDC, camshaft eccentric is pointing away from pump, easing pump installation.

At rear of CVCC head, attach distributor housing and thermostat cover. Three bolts secure distributor housing to head; two through thermostat cover. When installing distributor, make sure its edges fit in distributor-housing recess. Otherwise, thermostat cover will crack when bolts are torqued.

Thermostat—Make sure the lip inside the thermostat boss is clean, then install the thermostat. The spring side goes inside the manifold. Use RTV sealer on both sides of the gasket and position it over the thermostat. Install the thermostat housing and tighten its two bolts. If any intake-manifold fittings were removed, install them now.

Fuel Pump—Apply RTV sealer to both gaskets on the phenolic-resin fuel-pump spacer and slide the spacer onto the fuel-pump studs. Slip the pump onto the studs, add the two lock washers and nuts, and tighten the nuts evenly in several steps.

When cylinder 1 is at TDC as it is now, the fuel-pump eccentric on the camshaft is offset *away* from the fuel pump. This means you won't have to tighten the fuel-pump nuts against pump-return-spring force. If you find yourself fighting the return spring, cylinder 1 is probably not at TDC, or you may have forgotten the spacer. If all else fails, rotate the crankshaft to see if spring force is reduced. If it is, double-check camshaft installation.

Distributor—Installing the 1200 distributor is basically the same as for the CVCC. Follow the directions on page 134.

Carburetor—Using new gaskets, *and no sealer,* fit the carburetor spacer onto its manifold studs. Bolt on the carburetor, using a criss-cross pattern to tighten the four nuts and washers. Attach the fuel line between the carb and fuel pump and clamp each end.

Sparkplugs—Gap and install the sparkplugs. See the underhood sticker for specified plug type and gap. Use an extremely light coating of anti-seize compound on the plug threads. Don't get anti-seize on the first two threads. If it gets on an electrode, anti-seize compound will foul the plug.

Finally, dismount the head and clean the head bolts and studs as described for CVCC engines, page 136. Of course, if you want to install the engine with its head on, don't remove it. Then continue with "Small Parts," page 136.

FINAL HEAD ASSEMBLY—CVCC

Thermosensors—On CVCC engines, install the A and B thermosensors. The A thermosensor fits into the head. It's the one with the offset pins. The B thermosensor fits below the A sensor in cast-iron blocks, and its pins are centered. This sensor is mounted in the lower radiator tank on aluminum-block engines. Using a new O-ring, tighten each sensor with an adjustable wrench. Do not overtighten the sensors or you could strip out the aluminum threads in the head.

Distributor/Thermostat Housing—Clean the distributor/thermostat housing in solvent. Peel out the old O-rings from the housing and thermostat cover, then put in new ones. The gasket between the head and housing gets RTV sealer on both sides. Stick the gasket to the housing, then pop the housing in place on the head. The O-ring that seals the rear of the camshaft will hold the housing in place while you fit the thermostat.

If you removed the water-temperature sending unit or any ported-vacuum switch (PVS), replace them now. On all but '75 1488cc engines, the sending unit threads horizontally into the distributor housing, next to the thermostat. The sending unit in '75 1488cc engines installs vertically, under the thermostat. Use non-hardening sealer on the PVS threads and a new O-ring on the sending unit.

The spring side of the thermostat fits inside the distributor housing. Some engines use a thermostat with an air-bleed pin. Install this type of thermostat with the pin at the top, or 12 o'clock, position.

Pre-'80 1488s and all 1600s use a paper gasket and the O-ring to seal the thermostat cover in coolant. All 1751s and '80-and-later engines use a seal that fits over the thermostat and the O-ring. Apply a thin coat of RTV sealer over these parts and bolt on the thermostat cover. Two long bolts pass through the thermostat cover and into the head. A third, shorter bolt fits through the distributor housing to the right of the thermostat cover. On many engines, this bolt attaches a wire clip, so don't forget the clip. Remember, those bolts are going in aluminum—10 ft-lb torque is plenty.

Manifolds—If you have the special Honda manifold wrench, you can join the intake and exhaust manifolds on the bench and bolt them onto the head as a unit. Without the wrench,

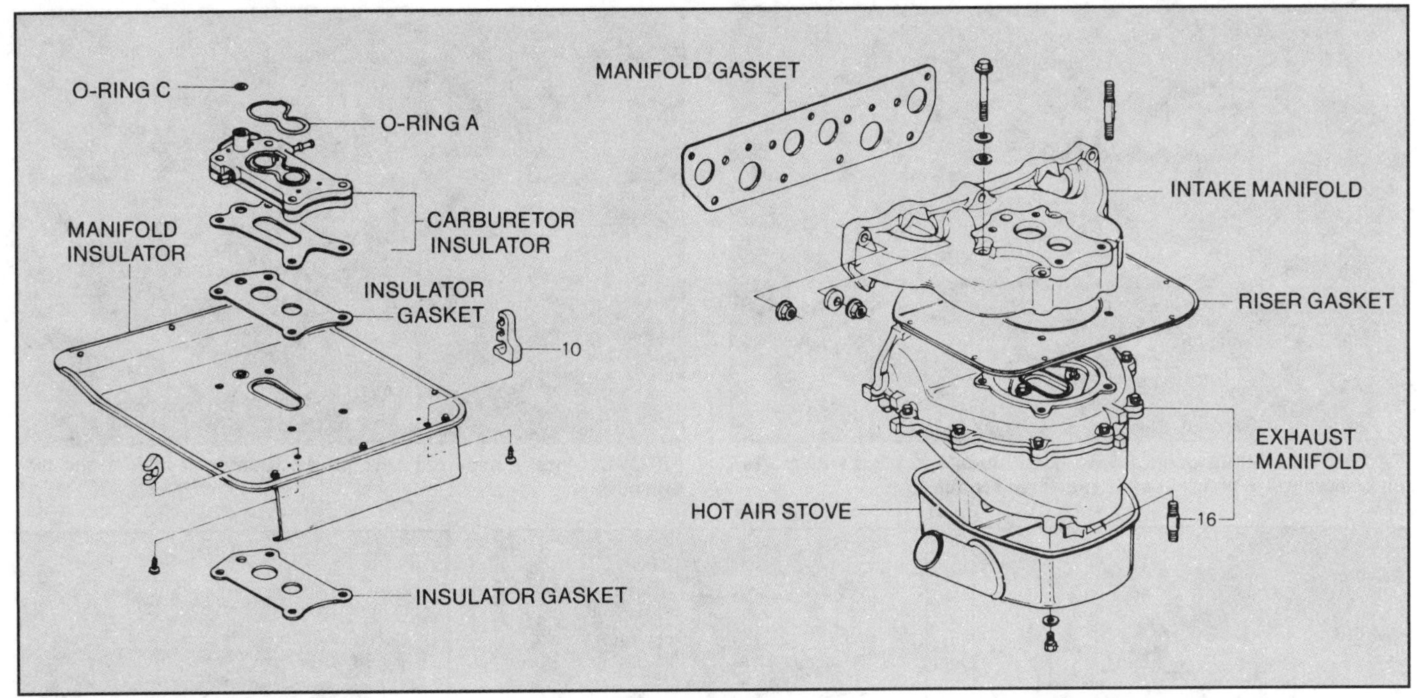

Early CVCC intake and exhaust manifolds; 1976 type shown, 1977—'79 type similar. Courtesy of American Honda Motor Co., Inc.

install them individually.

Before attempting to install CVCC-engine manifolds, make sure you have a new *riser gasket*. This asbestos sandwich-type gasket fits between the exhaust and intake manifolds. While you can use the old *heat shield*—the similar looking gasket between the intake manifold and carburetor—you *must* use a new heat-riser gasket.

Besides the gasket and manifolds, you need the two or four bolts that pass downward through the intake manifold into the exhaust manifold and the nuts and special thick washers that hold the manifolds to the head. The '75 1488cc engine uses 10 manifold-to-head nuts; all eight-port heads use 11 nuts; and all six-port heads use nine.

If you removed the EGR valve and anti-afterburn valve from a late-model manifold for cleaning, re-install them now. Use new gaskets and fit all support brackets.

Without using any sealer, slide the head-to-manifold gasket onto the studs. Now, slip on the intake manifold and start its center nut. If your engine uses exhaust-manifold-to-head seal rings, make sure they are positioned as shown, page 134. While holding the exhaust manifold against the head, install the special large dished washers—concave-side toward the head—and start all intake- and exhaust-manifold nuts.

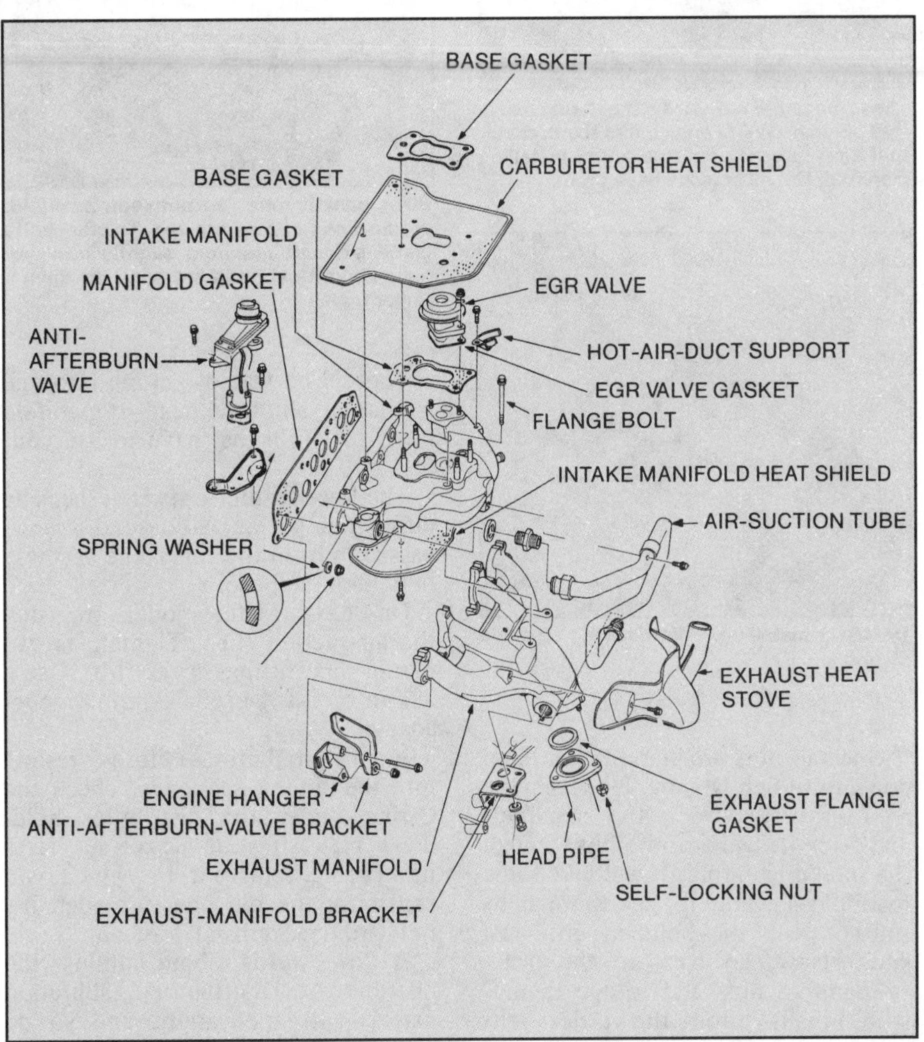

Exploded view of late CVCC intake and exhaust manifolds; '82 1751 shown. Courtesy of American Honda Motor Co., Inc.

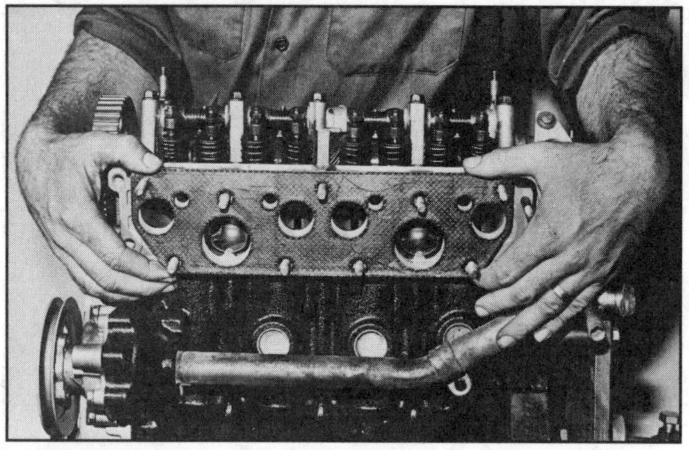

Fit CVCC manifold gasket onto head studs and align with ports and passages in head. Gasket sealer isn't required.

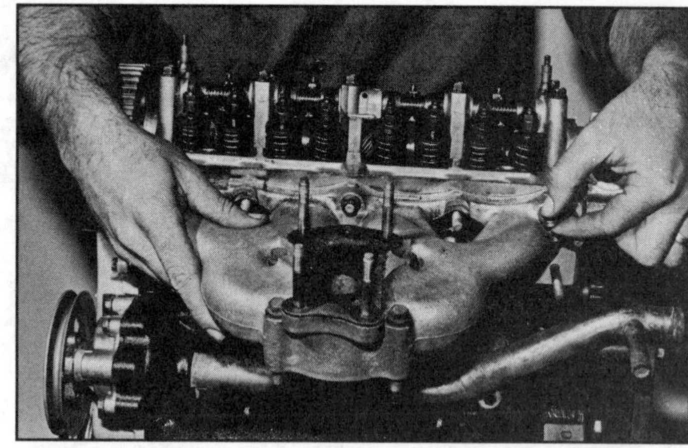

Fit CVCC intake manifold onto studs, then start middle and two end nuts.

These rings fall out easily. If exhaust manifold on your CVCC engine has them, check that they are not omitted when installing manifold. Not all engines have them.

Riser gasket must go between manifolds before bolts can be started. To start bolts, jostle exhaust manifold slightly until you have all vertical bolts started. Run them in finger tight.

Intake and exhaust manifolds *must* be free to move independently of each other when manifold-to-head nuts are tightened. Otherwise, manifold ears will break.

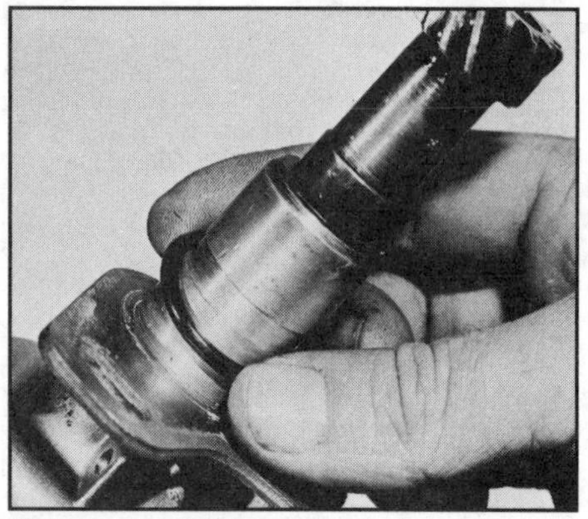

Before installing distributor, replace O-ring.

Once all nuts are started, run them down until lightly snug. Now, slip the riser gasket between the manifolds and drop the two or four bolts through the intake manifold. It will take some manifold jiggling to get these bolts started. Run the bolts in until just snug. Now, go back to the head-to-manifold nuts and torque them 7 ft-lb. Finally, torque the vertical bolts and then the head-to-manifold bolts both to 14—17 ft-lb.

There's no way of getting a torque wrench on all of the head-to-manifold nuts, so you'll just have to use your feel.

The lower center nuts are especially difficult to get at. A box-end wrench passed in from the side is the best way of tightening these.

On '82-and-later models, reinstall the air-suction tube. Tighten its two 32mm pipe fittings to 72 ft-lb. Use a 32mm or 1-1/4-in. flare-nut crowfoot adapter.

Time Distributor—Before getting into the timing sequence, check the O-ring under the distributor hold-down plate. If your gasket set has a new O-ring, replace it. In a pinch, you can reuse the old one, provided it's not hard, cracked and dried out.

A few words about timing the distributor: Distributor calibration varies with application, and so do Honda's timing procedures. There's a boggling array of procedures for high altitude, California, 48 states, manual transaxle, automatic transaxle, breaker-point ignition, electronic ignition and different model years. To make things simple, I give *one* general procedure for *all* distributors.

The engine should be at TDC, cylinder 1. If not, rotate the engine counterclockwise until the crankshaft and camshaft timing marks are at TDC. You did this during camshaft timing, page 129.

Next, determine that distributor rotor position is at cylinder 1. Look at the outside of the distributor cap. Near the base of the cap on most Hondas is a cast vertical line. This line marks cylinder-1 *tower*—female terminal for the sparkplug wire. Other cap designs feature a numeral 1 cast near the cylinder-1 tower, on top of the cap.

If your distributor cap doesn't have this cast line or numeral, you'll have to find cylinder-1 sparkplug wire on your own. Mount the distributor cap

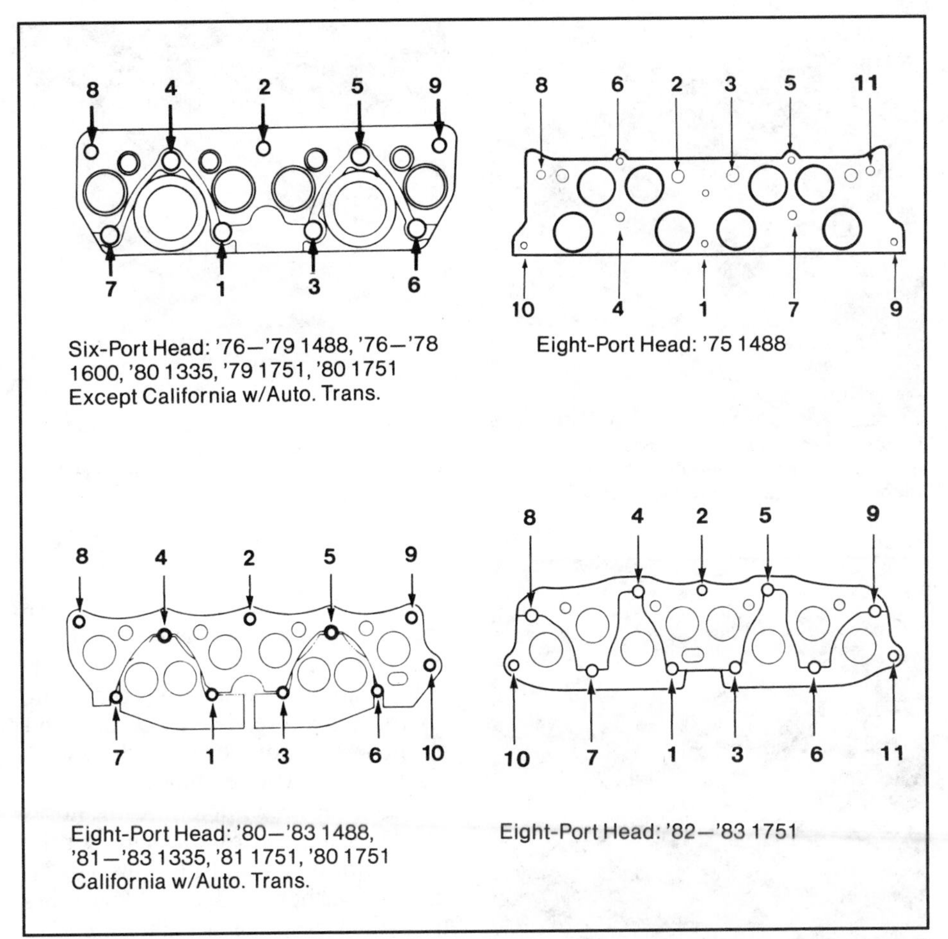

Final torque CVCC-manifold bolts 14-17 ft-lb following appropriate sequence. Courtesy of American Honda Motor Co., Inc.

Cast line on cap indicates cylinder-1 terminal. Some caps have a numeral 1 cast on top. Firing order is 1-3-4-2 and distributor rotation counterclockwise. As shown, number-1 lead is at front center, 3 to the right, 4 just visible behind 1, and 2 is at left. Lead in center goes to coil.

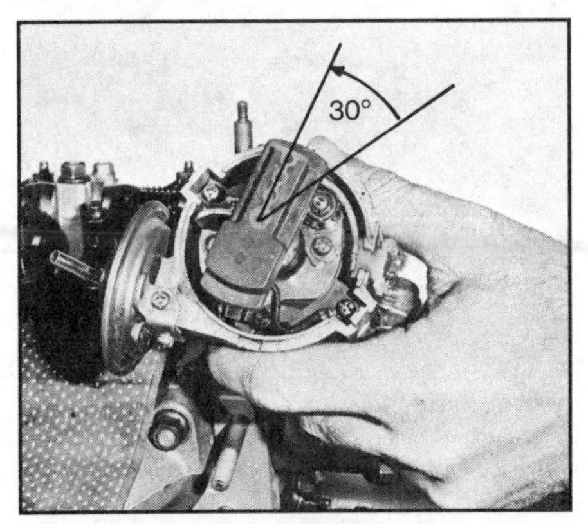

Distributor shaft will rotate counterclockwise about 30° during installation as helical gears engage. Text describes how to compensate for this movement.

on the distributor. Install the distributor so the hold-down bolt hole aligns with the eccentric opening in the distributor hold-down plate. For now, *don't worry about rotor position.*

If you are reusing the old sparkplug wires, lay them out so they fall toward their cylinders. With new wires, rely on the firing order. It's 1-3-4-2, numbered from the timing-belt end. The distributor rotates counterclockwise. When in doubt, refer the the firing-order illustrations on page 13. After determining which wire is for cylinder 1, remove the distributor.

Once you've determined cylinder-1 position, mark the distributor body with a piece of tape or felt-tip pen with the distributor cap in place. Then, remove the cap and align the rotor with the tape or mark.

Don't install the distributor yet. Because the distributor drive and driven gears are *helical*—curved, the distributor shaft will rotate approximately 30° as the gears engage. If you install the distributor with the rotor pointing toward cylinder-1 terminal, rotor position will be *30° advanced* after installation.

The solution, of course, is to retard the rotor 30° before installation. That means turning the rotor clockwise about two-thirds of its width.

Install the distributor, making sure the hold-down bolt holes in both distributor and head line up. When the distributor seats, the rotor should point at the tape or felt-tip mark—cylinder-1 position. If it doesn't, remove the distributor and try again. It's not unusual to get the distributor off one tooth the first time.

For 1335 or '80-and-later 1488 or 1751cc engines, there's another procedure. You can either use the steps just given, or the following Honda instructions:

Look at the distributor body and drive gear. At the base of the body you'll find a vertical line. It looks similar to the one cast in some distributor caps. There is a dot on the shank portion of the drive gear, just above the gear teeth. Turn the gear until the dot aligns with the vertical line on the body. Now, install the distributor. The rotor will automatically turn to cylinder-1 position. Of course, cylinder 1 must be at TDC when installing the distributor.

The distributor cap on these later engines should have either the usual vertical line or a circled numeral 1 cast in. If it's a numeral, it will be on top of the cap. This lets you quickly double-check distributor timing and takes the guesswork out of identifying ignition wires.

Sparkplugs—Gap and install a new set of sparkplugs. See the sticker under the hood for plug type and gap. If the old plugs had very few miles on

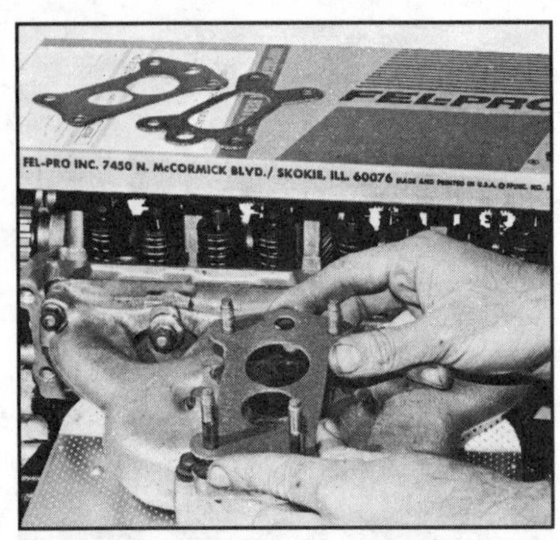

High-quality gaskets in complete sets ease engine assembly. As with remaining carburetor gaskets, this one needs no sealer.

Install heat shield and new carburetor gasket. On some models, heat shield attaches to intake manifold with screws.

Gasket and two O-rings fit atop carburetor insulator. Unusual, preformed piece is A O-ring. Round one is C O-ring.

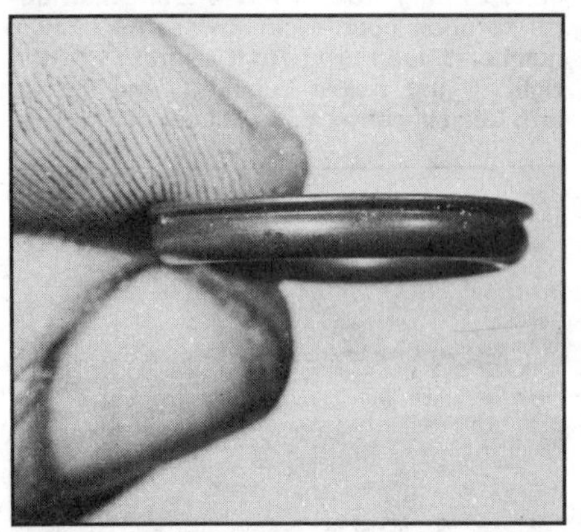

When installed correctly, C O-ring has small lip at top.

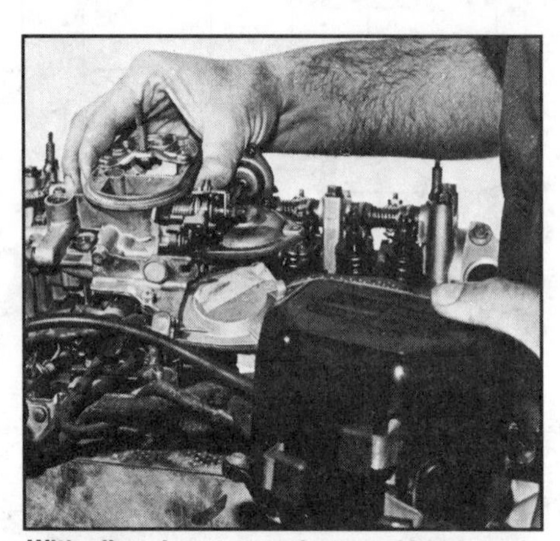

With all carburetor gaskets and insulator in place, carburetor and black box can be installed. Each carburetor stud gets a flat washer, then a wave washer—concave side down—and a nut. Torque nuts to 15 ft-lb.

Fit new O-ring to pump end of water pipe and push it into pump. Slip in pipe only far enough to install mounting bolt and align clamp with hole in block.

them, clean and reuse them. An engine usually runs better on fresh plugs, however. Use an extremely light coating of anti-seize compound on the plug threads. Keep the compound away from the electrode end.

Carburetor—The carburetor can be installed now, complete with hoses and emission-control box(es). Set the carburetor on its studs, followed by attaching washers and nuts. There are two washers per stud. Put the flat, thicker one on first, followed by the *wave* lock washer and nut. Tighten the four nuts in a criss-cross pattern.

Place the rocker cover on the head to keep out dirt. The head is now ready to dismount from the block, if you are going to install the engine without its head. Slip off the timing belt and remove the head bolts. Before lifting off the head, ready a place to set it down. Placing it across a metal drain pan or small trash can works well. Wherever you place the head, avoid placing it combustion-chamber-side down on a hard, flat surface.

SMALL PARTS—ALL ENGINES

Clean Head Bolts—If you haven't gotten around to cleaning the head bolts or aluminum-block engine studs, do so now. *Head bolts/studs must be clean.* Dirty fasteners will not torque correctly, and the head gasket and head could suffer. Lightly wire brush the bolt/stud threads and shank. Or, you can chase the bolt/stud threads with a 10 x 1.25mm die. Don't wire brush the bolt heads or you'll round them off. Instead, scrub bolt heads in solvent with a nylon brush.

Use non-hardening sealer on oil-pressure-sending-unit threads, then wrench it tight into block. On 1200s, sending unit mounts on opposite side of block.

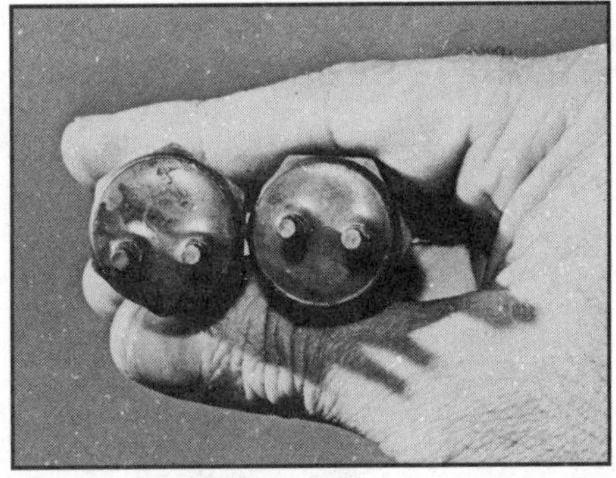

A and B thermosensors have different electrical-connector spacing. A is at left, B is on right.

On engines with two thermosensors, B thermosensor goes in block as shown and A thermosensor goes above it in head. Later CVCC engines have only head-mounted thermosensor. The 1170 and 1237 engines do not have thermosensors. The 1335 has an A thermosensor above oil filter. Use new O-rings when installing thermosensors.

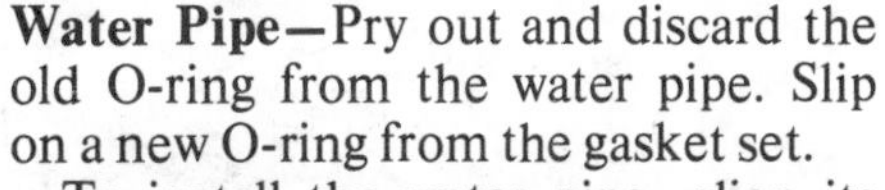

Water Pipe—Pry out and discard the old O-ring from the water pipe. Slip on a new O-ring from the gasket set.

To install the water pipe, align its mounting tang with the boss on the block, then push the pipe into the back of the water pump. Non-hardening sealer on the O-ring is OK, but it isn't necessary. There will be some resistance at first, then the pipe will pop into the hole. The O-ring fits into a detent. Tighten the mounting-tang bolt.

Oil-Pressure Sender—Lightly coat the oil-pressure sending-unit threads with non-hardening sealer. Then, start it in its hole, above the oil-filter boss on cast-iron blocks; low on the block under the intake manifold on aluminum blocks. Run it in finger tight and finish installation with an adjustable wrench.

Oil Filter—Before spinning on a new oil filter, fill it with oil. Open a can of 30W SE or SF oil and fill the filter halfway. Swirl the oil around in the filter until it is absorbed by the paper element. Pour in more oil until the paper is saturated and the filter begins to fill. Smear some oil on the filter seal, then quickly spin the filter into place. Note when the rubber seal contacts the block, then turn the filter another *two-thirds* of a turn.

Filling the oil filter helps speed engine priming. If it isn't in your way, slip in the dipstick. It will cover the dipstick-tube hole—and help keep dirt out of your fresh engine.

Engine Mounts—All engines have a front mount that attaches to the block. On iron-block engines, the mount goes to the left of the oil filter. On aluminum blocks, the front mount fits over the water pipe.

Bolt front engine mount to right side of block.

The left mount was installed with the cam-timing apparatus, so there's nothing to do there now. The setup is similar with the rear engine mount. Most rear mounts attach to the engine with only one bellhousing bolt—some use two. But, because the mount uses bellhousing bolts, you can't attach the mount now. It will go on during engine installation.

TORQUE SPECIFICATIONS

		ft-lb	kg-m
Auxiliary-valve collar	CVCC	54-61	7.5-8.5
Connecting rod	Except 1751	20	2.8
	1751	22-25	3.0-3.5
Exhaust Manifold-to-pipe	2-nut	40	5.5
	3-nut	36	5.0
	1200	14-18	1.9-2.5
Flywheel, driveplate	All	36	5.0
Front cover	All	6-9	0 8-1.2
Crankshaft pulley	'75-'80 CVCC	58-65	8.0-9.0
	'81-and-later CVCC	80	11.0
	All 1200	34-38	4.7-5.2
Head bolts	Iron blocks	44	6.0
	Late EB1, all EB2 & EB3, all 1335	37-42	5.1-5.9
	Early EB1 1019949 and earlier	30-35	4.2-4.8
Main-bearing-cap bolts	1170 1237 1335	29	4.0
	1488 1600	32-35	4.4-4.8
	1751	44-51	6.1-7.0
Manifolds	1170 1237	13-17	1.8-2.4
Manifold studs	CVCC	14-17	2.3
Manifold vertical bolts (See text for manifold torque sequence)	CVCC	14-17	2.3
Oil pan	Bolts	13	1.8
	Nuts	8	1.0
Oil pump	All	7-10	1.0-1.4
Rocker-arm assembly	Small bolts	10	1.4
	Large bolts	17	2.4
	1200, all	13-16	1.8-2.3
Water pump	All	10	1.4

Engine Installation 8

Installing and starting your engine is the last major operation of the rebuild. Everything in the parts pile should be used up or replaced when you are done. So, if there is any other work you want to do, such as rebuilding the radiator or starter motor, do it now. The remaining parts should be clean and neatly arranged.

One job that will save a lot of frustration is charging the battery. It's maddening to find when you're ready to fire up your new engine, that the battery that sat on the floor for a few weeks is dead. If you don't have a battery charger, charge the battery at a service station. Or, borrow a friend's charger.

Engine-Compartment Cleaning— Another important but often-neglected job is cleaning the engine compartment. It would be a shame to drop your new engine into a compartment with 100,000 miles of accumulated crud. Aesthetics aside, a clean engine compartment makes engine installation easier and safer. Any sort of degreaser works well, but aerosol engine cleaners are best. Don't approach this cleaning job haphazardly; you could harm the car's mechanicals, electrical system or cosmetics.

Take the time to mask parts that must remain dry—particularly electrical components. Engine cleaners contain strong detergents that can harm paint, so be careful not to splash it carelessly on the vehicle exterior. If you steam cleaned or pressure washed the engine and its compartment before pulling the engine, tidying up the compartment should be easy.

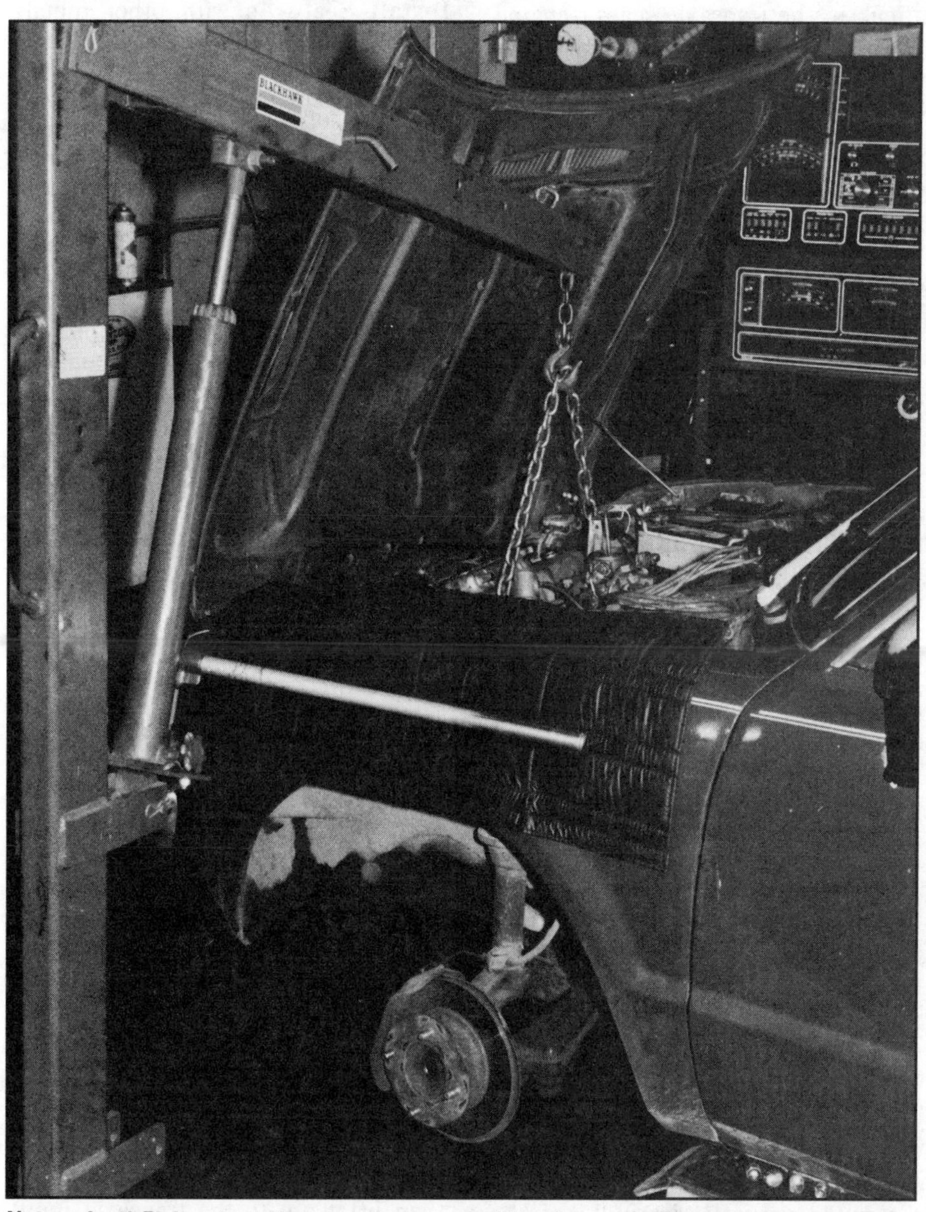

Not so fast! Before installing your freshly rebuilt engine, clean the engine compartment. Also, check condition of transaxle front seal and manual-transaxle clutch. Photo by Ron Sessions.

Another way to clean the engine compartment is at the car wash. It's a lot of work to tow a car to and from a car wash, but at least you leave the mess there and have the use of high-pressure hot water. You can also clean loose bolts, brackets and other hardware at the car wash. Don't forget to clean the underside of the hood.

When you're finished cleaning the engine compartment, get out those fender covers or blankets and tape them to the inner fenders. It would be a shame to scratch a fender in your eagerness to install your rebuilt engine.

Take Your Time—Before getting into the mechanics of installing the engine, remember, *don't rush the job.* It's only natural to be overeager when installing a new engine. It's the last part of a long and involved job, so wanting to hurry and see the results is natural.

You can work quickly, but remember, do each job *thoroughly* and *carefully.* It doesn't make sense to spend a lot of time and money pulling, tearing down, inspecting and assembling an engine, only to do a hurry-up, haphazard job of installing it. You may end up wasting your time and money. Work patiently. If you find yourself in too big a hurry, quit for the day and start again tomorrow.

TRANSAXLE FRONT SEAL

While the engine is out, inspect the bellhousing interior for oil. Check the front transaxle seal for leaks. A bellhousing that's gooey with oil indicates a leak in the transaxle seal or engine rear-main seal. Closely investigate the oil mess. You can probably tell which seal it came from. A good clue is the back of the flywheel or driveplate. If the engine side of this part is dry, the transaxle seal is leaking. Another giveaway is smell and color. Automatic transmission fluid (ATF) is reddish-pink when new, and reddish-brown when old; it smells like mineral oil. Motor oil, the factory fill for *manual transaxles* as well as engines, is brown and smells like petroleum.

If there is no transaxle leakage, great. Go on to the next step. If you do find leakage, there is good news and bad news. The good news is for automatic-transaxle owners: The seal in the automatic gearbox is cheap and easy to change. The bad news is for manual-transaxle owners: Removal and complete manual-transaxle disassembly is required to change the front seal. You know what I'm going to say next. If the seal is leaking, bite the bullet and have it changed. Letting it leak will only ruin the clutch and possibly the transaxle, besides making a mess in the driveway.

Transaxle work is beyond the scope of this book, so refer to the factory shop manual, a dealer or independent garage.

Automatic-Transaxle Front Seal— First step in changing the front seal is removing the torque converter. The converter is keyed to splines on the transaxle input (main) shaft and stator shaft. The converter is heavy because it's about half-full of fluid. Grasp the converter and pull it off the shaft. Don't tilt the converter neck down or ATF will pour out and make a mess.

After you have the converter clear of the engine compartment, tilt it over a bucket or drain pan and pour out the old ATF. Later, after you've installed the seal, replace the torque converter dry. You can then add *fresh ATF* and fill the converter through the dipstick tube.

With the converter out of the way, the seal will be visible. Pry out the old seal with a screwdriver, taking care not to gouge the stator shaft or aluminum transaxle housing. Work around the seal; don't pry in one spot only.

Remove all oil and grit from the seal bore with a paper towel and lacquer thinner. Lube the lip of the new seal with fresh ATF.

Install Seal—As with other metal-housed lip seals, the front automatic-transaxle seal must be started and driven in squarely. I doubt that you have a socket deep or large enough to fit over the transaxle input (main) shaft and stator support. If you don't, a pipe or steel tube will work. *If you're careful,* you can install the front seal with a hammer and small wood block.

To drive in the seal with the hammer-and-block method, hold the seal squarely over its bore, lip pointing toward the transaxle. Begin working it in with light taps around its periphery. Start with taps opposite each other: at 3 o'clock, 9 o'clock, 12 o'clock and 6 o'clock. After the seal is well started, tap it around its circumference. When the seal bottoms, you'll feel it; stop tapping.

After you've installed the seal, clean inside the bellhousing with lacquer thinner or other solvent. Don't get any solvent on the clutch-release bearing.

Install Converter—The converter is installed by slipping it onto the transaxle input (main) shaft, then indexing it with the stator shaft.

First, inspect the seal surface on the torque-converter neck for any roughness or varnish. If the surface is rough, give it the same 400-grit sandpaper treatment you gave the crankshaft journals. If it's just coated with varnish, clean it with solvent or lacquer thinner.

Next, apply some ATF to the converter-seal surface to lubricate the new seal. Grasp the converter firmly, then lift it up to the transaxle input shaft. Don't let the converter hang from the shaft. This may damage the seal or shaft. You should be able to install the converter with one steady motion. Make sure you mate the converter to both sets of splines—input shaft and stator shaft.

PREPARATION

With the chassis work out of the way, turn your attention to the engine and parts that bolt to it.

Clutch & Flywheel—With a manual transaxle, you must inspect the clutch before installing the engine. Now is the logical time to do such a job—while the engine is out.

Friction between the disc and pressure plate wears the disc and heats the pressure plate. Heat can cause hard spots to form on the pressure plate. They show up as blue-black discoloration. Heat also warps the pressure plate and fatigues the spring. On the other side of the disc, the flywheel gets a similar treatment. But because the flywheel has considerably more mass—weight—than the pressure plate, it can take more punishment before it's damaged.

Inspect Pressure Plate—Honda clutches are the diaphragm type, using a single, flat disc-type *Belleville* spring. The center of the spring has many radial cuts that create *fingers.* These fingers extend toward the center of the pressure plate and are bent rearward at a slight angle. When sighting across the spring fingers, they should be the same height. If the fingers aren't, the spring could be damaged. Replace the pressure plate.

If the pressure plate has been in service awhile, the release bearing will have worn a groove in the fingers. Replace the pressure plate if the grooves are more than 0.0012-in. (0.03mm) deep.

Look for cracks where the fingers meet the uncut portion of the spring. If you find any, the spring must be replaced. If the spring is OK, turn the pressure plate face up and lay a straightedge across the pressure ring. If warped, you'll see a definite concave shape beneath the straightedge. The gap will be largest at the inboard edge of the pressure-ring friction surface. A 0.001-in. (0.025mm) bend is bad; 0.002 in. (0.05mm) definitely means replacement. Inspect the pressure plate for hard spots and *heat checks,* or cracks. Also, look for grooves cut into the pressure plate by clutch-disc rivets. If the disc was worn down to its rivet heads, the pressure plate and flywheel may be grooved.

You don't have much of a choice when it comes to servicing the pressure plate. Light scratches or scoring can be polished out with 500- or 600-grit emery paper. If there are surface cracks, the pressure plate must be replaced.

Replacing a pressure plate rather than resurfacing it is a smart move for several reasons. First, the pressure plate will be full thickness and better able to absorb heat. Second, the spring will be new and not fatigued like your old one, so you'll get maximum capacity from the clutch. Third, other internal parts will be new, not partially worn. Fourth, the rebuilt engine should have more power than when you pulled it out. This means the clutch must handle more torque. Finally, there are price and convenience considerations. Once your freshly rebuilt engine is installed, you don't want to have to pull it just to replace the blasted clutch!

Disc Inspection—The major indicator of disc condition is the thickness of its facings. The thinner they are, the less life the disc has. Real problems begin when the disc is worn so badly that the rivets stand flush with or above the surface of the friction material. Then, both pressure plate and flywheel will be grooved from rivet contact. You may have already discovered this from inspecting the pressure plate. The other disc-killer is oil. If the disc is oil-soaked, throw it away and start with a new one.

Measure disc overall thickness with a micrometer or vernier caliper. A new disc measures 0.327—0.354 in. (8.3—9.0mm) and has a 0.232—0.260-in. (5.9—6.6mm) service limit. A disc worn to anything approaching 0.232 in. should be replaced. You *do not* have to compress the disc to get an accurate thickness measurement on Hondas.

Another disc measurement is rivet depth. Use a depth gage to measure from the lining to a rivet. New, this distance is 0.047—0.059 in. (1.2—1.5mm) and the service limit is 0.009 in. (0.2mm). It all depends on how long you want the clutch to last before you must pull the transaxle to service it. If the disc is half worn or more, don't give it a second thought. Replace it.

Flywheel Inspection—Inspect the flywheel just as you did the pressure plate. One problem you don't have to be concerned with is warpage. A flywheel's large mass can absorb and dissipate more heat than the disc can generate before it self-destructs. However, look for grooving from rivets and blue-black hot spots.

Heat checking can be a problem with flywheels. Normal heat checking can be cured by light resurfacing, but cracks radiating outward from the center mean the flywheel *must* be replaced. Usually, these cracks *cannot* be removed by resurfacing. Such a flywheel is a potential hand grenade—it could explode at high rpm, causing serious damage to the vehicle and injury to passengers and bystanders.

Flywheel suffers from hot spots—dark areas—and radial heat checks. Flywheel was reconditioned by grinding.

If you have the flywheel resurfaced, use a grinder, not a lathe. The lathe cutting tool will literally jump over the hot spots, which are harder than the parent material, resulting in a bumpy flywheel surface. Grinding makes the flywheel perfectly flat.

If your flywheel is in good condition—no cracks, grooves or hot spots—freshen its working surface with 500- or 600-grit emery paper. This will remove bonding resin that was transferred from the disc to the flywheel. After sanding the flywheel, wipe it with non-petroleum solvent, such as lacquer thinner or alcohol, to remove any residual oil. Give the pressure plate and disc the same treatment.

Pilot Bearing—Unlike most other engines, the pilot bearing *does not* go in the end of the crankshaft. All Honda engines except the 1751 use a roller-type pilot bearing in the center of the flywheel or driveplate. The 1751 engine has no pilot bearing at all.

If originally equipped with one, install a new *pilot bearing* now. Just tap out the old and tap in the new. The old bearing makes a great driving tool for the new bearing.

Clutch-Release Bearing—There are two ways of looking at what to do with the release bearing, or *throwout bearing,* as it's sometimes called. You could take the approach that every time the clutch needs servicing, the release bearing should be replaced. Or, you could simply change it now because the engine is out and it is easily done.

Sealed pilot bearing installs in center of flywheel or driveplate on all engines except 1751. The 1751 uses *no pilot bearing.* If bearing feels rough or sticky, replace it. Tap out old bearing and use it to drive in replacement.

Either way you look at it, you should change the bearing now because of the labor involved in replacing this fairly inexpensive part later.

The bearing rides on a holder, which in turn, is concentric with the input shaft. On some models, the release bearing is cable actuated. The cable pulls up on a shaft-mounted lever outside the transaxle. This shaft has two ears to which the release-bearing assembly is clipped. By removing the spring from the release arm on the bellhousing's exterior, and a bolt inside the bellhousing, the bearing assembly will slide off. The two clips have to be lightly pried from small holes in the bearing holder.

Other models use hydraulic clutch actuation. Here, you merely free two spring clips and pull the bearing assembly off. It will take a little wiggling to free the assembly from the release arm.

Once the bearing assembly is out, tap the bearing holder out of the bearing ID. Tap the holder into the new bearing and mount it back in the bellhousing. This all sounds complicated, but is quite simple and goes quickly.

Flywheel/Driveplate Installation—Check the crankshaft flywheel flange for nicks or other imperfections that could cause the flywheel or driveplate to sit slightly skewed. Smooth any high spots with a file. Be careful not to gouge the flywheel in the process. A flywheel must be square with the crank—no wobble as it rotates—or it will have bad effects on transaxle internals, the rear-main bearing and

With a little wiggling, release bearing and holder unclip from this style release arm. Holder will tap from old bearing and into new one. Note configuration of spring clips before removing bearing.

Universal clutch-alignment tool keeps clutch disc centered while pressure-plate bolts are torqued. Disc must be centered or engine and transaxle will not mate.

Always use new exhaust-flange gasket for gas-tight seal.

bearing and crank seal.

Position the flywheel or driveplate so it engages the crankshaft dowel. Remember, 1200s have *no* dowel.

If you're fitting a driveplate, make sure it dishes *away* from the block. Don't forget the special washer that fits under each of the six bolts. With the drive plate or flywheel on the dowel, tighten the bolts in a star pattern to 36 ft-lb.

Install Clutch—Fitting the clutch to the flywheel can be a handful. Besides the pressure plate, disc and eight bolts, you'll also need an alignment tool. Use the tool to align the disc to the flywheel pilot bearing, if so equipped, so the input-shaft splines slide straight through into the bearing during engine installation.

Although I've lined up a clutch disc by eye in an emergency and had it work, I don't recommend that you do it. Mating an engine and transaxle is tough enough, even with a perfectly aligned disc. Doing it by eye is asking for trouble.

If you have a Honda transaxle mainshaft lying around the garage, use it as an alignment tool. If you don't have one, the local parts store has various alignment tools. They range from inexpensive plastic or wooden ones to more expensive universal tools. An inexpensive one will work fine.

Start clutch installation by cleaning the disc. Wipe it down with lacquer thinner to make sure it's clean. The disc may look the same on both sides, but it's not. The side with the long hub-center projection, damping springs and retaining plate must be installed *away* from the engine, *toward* the pressure plate and transaxle. If your install it backward, as I did once, you'll have disc-to-flywheel interference in short order.

Lay the disc on the pressure plate, hub-side down, and lift the two to the flywheel. Use one hand to hold the pressure plate and disc against the flywheel. With your other hand, insert the alignment tool through the center of the disc and into the pilot bearing, if so equipped.

Rotate the pressure plate until it is aligned with the dowels on the flywheel.

The alignment tool should support itself and the disc while you start the pressure-plate bolts. Run in the bolts finger-tight. Lift up on the end of the alignment tool to center the disc, then *tighten the pressure-plate bolts gradually,* no more than two turns at a time. Otherwise, you may bend the pressure-plate cover.

Check that the alignment tool slides in and out of the disc and pilot bearing without hanging up. If it does, loosen the pressure plate and center the disc again. If, after retightening the bolts, the alignment tool moves in and out freely, remove it. You're now finished at the rear of the engine.

Exhaust Seal—An item too easily forgotten is the metal/asbestos sandwich-type exhaust gasket. This gasket sits in a groove in the A-pipe mating flange. If you haven't done so already, pry out the old one with a small screwdriver and press in the new one by hand.

Position Crankshaft—On manual-transaxle applications, make sure the crankshaft is at TDC cylinder 1. This will speed camshaft timing later. On automatic-transaxle engines, the crank must be rotated to bolt up the torque converter, so there is no sense positioning it now.

ENGINE INSTALLATION

Prepare for engine installation by bringing all the pieces together. You'll need a cherry picker or chain hoist, floor jack, jack stands and the usual hand tools. It's also best to have a friend help with engine installation. One person can work the cherry picker and the other the floor jack. It's safer—and much easier.

Remove the hood from the car, taking care not to mix up any shims. If the lifting equipment has sufficient travel, raise the car and position it on jack stands. Block the rear wheels and set the parking brake. Use the lowest possible jack-stand setting. You need just enough space to comfortably reach the crossmember and exhaust connections from underneath the chassis.

Place a wood block on the floor-jack pad and slide the jack under the transaxle. Come in from behind the right front wheel. This way the jack

CVCC timing marks are visible through hole in block, at front edge of bellhousing; 1200 engine timing marks are on crankshaft pulley. I repaint timing marks so they are more visible.

Myron took this picture from the roof so everything looks backward. Standing over left fender and pulling block toward you will allow engine to clear hoses and bellhousing. Have helper work cherry picker while you guide engine. Keep your hands high on block in case it slips. Impatience got better of me when I skipped covering block deck with cardboard to protect deck and bores from damage, dirt and water. One deep nick and I'd have some tedious filing to do.

will not interfere with the cherry picker or be constantly underfoot when working around the engine compartment. Raise the transaxle somewhat higher than its normal installed position. A slight bellhousing-up attitude helps.

Remove the crossmember and any transaxle supports you may have added. Also remove the bellhousing dust cover and trans-mount cup.

Head-On Engine Installation—If you decide to install the engine complete with cylinder head and manifolds, keep a few things in mind. First, CVCC engine balance is affected by the weight of the manifolds hanging off the fire-wall side of the head. Try attaching the lifting chain to points in line with the head/manifold mating surfaces. Test these lift points by observing the angle of the oil-pan bottom. When hanging at the correct attitude for installation, the CVCC-engine oil-pan bottom will parallel the ground.

Because aluminum-block 1200s have crossflow manifolds, they are better balanced. Good lift points for these engines are the distributor hold-down bolt hole and the flywheel-end torque-rod mounting-bracket bolt hole.

You should have at least one helper when installing a complete Honda engine. It takes more than one person to carefully steer a complete engine into the compartment. Although it may be tempting to stick a hand under the engine to adjust something during installation, *don't do it!* That's an easy way to lose a finger. Use a 2x4, long breaker bar, crowbar or pull the engine out and try again.

Get the exhaust pipe and manifold close during installation, or other things will get in your way. Read the section on joining the exhaust, page 148, before engine installation.

Just because the short block and head are together, don't be lazy. Pull the oil-pump drive shaft so you can manually pressurize the oiling system before starting the engine.

Head-Off Engine Installation—What follows is the procedure for installing the engine minus its head. This method provides much more working space around components such as the master cylinder and booster, and makes clutch disc/transaxle input-shaft mating much easier.

Lift Engine—Attach the lifting chain diagonally across the block. You can use two head bolts and large washers passed through the chain and threaded into the deck, as I did. Or better yet, use two shorter bolts—10mm shank X 1.25 thread pitch X 50mm long. The shorter bolts will capture the lift-chain links between the bolt head and block deck. Why risk bending the head bolts if you don't have to? Either way, don't make the lifting chain overly long or the lifting hoist will run out of travel.

Fashion a cardboard cover to place over the block deck. This will keep dirt out of the engine, and protect against scratches and dings from the lifting chain. Tape the cover to the block.

Raise the block off the floor and adjust the angle at which it hangs. The block should hang level, or slightly lower at the bellhousing end. The deck tilts toward the radiator in the installed position, so if you can get the block to hang that way now, all the better. If not, adjust the block as it settles onto its mounts.

Lower Engine—Position the block over the engine compartment and lower it slowly. Watch for wires and cables. They tend to catch on engine pulleys and hang things up.

When mating an engine to an automatic transaxle, the block can be lowered almost vertically onto its mounts. You need only about 1 in. of clearance to keep the torque converter and bellhousing clear of the driveplate. So, just lower the engine, get the trans and block mated, run in

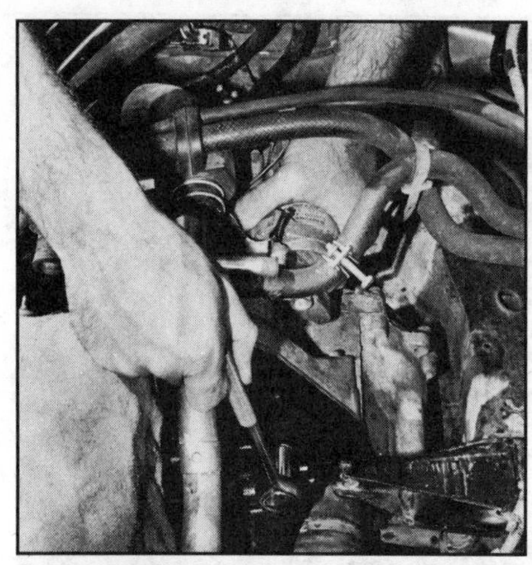

Starter bolts, including long, "backward" one are also bellhousing bolts. Rear engine mount has already been connected. Front engine mount can also go on. So can left mount, if no alternator or A/C compressor mounts there.

the bellhousing bolts and lower the block onto its mounts.

Manual-transaxle engines are more difficult. The input shaft must clear the pressure plate, then, at just the precise height, be pushed into engagement with the clutch-disc splines. On these applications, the engine should be about 6 in. from the transaxle so it will clear the input shaft.

With a manual transaxle, lower the engine until it's above the transaxle. Then increase the downward angle of the flywheel end. Lower the block some more, guiding the clutch over the input shaft. *Do not* put your hands between the block and transaxle. Serious injury could result if the two pieces slam together—especially if your helper is trying to push them together. You could crush a finger or hand.

If you have trouble getting the engine and transaxle together, remember that you can vary transaxle height slightly with the floor jack. Keep adjustments within reason. CV joints cannot tolerate severe angles.

Several things can keep an engine and transaxle from mating. If you forgot to remove the crossmember, the transaxle mount will block the driveplate or flywheel. You'll grunt and groan until you discover your mistake and remove the crossmember. If the manual-transaxle input shaft misses the center hole in the clutch disc and catches against the disc's center section, pull the engine away and try again.

Although unlikely with the small splines on the input shaft and clutch disc, another problem could be tooth-to-tooth spline contact between the two. Rotate the crankshaft slightly or engage fourth gear and turn the front wheels to rotate the splines into engagement.

The most common problem is lowering the block at too great an angle from horizontal. Sight down the space between the engine and transaxle. If the gap is equal top and bottom, the angle is OK. But if the gap is closed at the top, the angle is too great. A large angle will prevent the input shaft from slipping through the clutch disc and into the pilot bearing or flywheel.

When the block and transaxle are flush, begin installing the bellhousing bolts. Three bolts go in from the front side of the engine. On the back side, two bellhousing bolts double as starter-mounting bolts. The two bolts below them double as engine-mount bolts. Go ahead and install the starter and rear engine mount now. Sometimes, one of the starter bolts is "backward"—enters the bellhousing from the block side. There are no bolts on the bellhousing bottom.

On pre-1980 Civics and pre-1982 Accords only, after all bolts, starter and rear engine mount are installed, lower the engine/transaxle unit onto its mounts. The front and rear mounts use two nuts atop a flat plate that acts as a common washer. Additionally, there's a ground strap attached to each mount. Position the plate and ground strap, then thread the two nuts into the front mount. Repeat for the rear mount.

On all Preludes, '80-and-later Civics and '82-and-later Accords, the crossmember must be installed *before* lowering the engine/transaxle unit.

If no accessory mounts in front of the camshaft sprocket, bolt up the left engine mount. Use a punch or screwdriver to pry the chassis end of the mount out of the recess in the left fender. Run in the two small hex bolts through the two mount halves, then tighten the large hex bolt at the fender.

Driveplate—With an automatic transaxle, bolt the driveplate and torque converter together now. To install the eight bolts, rotate the crankshaft with a socket and ratchet on the front-pulley bolt. Access is through a hole in the left-front inner fender. Because the Honda engine is so small, you can reach both the crank-pulley and driveplate bolts while under the car. Or, have a helper turn the crank.

Dust Cover—Install the dust cover/splash shield. The cover attaches to the oil pan with two oil-pan bolts. Hold the cover in place so you can find which ones they are. Remove the two oil-pan bolts, position the dust cover and run the bolts back in. Now, position the metal cup or rubber crossmember mount against the dust cover and run its three bolts through the cover and into the bellhousing.

Lower Torque Rod—On all Preludes and 1980-and-later Civics, install the front, lower torque rod with two eye bolts between the body and bellhousing mount.

Crossmember—Raise the crossmember into position and install its four or six bolts.

On all Preludes, '80-and-later Civics, and '82-and-later Accords, the crossmember has already been installed. Lower the engine onto it. Snug the rubber engine-mount-to-crossmember stud nuts. Then, bolt up the left engine mount, unless an accessory mounts in front of it.

Remove Chain Hoist—Once the engine is bolted to its mounts, remove the lift chain.

A/C Compressor—If the A/C compressor mounts to the front side of the block, install it now. Once the head is installed, access is restricted.

Bolt the compressor lower mount to the block—two bolts. Unwire the compressor from the chassis. Inspect the adjusting mechanism. Note the long hex bolt that passes through a spacer. Also threaded through this spacer on the finished installation are two special, long-hexed lower mounting bolts. Get a good mental picture of how this adjuster works, because you'll have to bolt it together blind. The compressor blocks your view of the adjuster and two lower mounting bolts. When installing, orient the mounting spacer—actually a cylinder with threaded ends—with its ends equidistant from the bracket ends.

Once you understand how the ad-

On automatic-transaxle cars, bolt driveplate and torque converter together before installing dust cover. On manual-transaxle cars, bolt on dust cover. Beware of overtorquing dust-cover bolts—they thread into aluminum.

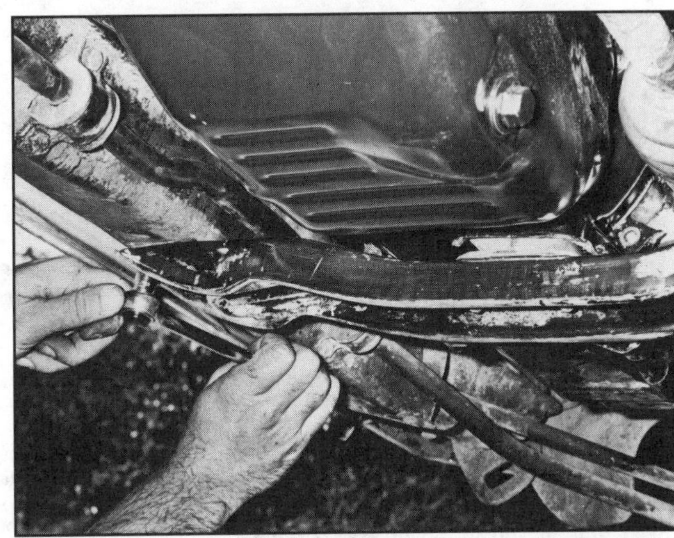

Crossmember mounts with six bolts on early models and four bolts on later ones. On '80-and-later models, transaxle mount must be bolted together, unlike early Accord shown.

juster works, move the compressor into position and thread two lower mounting bolts into the adjuster. These two lower bolts are special, long hex pieces. Now, install the upper bracket with its two bolts. Leave the A/C compressor belt off for now. The alternator belt must go on first.

Alternator—On 1170/1237 engines, attach the alternator lower bracket and the upper arm. Position the alternator and pass the lower bolt through the alternator and bracket. Then, run the washer, lock washer and nut onto the bolt finger tight. Now, install the upper mounting/adjusting bolt with its washer finger tight.

Drape on the fan belt, pull the alternator away from the block until the belt is tight and tighten all bolts. Adjust belt tension so it deflects 1/2 in., midway between the water-pump and alternator pulleys. Finish the installation by making the electrical connections. Follow your tape "flags" and notes.

On cast-iron-block engines, install the alternator *after* the cylinder head.

Power Steering—On cars so equipped, install the power-steering pump. The pump's lower mounting bolt threads into a bracket near the pan rail, below the water pump. If you haven't done so already, install that bracket now. The pump's upper attachment is a curved arm bolted to the water pump. Install the upper and lower mounts and drape on the drive belt. Adjust belt tension so deflection on the longest span is no more than 1/2 in.

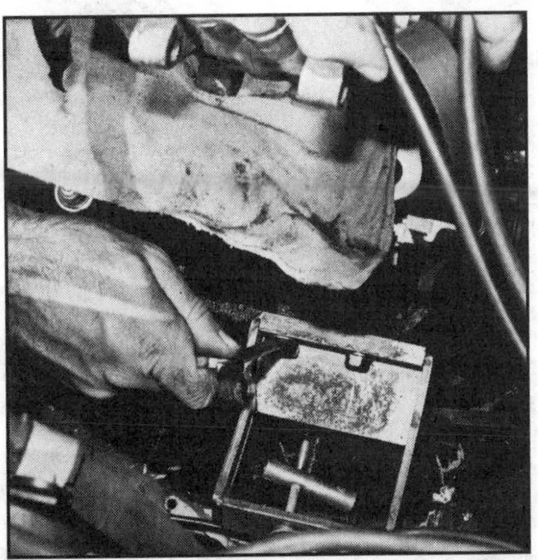

When mounting this style A/C-compressor bracket, make sure longer portion of round bar faces left fender. Bottom compressor bolts thread into this bar, so it must reach both bolts.

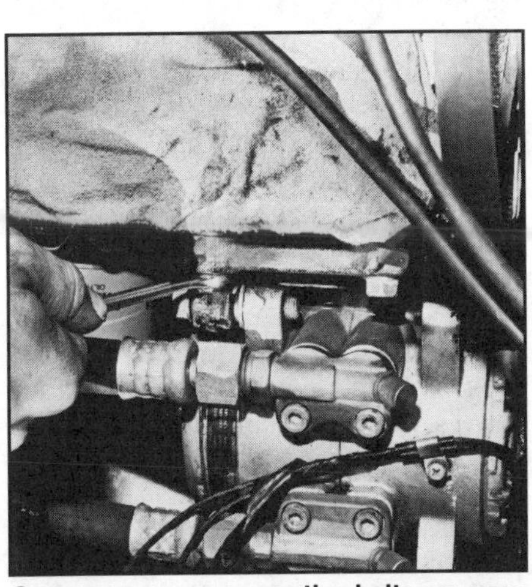

Compressor upper mounting bolts are easy to install. Get these top bolts started, then bottom ones. After all bolts are started, tighten them.

INSTALL CYLINDER HEAD

Note: Disregard this section if you've already installed the cylinder head. Instead, go to "Exhaust Pipe and Manifold," page 148.

Prepare for cylinder-head installation by removing the protective cardboard and cleaning the block deck. Use paper towels to wipe the deck. Paper lint dissolves in oil and will not clog the lubrication circuit—cloth lint will cause *big* problems. A little lacquer thinner or other, non-petroleum-based, fast-evaporating solvent is helpful. Also check for anything that may have gotten into the cylinders and remove it.

When you think it's clean, run your palm across the deck—assuming your palm is clean. Feel for grit and watch for smearing grease. Once the deck is spotless, give the head mating surface the same treatment.

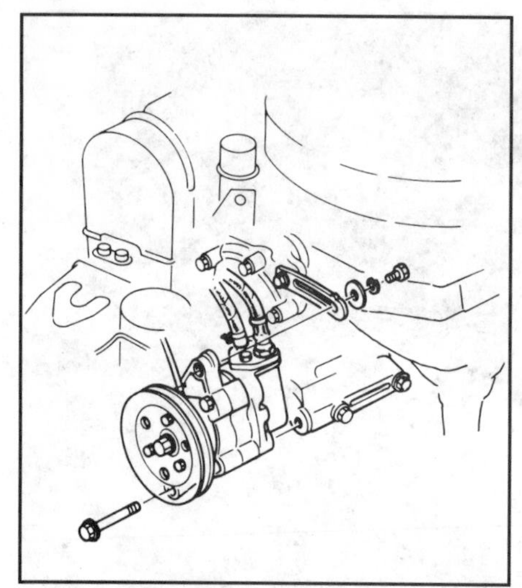

Power-steering-pump mounting. Courtesy of American Honda Motor Co., Inc.

Fel-Pro Print-O-Seal head gasket (top) and Honda OEM moly-coated gasket (bottom): Print-O-Seal design features extra sealing power and works great in all Honda applications except aluminum-head/iron-block. Dissimilar metals expand at different rates and tear sticky gaskets. Moly-coated Honda gasket allows relative head and block movement without damaging gasket.

After wiping deck with clean hand to check for dirt and nicks, position head gasket over dowels without bending or cutting it. Use no sealer or cement. Gasket must be dry to work properly.

Head Gasket—Inspect the head gasket for nicks and cuts before laying it on the deck. Look for a TOP or FRONT designation on the gasket for orientation. If there is no such wording, make sure the manufacturer's trademark appears on *top.* The gasket should lie flat without bends or wrinkles. Replace a kinked gasket with another.

Double-check for two dowels in the deck *or* cylinder head. If there are no dowels, they probably were removed at the machine shop. *Don't attempt head installation without them.* Without dowels, the head and its gasket may not locate properly, leading to head-gasket-sealing problems.

If you've been working alone, get a helper for head installation, especially on CVCC engines. Unless you're an Olympic weight lifter, there is no way you can carefully set the heavy, poorly balanced CVCC head-and-manifold assembly on the block by yourself. The aluminum-engine head-and-manifold assembly is lighter and better balanced, but is still best handled by two people.

With the car hood removed, grasp the head and manifolds at one end while your helper does the same at the other. Lift the head and position it over the block. You and your helper will be facing each other across the front fenders. Now, carefully lower the head onto the block. On aluminum-block engines, the studs will guide the head into place. But, on cast-iron blocks, it's up to you to get head and block aligned and into engagement with the dowels without sliding the head around on the deck. Moving the head while resting against the head gasket may cause gasket trouble later.

Also align the exhaust manifold and pipe during head installation, especially on CVCC engines. On these models, Honda uses an unconventional, flexible header pipe. This "stainless-steel bamboo" allows you to move the pipe around quite a bit. So, if you don't get the pipe and manifold mated on these models while setting the head on the block, don't worry. Just slide under the engine and guide the pipe into position. Verify that the pipe and manifold are mated. Take the two large manifold-to-pipe nuts with you and start them on the manifold studs after giving each stud a dab of anti-seize. By threading on the nuts, the pipe won't slip off.

Head Bolts or Stud Nuts—Wiggle out from under the car and get out the head bolts or nuts and washers. On iron-block engines, slip a washer onto each head bolt. Run them into the head with a 10mm, 12-point socket. Torque the bolts in 10-ft-lb increments, following the sequence shown. Final head-bolt torque for all iron-block engines is 44 ft-lb.

It takes two people to set cylinder head/manifold assembly onto block. Head must be set straight down on gasket—sliding head into engagement with dowels will damage it. Once you feel dowels engage head, lower assembly into place. *Keep your fingers from between head and block!*

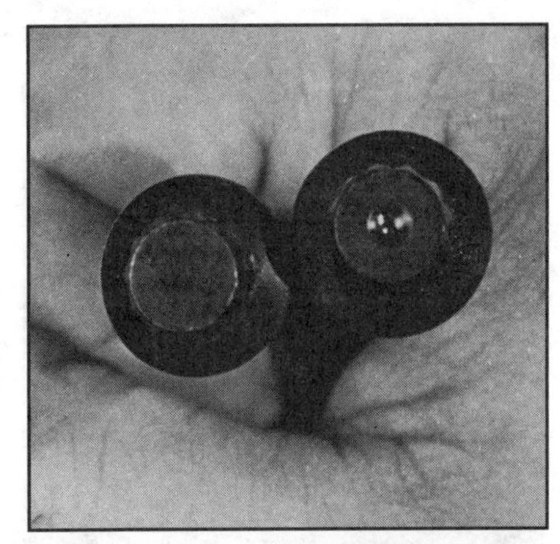

Dimple in bolt head identifies short cylinder-head bolt. It's used in 1975—'79 1488 and 1600 heads. Short bolt goes in sparkplug side, extreme distributor-end bolt hole. If you didn't keep head bolts in order at disassembly, you'll be glad this dimple is there.

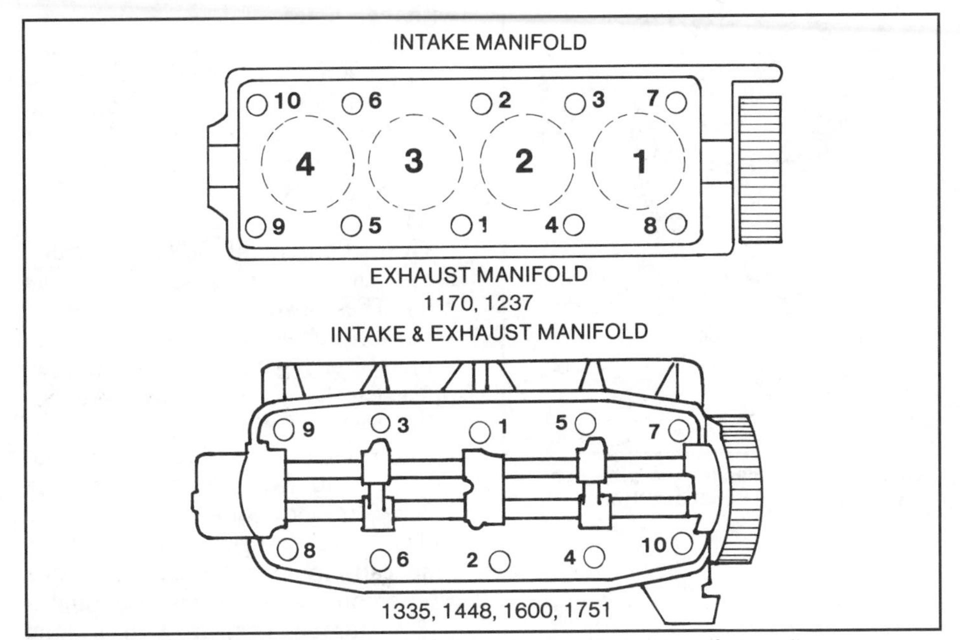

Cylinder-head torque sequence must be followed to ensure head-gasket sealing.

Before torquing cylinder-head bolts, double-check that head is sitting squarely on block and exhaust pipe is mated with manifold. Tighten head bolts in 10-ft-lb increments to specifications. See text for final torque specifications.

On aluminum-block engines, snug all head nuts and bolts with a 14mm hex socket. Then, torque in the sequence shown 37—42 ft-lb, except for engines EB1 1019949 and lower. These get only 30—35 ft-lb, unless you took my advice and replaced the early bolts with the newer ones. New bolts and all stud nuts get 37—42 ft-lb.

It takes several minutes to go through the head-bolt torquing sequence, so don't rush the job. After reaching final torque, let the bolts, studs and head gasket relax, or *normalize,* for a couple of minutes while you do something else. Then, come back and retorque them to make sure none of the bolts loosened.

Timing Belt—Double-check that both cylinder-1 valves and piston are at TDC, page 129, then install the timing belt. Pull the belt up from the crank sprocket as far as it will go and spread it with your thumbs. Slip the belt onto the cam sprocket, starting with the right side. Try to get all slack out of the left side of the timing belt. There should be slack in the right side because that's where the timing-belt tensioner is.

Take up the slack by adjusting the tensioner. Both the tensioner pivot and adjusting bolts should be finger-tight only. Double-check by reaching down the front of the timing-belt lower cover where the bolts stick out. Rotate the crank counterclockwise one-quarter turn or more and hold it there. This pulls the belt in the direc-

After double-checking that crank and camshaft are at TDC cylinder 1, loop timing belt around cam pulley and push it on.

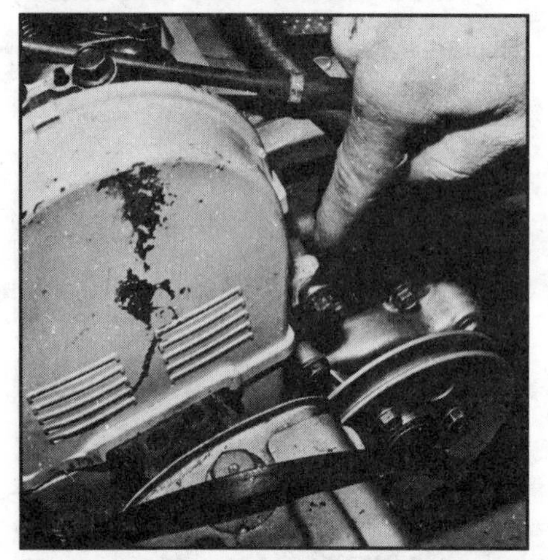
Fit two loose gasket ends from timing-belt lower cover into upper cover and bolt to head.

tion of rotation and slackens the belt's right side. Finish adjustment by tightening the adjusting bolt—the lower or right bolt—then the pivot bolt—the upper or left bolt. The tensioner automatically takes up slack during the adjustment procedure because it's spring-loaded. Later on, if you think the belt is too loose, repeat the adjustment procedure.

Install the timing-belt upper cover. Slip it over the cam sprocket and bolt it to the head. This will secure those two loose gasket ends sprouting from the timing-belt lower cover.

Exhaust Pipe & Manifold—Bolting these two pieces together is like paying taxes. It's no fun and you can't avoid it. So, let's get it over with.

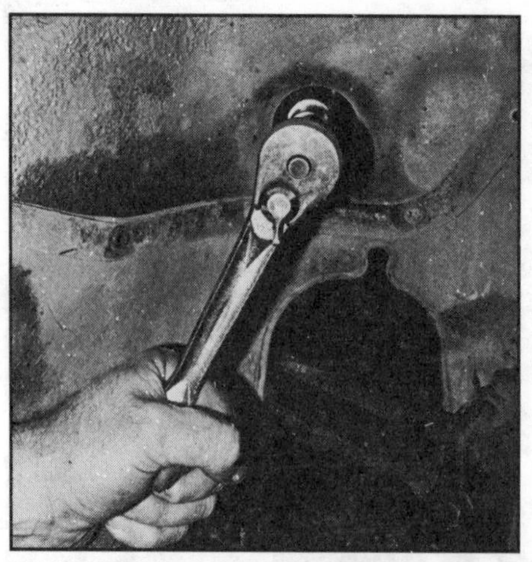
Setting timing-belt tensioner requires rotating engine counterclockwise a quarter turn or more. Crankshaft-pulley nut is accessible through this hole in left front inner fender. Don't forget to remove socket wrench before starting engine!

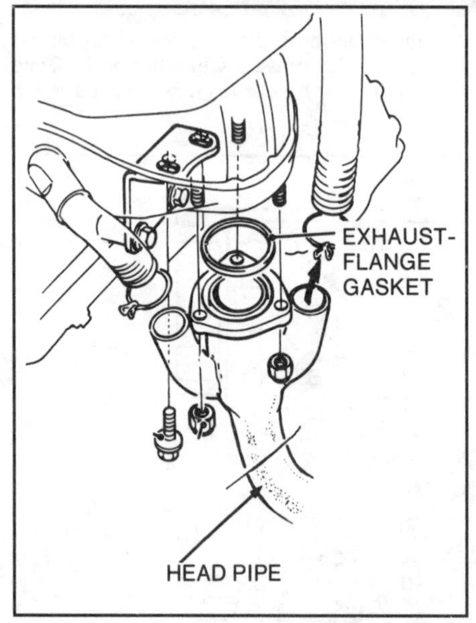

Typical exhaust-manifold-to-pipe connection. Courtesy of American Honda Motor Co., Inc.

1170, 1237—For non-CVCC engines, connecting the header pipe is no nightmare. Align the exhaust pipe over the manifold studs, give the stud threads a dab of anti-seize compound and run on the nuts. Before tightening these nuts, attach the pipe-to-manifold bracket to the base of the block. Now tighten all nuts and bolts—torque manifold-to-pipe nuts 14–18 ft-lb.

CVCC—Get *all* the bolts started before tightening them. If you start tightening them before all are in, you

Obviously, one drawback to installing engine minus head is difficulty of adjusting timing-belt tensioner later in tight quarters. While applying force to crankshaft-pulley bolt in direction of engine rotation (counterclockwise), tighten adjusting bolt, then pivot bolt.

won't be able to start the final few bolts. Then you'll have to backtrack, loosening bolts until there's enough play in the pipe to line up the last bolts with their holes.

The two or three manifold-to-pipe nuts are already installed finger tight. Start two bracket-to-block bolts and two bracket-to-bracket bolts. Also put anti-seize compound on the manifold-to-pipe threads if you didn't the first time.

Now, tighten all bolts and nuts. First, tighten the manifold-to-pipe nuts. These two or three nuts take a lot of torque because they provide the primary sealing pressure. And, they must seat the steel/asbestos gasket between them. Torque for the two-nut version is 40 ft-lb; three-nut connections get 36 ft-lb. Then tighten the bracket-to-block bolts.

The manifold-to-pipe connection is practically impossible to get a torque wrench on. On some models, you can use a torque wrench with a long extension; you'll just have to guess on the rest. Keep in mind that with an average-size 17mm wrench, a 40-ft-lb pull is right on the "grunt threshold." Tighten the nuts in a *crisscross pattern.* Go back and forth until the nuts will tighten no more, even with considerable effort.

Finally, tighten the bracket-to-bracket connections. If you're working on a Prelude or late Accord, don't forget to install the two splash guards under the engine and transaxle—three bolts each. You can now lower the car, if you wish. I

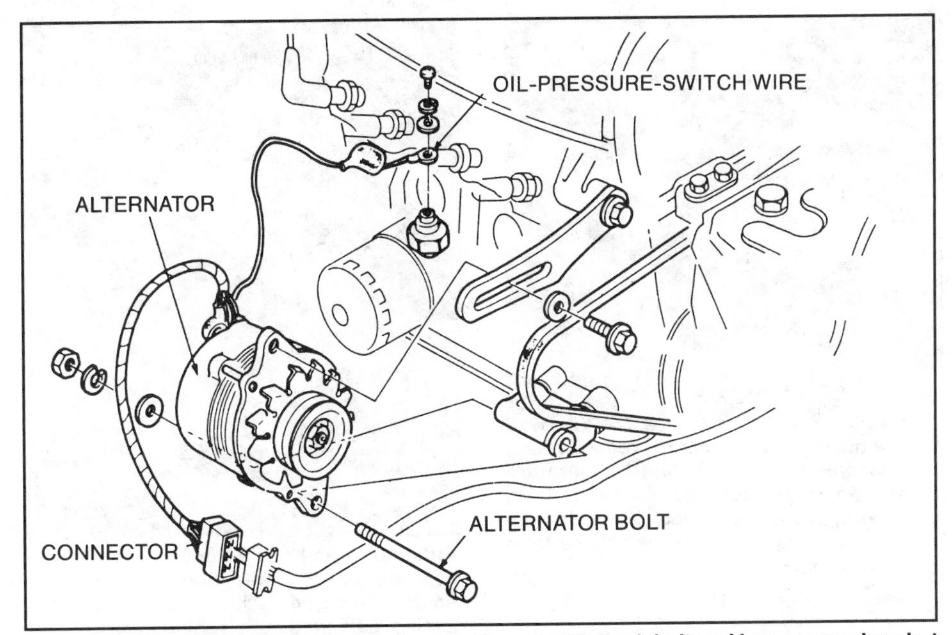

Alternator mounting—non-A/C cars: Leave alternator through bolt and inner, upper bracket bolt loose while adjusting belt tension. Tighten adjusting bolt first, then upper-bracket and through bolts. Courtesy of American Honda Motor Co., Inc.

Belt guard is an important part of this style alternator mount. It can keep neckties and other foreign objects from getting wrapped around rotating pulley. Don't leave it off!

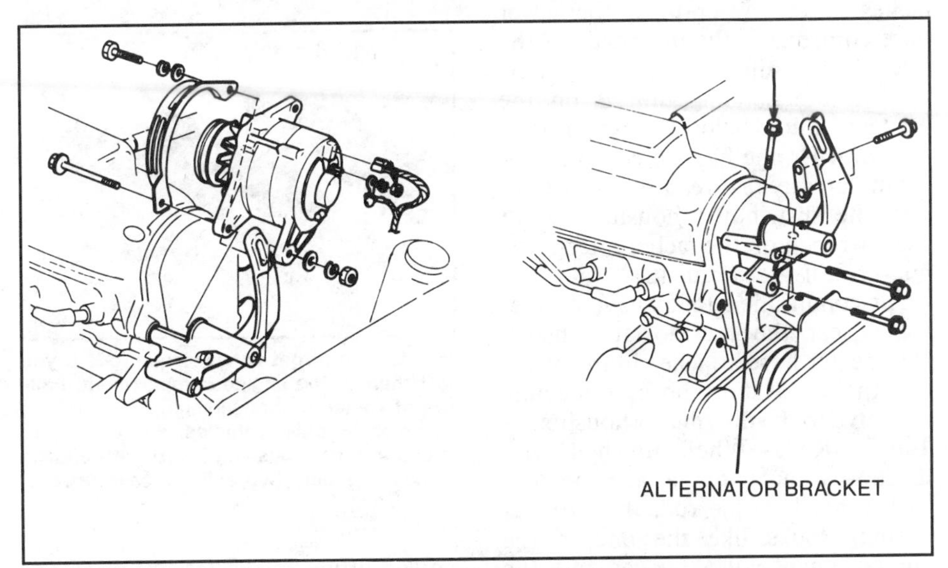

Early alternator mounting—A/C cars: With this type of alternator mounting, make sure you follow this installation order: timing-belt upper cover, drive belt, left engine mount, alternator bracket, alternator. This mounting is similar to late A/C-compressor-pump mounting. Courtesy of American Honda Motor Co., Inc.

prefer to keep the car raised until right before I start it. It keeps the engine a little higher off the ground so I don't have to lean over so far—easier on this six-footer's back.

Alternator—On all iron-block engines, the next step is alternator installation.

The standard alternator location is low, alongside the front side of the block. Install the lower alternator bracket to the block with its two bolts. Set the alternator on the lower bracket and install the bottom through bolt. Don't tighten the through-bolt nut, yet. Instead, bolt the upper alternator arm to the head, swing the alternator out and thread in the alternator adjusting bolt through the slot in the upper arm. Slide the alternator drive belt over the crank, water-pump and alternator pulleys. Now, pull the alternator away from the block until the belt deflects 1/2 in. at midspan. Tighten the through and adjusting bolts.

The other type of iron-block alternator mounting faces the camshaft sprocket. Its mount uses a large wishbone-shaped bracket attached to the head and left engine mount. To install a high-mounted alternator, first fish the drive belt onto the crankshaft pulley and drape it over the timing-belt upper cover. Next, pull the left engine mount from its hole in the fender. Set the alternator bracket in place on top of the left engine mount. Start all six bracket bolts. Two bolts pass downward through the alternator bracket and into the left engine mount. Two bolts thread into the head next to the manifolds, while the last pair bolt into the head on the other side. These last two bolts run parallel with the crankshaft. Once all bolts are started, tighten them. This will connect the left engine mount, so when you are done with six bracket bolts, tighten the large, left engine-mount bolt on the fender.

Lay the high-mounted alternator on its bracket and install the through bolt and belt guard. Lay the drive belt into the alternator pulley, then swing the belt guard into position. Pass the adjusting bolt through the belt guard and thread it into the alternator. Adjust belt tension and tighten the through and adjusting bolts. Finally, connect the electrical leads.

A/C Compressor—For A/C compressors that bolt to the cylinder head, use the same procedure for installing the high-mounted alternator. Those crafty people at Honda figured out a way to use the same wishbone-type mount for a high-mounted alternator *or* A/C compressor. This mount attaches to the left engine mount and both sides of the head. So, when you're using the high-mounted *alternator* installation procedure detailed previously, mentally substitute the word *A/C compressor,* and you've got it.

Air Pump—On 1975—'79 1237cc

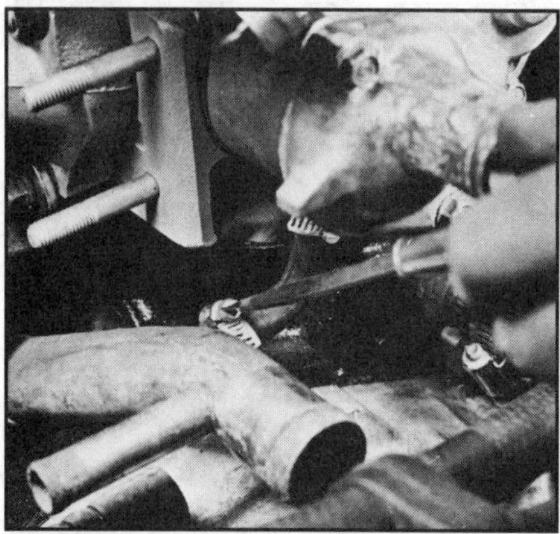
A new, pliable CVCC coolant-bypass hose is a lot easier to install than reusing old, hardened hose.

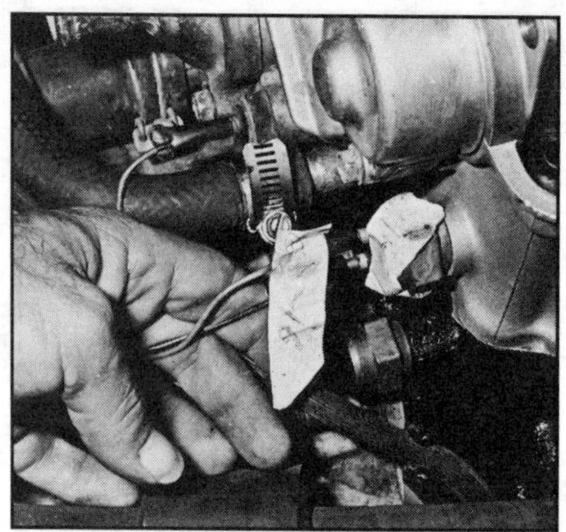
Follow numbering system for reconnecting thermosensor and coolant-temperature leads. Markings were good and dark to start with, but faded rapidly. Black felt-tip marks take longer to fade than those from light-color markers.

Although it looks like I'm strangling a boa constrictor, hose installation must be done gently. Soldered radiator connections are easily damaged by yanking on hoses. Loosen hose clamps completely, using twisting motion to install hoses. Do not overtighten hose clamps.

It's easy to forget about torque rods if you left them in the engine compartment. Failure of voided rubber-bushing insert, left, is common on older vehicles. You can try to press in a new bushing insert with a large bench vice, but a hydraulic press is better.

engines, install the air-injection pump. Bolt the upper bracket to the head, above the water pump. Hold the air pump in place and insert the through bolt through the pump and upper bracket. With the pump hanging from the through bolt, install the adjusting bolt through the adjusting arm sprouting up from the water pump, and into the air pump. Use the lever pin at the rear of the pump to pull the drive belt tight. Don't lever against the pump body or you'll damage it. When correctly adjusted, the drive belt can be deflected about 1/2 in. midway on its longest span. Finish air-pump installation by connecting the supply, bypass and bellows hoses.

Hoses & Wires—At the bellhousing end of the engine, connect the bypass, radiator and heater hoses. Use new hoses if the old ones are two years old or more.

On CVCC engines, you'll need a new water-pump bypass hose just so you can bend it into position between the water pipe and distributor housing. The lower radiator hose connects with the water pipe; the upper hose with the thermostat housing. The heater hoses connect to the water pipe and cylinder head, immediately below the distributor housing. Non-hardening sealer spread inside the hoses will greatly speed installation and help seal the connections.

Follow your numbering system to connect the water-temperature and oil-pressure-sending-unit leads. On CVCC engines, connect the two thermosensor leads.

On 1200s, the water hose-and-wire arrangement is a little different. The upper heater hose at the fire wall attaches to the water pipe; the lower hose connects to the underside of the intake manifold. The water-temperature sending unit is on the intake manifold below the carburetor.

If your Honda has a mechanical tachometer drive, thread it finger-tight into the distributor housing. Don't use a wrench on the tachometer-drive nut—you'll overtorque it.

Air-conditioned 1335s have a water valve near the thermostat housing. One heater hose attaches to this valve and there's a short hose connecting the valve to the distributor housing.

Torque Rods—When finished with the hoses and wires, attach the fire-wall-to-head torque rod. If its rubber bushing looks like the one in the photos, change it. You can buy the bushing separately, but a hydraulic press is needed to install it. Because the torque-rod bushing is *voided*—asymmetrically shaped to tune out vibrations—mark the position of the old bushing before removal, and install the new bushing in the same position. If your Honda has a rear lower or front lower torque arm, install it now.

Oil Cooler—If equipped with an engine-oil cooler, install its hoses now. As you stand in front of the engine compartment, the oil cooler is to the right of the radiator. The connections are: outlet at bottom, inlet at top. Clamp the outlet hose to the cooler's bottom fitting, and to the right fitting under the oil filter on the block. The inlet hose runs from the upper cooler fitting to the left fitting under the oil filter. Double-check that all connections are clamped.

Distributor—Attach only the distributor primary coil wire for now.

Black Box(es)—Hang the emission-control box or boxes on the fire wall. Join any electrical connections at the "black box(es)."

Choke & Throttle Cables—Both cables attach the same way. First, lay the cable in its approximate installed position. Unthread the nut closest to the cable end. Guide the cable and cable housing through the cable bracket until the second nut stops the cable housing. Thread the first nut back onto the cable housing and tighten it against the bracket. Hold the second nut with another wrench.

For choke cables, you'll need a

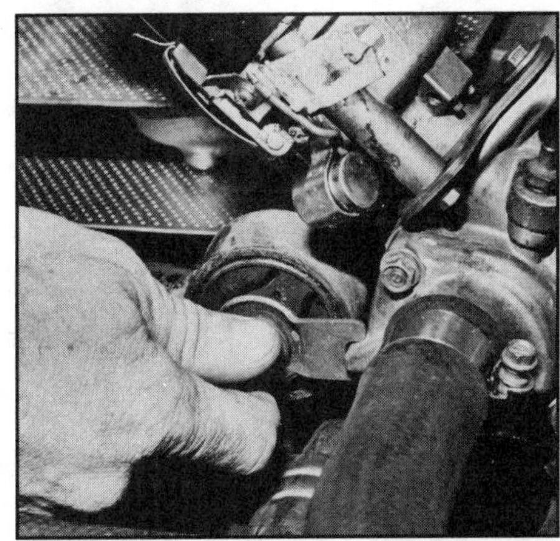

Strange washer/bracket on upper torque rod rests against knob on thermostat cover. This keeps rubber bushing from twisting when rod-eye bolt is tightened.

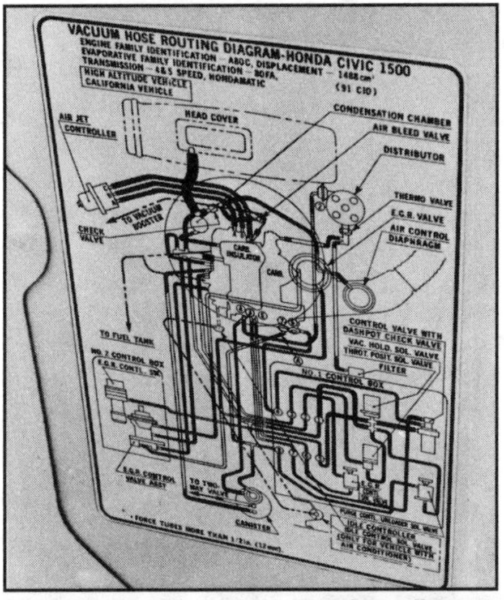

Later cars have hose-routing diagram under hood. It can help sort out things if your numbering system has you confused.

Cracked vacuum hoses can cause driveability problems. If end of hose is cracked, trim off bad end and reconnect it if there's enough slack. Otherwise, replace hose.

Fill cooling system until air-free coolant flows from air bleed. Air bleed opens and closes the same as a brake bleeder. On 1200s, air bleed is on intake manifold.

Oil poured down oil-pump drive-shaft hole goes directly into pan. Pouring some oil over rocker-arm assembly helps with initial lubrication, but makes it more difficult to tell when oil system pressurizes.

10mm wrench, throttle cables, 12mm. Thread the cable end onto the choke- or throttle-actuating arm. The cable end has a small cylinder permanently attached. Pull the actuating arm back to gain some cable slack, slide the cylinder through the hole in the actuating arm, lay the cable in the actuating arm and return the arm to its at-rest position.

Fuel & Vacuum Hoses—While you're at the carburetor, attach the fuel and vacuum hoses. This is where you put your numbering system to the test. I'll list some of the major connections for you, but there are too many possible hose routings to list all of them.

Fuel connections are easy: On CVCC engines, join the two fuel lines at the fuel-line junction leading out of the fire wall. On 1200s, clamp the fuel line from the fire wall to the fuel pump. The pump-to-carburetor line should already be attached.

The large vacuum hose running from the brake booster connects to the intake manifold. On 1200s, the fitting is directly below the fuel pump; CVCCs hide it below the carburetor. Some of the more common emission-control vacuum lines run to the start-control solenoid on the driver's side of the fire wall, the charcoal canister near the left fender and the PCV connection atop the rocker cover. You'll have to wait until the rocker cover is installed before making that connection. Don't forget to hook up the distributor vacuum-advance hose. It typically sprouts out of the black box.

Battery—Clean all battery cables and posts, then install the battery. If you didn't charge the battery during storage, charge it now. With tight clearances, the starter will need every ounce of energy the battery can supply.

Add Coolant—Here's a good test of your optimism. If you're confident the new engine will run correctly, fill the cooling system with a 50%-water/50%-antifreeze mix. If you are naturally cautious, fill the system with plain water. Then, a coolant leak won't have you crying over spilled antifreeze.

Filling the Honda cooling system is easy because of the nifty air bleed. On other engines, air gets trapped behind the closed thermostat when adding coolant to a cold engine. Then, as the engine is started and begins to warm, only air touches the thermostat. The thermostat thinks the coolant is still cold and stays shut. But the engine continues to heat the small amount of coolant. Sometimes, increased coolant pressure will force hot coolant against the thermostat and open it. Other times, revving the engine slightly will force coolant against the thermostat due to increased water-pump action. An alert mechanic will see the pegged water-temperature gage, shut the engine off and manually bleed air out by removing the thermostat and adding water behind it. In all cases, there's a risk of severely overheating the engine, resulting in broken rings and scored cylinder walls.

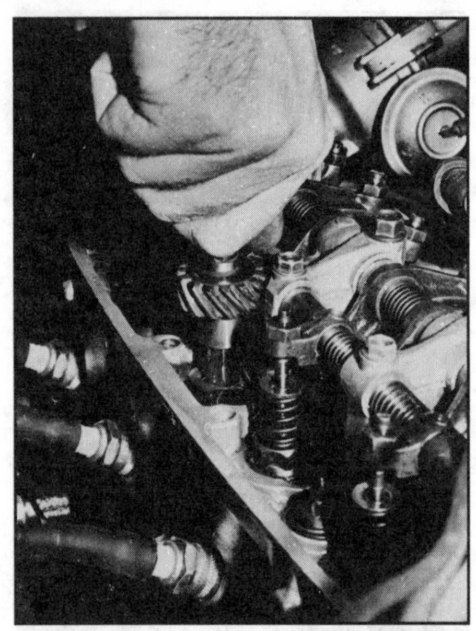

Building oil pressure manually: Upside-down driven gear makes a good handle. Rotate gear counterclockwise until oil flows into rocker-arm area. Drill motors and speed handles are faster, but not necessary.

Moving rocker-arm spacer slightly might be necessary to install oil-pump drive gear. Always use new driven gear with new camshaft.

Without dropping dowels inside engine, install drive-gear cover. Torque cover bolts 6—9 ft-lb.

Rocker cover fits over edge of timing-belt upper cover. Both hold-down studs provide mounting for wire or cable standoffs. Don't forget to attach ground cable to front—timing-belt end—hold-down stud.

With Honda's air bleed, you don't have to worry about these problems. Just open the air bleed several turns, and add coolant at the radiator until it flows without air bubbles from the bleed. The cooling system is purged. Close the bleed and top off the system. Mop up the coolant spilled from the bleed, install the radiator cap and you're finished filling the system.

Cooling-system capacity is 4.4 quarts on 1200s and 1335s; 4.8 quarts on early 1488s and 1600s; 6 quarts on '80 and later 1488s; and 7 quarts on 1751s. These figures are for completely dry cooling systems, heater core included.

Therefore, somewhere between 2 and 3.5 quarts of antifreeze will give the desired 50/50 mix. If you add plain water now, drain some of it after you've run the engine. Refill it with antifreeze.

Add Oil—Fill the crankcase with oil by pouring it down the oil-pump-drive-shaft passage. Use a major brand, SE- or SF-grade oil. In temperate climates, use 30W. Cold climates with sustained temperatures under 20F require 5W-20 or 5W-30 oil. In hot climates, where it's typically 80F or more, use 20W-50. You'll need 3.6 quarts to fill a dry engine, except for 1200s; they get 3.2 quarts. Subtract the amount of oil added to the filter during engine assembly.

Build Oil Pressure—Before installing the rocker cover, it's a good idea to prime the oil pump so pressure builds quickly when the engine is started. Insert the oil-pump drive shaft into its passage and engage the slot in the oil-pump shaft. Place the oil-pump drive gear *upside down* on the drive shaft. The pin through the shaft will engage the slot in the oil-pump gear in this upside-down position.

Now, turn the pump gear and shaft counterclockwise by hand until oil flows into the rocker-arm area. There is no need to turn the shaft with an electric drill or speed handle. However, a good idea is to rotate the crankshaft while turning the oil pump. This ensures oil flow through all crankshaft oil passages and rod bearings. Have a helper turn the crankshaft with a ratchet/socket combination through the hole in the left fender while you turn the pump gear.

After oil flows into the rocker-arm area, the engine is primed. Pull off the pump gear, turn it right-side up and drop it into position on the pump drive shaft. Squirt or pour some oil on the pump gear. Then, lay the oil-pump-drive-gear cover over the gear and bolt it into position. Watch for two dowels in the cover during installation. They tend to fall out, and you sure don't want to fish one out of the oil pan after it drops through the drive-shaft passage.

With the gear cover in place, install the rocker cover with a new gasket. The rocker cover fits over the timing-belt upper cover and is retained by two acorn nuts. Drop in the two rubber cones, cover them with the cupped washers, fit any ignition-wire looms or cable guides, then tighten the acorn nuts. Slip on the breather hose.

Start Engine—So you can watch for fuel and vacuum leaks, leave the air cleaner off for initial engine start-up—if it's not windy. Otherwise, set the air cleaner in place. Don't bolt it down. You'll have to remove it to check valve clearance.

Time Engine—Loosen the distributor

After engine develops oil pressure from cranking engine, connect coil negative lead (shown) or plastic connector on electronic ignition.

Check timing during initial engine run-in. Try not to let engine idle slowly for more than a couple of seconds to set timing. Fast idle is required for cam break-in.

hold-down bolt so you can *just* turn the distributor. Hook up a timing light. Inspect the wires leading to the coolant-temperature sensor in the radiator tank. Make sure they are connected to the sensor. If the fan or fans were not operating before the engine rebuild, jump the two sensor leads and tape them together. When the ignition switch is on, the fan(s) should run. If the fan(s) still don't run, you *must* keep an eye on the coolant-temperature gauge. An idling engine *will* overheat without the fan(s). In a pinch, use a common house fan. One of these blowing through the grille will work almost as well as the car's fan (s).

Crank the engine until the oil-pressure light goes out or oil-pressure gage indicates pressure. Connect the distributor-ground lead or electronic-ignition connector. Start the engine and run it at fast idle—about 1500 rpm—for 15 minutes. Watch for leaking fluids and keep your eye on coolant temperature.

After making sure there are no leaks or funny noises, set ignition timing. Follow the underhood sticker's recommended ignition-timing specification. If you can't find the sticker, use these rough guidelines: 1973—'77 1170 and 1237, 5° BTDC; 1978—'79 1237, 2° BTDC; 1975—'79 1488 and 1600, 2° BTDC; All 1751, 1335 and '80-and-later 1488, TDC to 5° BTDC. Remember, these are rough approximations. For a thorough engine tuneup, I highly recommend taking your car to a professional tuneup shop—one with an exhaust analyzer and scope. Although setting dwell and timing is no big deal, ignition wires, cap and rotor should be checked. And the CVCC-engine three-barrel carburetor is practically impossible to adjust without special tools.

On 1200s, timing is set at the *crankshaft pulley.* The timing pointer is a raised line on the timing-belt lower cover. There are two marks in the crank pulley—5° BTDC and TDC. The pulley rotates counterclockwise, so the 5° BTDC mark is on the left.

All other engines are timed through a hole in the *bellhousing.* Through this hole, the timing marks on the flywheel or driveplate can be aligned with a stationary pointer. Typically, there will be a white T that merely means you are looking at the timing marks. A white line is usually TDC and a red line is usually where timing should be set. I say usually because sometimes there is a blue, yellow or other-color painted line instead of the red one. If there is only one line, usually white, set timing on it. If there are both red and white lines, set timing on the red. If possible, refer to the car's underhood emission sticker.

When timing is set, tighten the distributor hold-down bolt and disconnect the timing light.

Post Run-in Checks—After 15 minutes, shut off the engine. Check coolant level at the translucent overflow reservoir. Do *not* open the radiator cap on a hot engine—you may be scalded by hot coolant! Take several minutes to inspect the engine for leaks.

Now, let the engine cool so you can check the valve adjustment. It will take at least one-half hour before the engine is cool enough for this job, so busy yourself with picking up tools and cleaning up the area. It's also a good time to take a break, and show off your new engine to all the friends who stop by.

Once the cylinder head has cooled to the touch—only enough heat to determine that the engine was run recently—adjust the valves. Follow the valve-adjustment instructions, page 130. Retorquing the cylinder head with Honda's moly-coated head gasket or with certain aftermarket head gaskets, like Fel-Pro's Print-O-Seal, is not necessary. Check the instructions that came with the head-gasket set.

Of course, you'll have to remove and replace the rocker cover. When you set the rocker cover on the head, wipe off the mating surface of the cover gasket. Remember that the heat remaining in the engine will close valve clearance slightly—maybe 0.0005 in. (0.01mm). So, the valves may feel slightly tighter—that's OK.

New air-cleaner element provides maximum protection for your new engine. Later air-cleaner housings have four bolts *inside* housing. Two hold housing to carburetor; other two retain bundles of vacuum hoses to bottom of housing.

Old wear lines provide an excellent guide for exact hinge placement. Snug bolts and check hood fit. Then tighten hinge bolts.

Just make sure none are very loose or tight.

If your car has cruise control with a large vacuum canister, mount the unit after installing the rocker cover. The canister stud extends into a bracket at the radiator. The rod end uses washers and cotter keys to mount to the throttle-actuating arm.

Air Cleaner—With the torquing and valve-adjusting chores out of the way, install the air cleaner. Use a new rubber gasket between the air-cleaner housing and carburetor. Make all vacuum-hose connections, set the bottom air-cleaner housing in place and start all bolts through the mounting brackets. Don't forget the two or three bolts and two nuts inside '80-and-later air cleaners. Check that the rubber gaskets inside the air-cleaner housing are firmly glued down. If they are loose or swollen, pull them up, cut out enough material to take up the slack, then glue them back down. 3M's Super Weatherstrip Adhesive is perfect for this job.

Install a new filter element and the air-cleaner cover. Check that the element is the right height, especially on 1977-and-later 1488s and all 1600s. The 1975—'76 element was shorter than later versions. If it is too tall, you won't be able to install the air-cleaner cover. But a short element may go unnoticed and the engine will inhale unfiltered air. Visually check for a short element with the lid off. Once the cover is in place, tighten all air-cleaner-housing bolts. Connect any fresh- and hot-air ducting.

Hood—Have someone help set the hood in place, then bolt down the hinges. Because the hood is front-hinged, it can be adjusted in the safety-latch position. Don't neglect any shims or the hood will not fit properly. Follow your reference marks to align the hood. This job goes quickest when you install all hinge bolts and snug them. Gently bump the hood into position and tighten the bolts. Replace the grille, headlamp doors or windshield-washer bag, if removed to gain access to the hood bolts.

Read Chapter 9, Tuneup, before driving your Honda car. Following break-in and tuneup suggestions given there will add many miles to your engine's longevity.

Tuneup 9

To get the most out of your rebuilt engine, you must break it in and tune it. Engine break-in sets an engine's wear pattern for life. A professional tuneup helps ensure that the engine operates at peak efficiency—with good power, good fuel mileage and low exhaust emissions.

Break-in—Engine break-in is the subject of many tall tales. You may have heard it takes 2000 miles to correctly break-in an engine or that the best thing you can do is "baby" the engine. A lot of these stories stem from long break-in times mandated by new-car manufacturers. These long—2000-mile—break-in periods are based more on a desire to minimize warranty costs than on real engine needs.

In today's engines, most parts need only seconds to break-in. If the rod and main bearings haven't bedded in within five minutes, for example, they may never bed in. The big exception is piston rings. The moly rings used in Honda cars should completely break-in by 300 miles. If they haven't done so by 500 miles, something is wrong. Either the gaps are incorrect, cylinders are tapered, a ring broke during installation or something of the sort. Therefore, your Honda engine should be fully broken-in by 300 miles. If the engine still has oil-consumption, overheating or power problems at 500 miles, they are not break-in related.

The other break-in consideration is *how* to break-in the engine. Again, most people are led by new-car recommendations to go especially easy on a new engine. It's not the worst advice, but the best way to operate a new engine is very close to normal practice. Things to avoid are sustained operation at constant rpm, lugging, short trips and lack of maintenance.

If the engine is run at a constant rpm, the rings will wear for those conditions only. Constant-rpm operation will also prolong ring break-in time. Sustained high-speed operation is especially harmful because it over-stresses the engine before it is fully broken in.

Lugging is bad because it puts a high load on bottom-end parts—main bearings in particular. It can also heavily load the rings.

When a car is driven only on short trips, it may not warm to operating temperature. Because a rich air/fuel mixture is required for cold-engine driveability, sulphuric acid—a byproduct of burning sulphur fuels—forms rapidly in the oil. The corrosive acid can cause a lot of "dry starts" because the oil has run off of the bearings.

By lack of maintenance, I mean driving the new engine for a month without ever looking under the hood. You might get away with that under normal conditions, but immediately after taking the engine apart and bolting it back together, there's a chance something might come loose or leak.

So, during break-in, you should: vary engine speed, vary engine load and keep an eye on engine operation. Varying engine speed helps the rings break-in under varying conditions. The same is true of varying engine load. This is the biggest single factor in rapid and correct ring break-in. During the first road test, accelerate at one-half or three-quarter throttle to 55 mph, then close the throttle and decelerate to 20 mph. Repeat the proc-

Professional tuneup mechanics have knowledge and tools to give your Honda the best possible tune. This translates into better performance, economy and engine longevity.

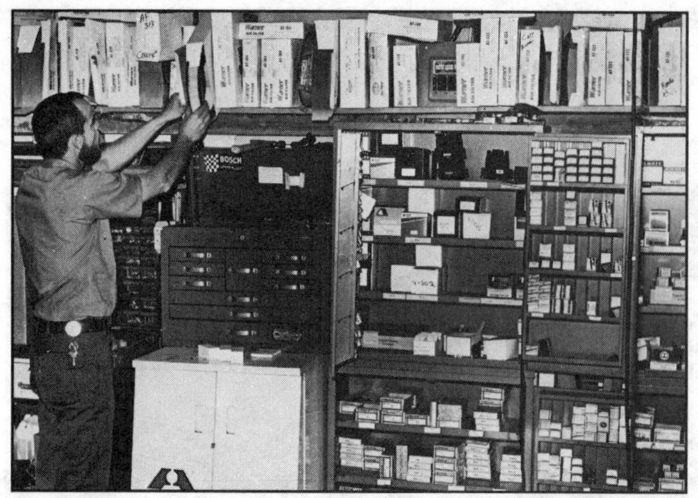

Many tuneup shops also have extensive parts supply. It's another reason why they can tune an engine so quickly.

Tuneup man discovered that project car's rubber seals in air-cleaner housing were loose and swollen oversize. Cutting a section out of seals and regluing seal to housing with weatherstrip adhesive cured problem. Abrasive paper prepped surface to improve bond.

George at Tune-N-Go pointed out difference between air-cleaner elements. He's seen many Hondas fitted with short filter when they needed taller version. Gap between cleaner-housing lid and element allows unfiltered air to enter engine.

Lambda-linkage adjustment screw, otherwise known as *auxiliary throttle-linkage adjustment*, synchronizes movement of auxiliary throttle with main throttle. *Leave this screw alone.* It has a cap over it to prevent tampering. Adjustment is critical and requires a special dial indicator. Even pros don't like this one—so don't touch it.

ess five times in succession. During acceleration, half the ring—thrust side of the piston—is stressed; during deceleration the other half takes its medicine. After this procedure, drive normally and follow the break-in guidelines stated earlier.

Keeping an eye on engine operation is self-explanatory—the sooner you spot trouble, the quicker it can be fixed with less chance of damage. Be prepared to readjust engine idle several times. As friction drops with increasing miles, idle speed increases.

The oil and filter should be changed at 25 and 300 miles to remove any leftover machining grit, dirt, lint or metal filings as parts bed in. Thereafter, your Honda engine will last longer with 4000-mile-or-shorter oil-and-filter-change intervals. Frequent oil changes are about the best thing you can do for a hard-working, high-revving Honda engine.

TUNEUP

As soon as your engine is fully broken-in, take it to a professional tuneup shop. You need a shop with an exhaust-gas analyzer and engine analyzer—a *scope*.

You might balk at the idea of taking your car to a tuneup specialist. After all, the engine is fresh, with new sparkplugs, air filter and maybe points. Well, if for no other reason, you should take your car to a pro just to have someone else look at your work. There might be some incorrectly installed vacuum hoses or a loose bolt. The pro sees a lot of cars and should find these problems fairly easily. It's not that you shouldn't trust your work, just that another eye never hurts.

Even more importantly, the pro and his equipment can tune your car better than you. Unfortunately, modern cars are difficult to tune. The pro has the tools and knowledge needed to get your Honda to run its best. Your rebuilt engine has better compression and will require carburetion adjustments. The ignition system can probably benefit from some fine tuning, also.

Because you've installed many new parts, brought compression to original specs and adjusted the valves, the tuneup shouldn't cost much. And, it shouldn't take all day, either. This is also a great time to talk to a pro about your car. You might pick up some information that will save you a lot of money.

Finally, a tuneup and smog certification may be required to keep your car licensed. This is a great time, mechanically speaking, to *desmog* your car. If you live in an area where periodic tailpipe-emission checks are required, make sure the inspection coincides with the registration deadline. Even if you don't go through a smog-equipment check, you'll be doing your part to keep the air clean.

Appendix

METRIC/CUSTOMARY-UNIT EQUIVALENTS

Multiply:	by:	to get:	Multiply:	by:	to get:
LINEAR					
inches	X 25.4	= millimeters(mm)		X 0.03937	= inches
feet	X 0.3048	= meters (m)		X 3.281	= feet
miles	X 1.6093	= kilometers (km)		X 0.6214	= miles
AREA					
inches^2	X 645.16	= millimeters^2(mm^2)		X 0.00155	= inches^2
feet^2	X 0.0929	= meters^2(m^2)		X 10.764	= feet^2
VOLUME					
inches^3	X 16387	= millimeters^3(mm^3)		X 0.000061	= inches^3
inches^3	X 0.01639	= liters (l)		X 61.024	= inches^3
quarts	X 0.94635	= liters (l)		X 1.0567	= quarts
gallons	X 3.7854	= liters (l)		X 0.2642	= gallons
feet^3	X 28.317	= liters (l)		X 0.03531	= feet^3
feet^3	X 0.02832	= meters^3(m^3)		X 35.315	= feet^3
MASS					
pounds (av)	X 0.4536	= kilograms (kg)		X 2.2046	= pounds (av)
FORCE					
pounds—f(av)	X 4.448	= newtons (N)		X 0.2248	= pounds—f(av)
kilograms—f	X 9.807	= newtons (N)		X 0.10197	= kilograms—f

TEMPERATURE

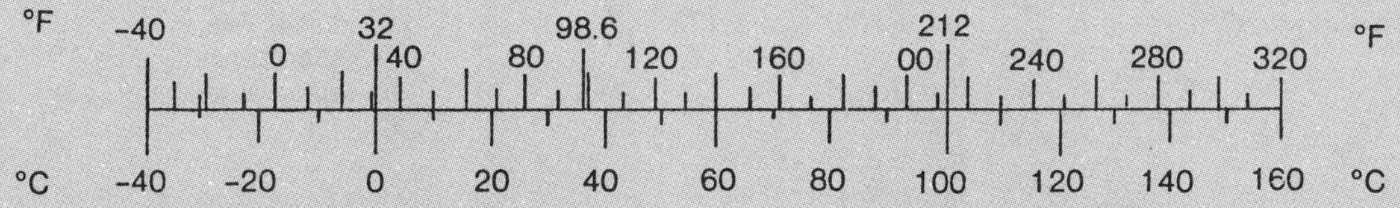

°F: −40, 0, 32, 40, 80, 98.6, 120, 160, 200, 212, 240, 280, 320 °F

°C: −40, −20, 0, 20, 40, 60, 80, 100, 120, 140, 160 °C

Degrees Celsius (C) = 0.556 (F - 32)

Degrees Fahrenheit (F) = (1.8C) + 32

Multiply:	by:	to get:	Multiply:	by:	to get:
PRESSURE OR STRESS					
inches Hg (60F)	X 3.377	= kilopascals (kPa)		X 0.2961	= inches Hg
pounds/sq in.	X 6.895	= kilopascals (kPa)		X 0.145	= pounds/sq in
POWER					
horsepower	X 0.746	= kilowatts (kW)		X 1.34	= horsepower
TORQUE					
pound-inches	X 0.11298	= newton-meters (N-m)		X 8.851	= pound-inches
pound-feet	X 1.3558	= newton-meters (N-m)		X 0.7376	= pound-feet
pound-inches	X 0.0115	= kilogram-meters (Kg-M)		X 87	= pound-feet
pound-feet	X 0.138	= kilogram-meters (Kg-M)		X 7.25	= pound-feet
VELOCITY					
miles/hour	X 1.6093	= kilometers/hour(km/h)		X 0.6214	= miles/hour
kilometers/hr	X 0.27778	= meters/sec (m/s)		X 3.600	= kilometers/hr

COMMON METRIC PREFIXES

mega (M)	= 1,000,000	or	10^6	centi (c)	= 0.01	or	10^{-2}
kilo (k)	= 1,000	or	10^3	milli (m)	= 0.001	or	10^{-3}
hecto (h)	= 100	or	10^2	micro (μ)	= 0.000,001	or	10^{-6}

Conversion chart courtesy of Ford Motor Company.

Index